D1446716

R00219 67762

CHICAGO PUBLIC LIBRARY
HAROLD WASHINGTON LIBRARY CENTER

R0021967762

REF
QD
923
.I57   Introduction to
       liquid crystals
Cop. 1

DATE DUE

REF
QD
923
.I57
cop. 1

FORM 125 M

BUSINESS/SCIENCE/TECHNOLOGY DIVISION

The Chicago Public Library

SEP 1 1977

Received

INTRODUCTION TO LIQUID CRYSTALS

INTRODUCTION TO LIQUID CRYSTALS

Edited by
E. B. Priestley
Peter J. Wojtowicz
Ping Sheng
RCA Laboratories
Princeton, New Jersey

PLENUM PRESS · NEW YORK AND LONDON

Library of Congress Cataloging in Publication Data

Main entry under title:

Introduction to liquid crystals.

Includes bibliographical references and index.
1. Liquid crystals. I. Priestley, E. B., 1943- II. Wojtowicz, Peter J., 1931-
III. Sheng, Ping, 1946-
QD923.I57   548'.9   75-34195
ISBN 0-306-30858-4

REF
QD
923
.I57
cop. 1

© 1974, 1975 RCA Laboratories
Princeton, New Jersey

Plenum Press, New York is a division of Plenum Publishing Corporation
227 West 17th Street, New York, N.Y. 10011

United Kingdom edition published by Plenum Press, London
A Division of Plenum Publishing Company, Ltd.
Davis House (4th Floor), 8 Scrubs Lane, Harlesden, London, NW10 6SE, England

All rights reserved

No part of this book may be reproduced, stored in a retrieval system, or transmitted,
in any form or by any means, electronic, mechanical, photocopying, microfilming,
recording or otherwise, without written permission from the Publisher

Printed in the United States of America

Preface

The existence of liquid crystals has been known for nearly a century; yet it is only in the last ten years that their unique optical, electrical, electro-optic, and thermal properties have been exploited to any significant extent in such technological applications as digital displays and thermography. Digital watches equipped with liquid-crystal displays (LCD's) have recently made their debut in the electronic watch market, and the large-scale use of LCD's in a variety of other applications requiring reliable, low-power digital displays is imminent. There is good reason to believe that liquid crystals will be the first electro-optic materials to find widespread commercial use. Apart from applications, liquid crystals are unique among the phases of matter. Lurking beneath their garish display of color and texture is a great complexity of physical and chemical interaction that is only now beginning to unfold in the face of a decade-old resurgence in all aspects of liquid-crystal research. RCA Laboratories has participated in this resurgence from its beginning in the early 1960's and at present maintains active liquid-crystal programs both in basic research and in device engineering.

In view of the widespread interest in liquid crystals at RCA Laboratories, an in-house weekly seminar devoted to the subject of liquid crystals was organized in the fall of 1973. The resulting lectures were subsequently published in three issues of the *RCA Review* and, with the incorporation of much additional material, eventually grew into the present volume.

The book is intended as a tutorial introduction to the science and technology of liquid crystals. We believe it will serve as a useful primer for those interested in the physics of liquid crystals and those using or contemplating the use of liquid crystals in practical devices. The book is not meant to be a review of the entire field of liquid crystals, and no attempt has been made to include exhaustive compilations of literature references. Emphasis has been given to areas generally ignored in other texts; specific topics emphasized include the statistical mechanics of the molecular theory and various aspects of device fabrication.

The eighteen chapters in the present volume can be divided into four groups. Chapters 1 and 2 give a brief introduction to the structural and chemical properties of liquid crystals. The next eight chapters develop both the microscopic statistical theories (Chapters 3–7) and the macroscopic continuum theories (Chapters 8–10) of liquid crystals. Chapters 11 through 17 treat various aspects of device applications and related considerations, including packaging; optical, electro-optic, and electro-chemical effects; addressing techniques; and the use of liquid crystals in optical waveguides. The final chapter presents an overview of lyotropic liquid crystals with particular emphasis on the crucial role of the hydrophobic effect in their stability.

Important contributions to the success of this venture were made by several individuals. We wish to express our sincere appreciation for their kind efforts. Roger W. Cohen made the initial suggestion for the seminar series and has given continued encouragement and support during the preparation of the book. George D. Cody was most instrumental in arranging for the publication of the lectures in the *RCA Review* and had the foresight to suggest that they be incorporated in the present volume. Ralph F. Ciafone provided outstanding editorial support during publication of the lectures in the *RCA Review* and also assumed substantial editorial responsibility for many of the tasks involved in putting the lectures together as a book. Mrs. Dorothy C. Beres was especially helpful in typing the manuscripts for the majority of the chapters. Thanks are also due each lecturer-author whose contributions were invaluable in making the seminar series and this resulting volume complete.

E. B. Priestley
Peter J. Wojtowicz
Ping Sheng

Contents

Chapter 1
Liquid Crystal Mesophases ● E. B. Priestley

1. Mesophases 1
1.1 Disordered Crystal Mesophases 2
1.2 Ordered Fluid Mesophases 2
2. Types of Liquid Crystals 3
2.1 Thermotropic Liquid Crystals 3
2.2 Lyotropic Liquid Crystals 3
3. Classification According to Molecular Order 4
3.1 Nematic Order 4
3.2 Cholesteric Order 5
3.3 Smectic Order 7
4. Polymorphism in Thermotropic Liquid Crystals 9
5. Molecular Structure of Thermotropic Mesogens 10
6. Properties of Ordered Fluid Mesophases 12

Chapter 2
Structure – Property Relationships in Thermotropic Organic Liquid Crystals ● Aaron W. Levine

1. Introduction 15
2. Organic Mesophases 16
3. General Structural Features of Mesogens 17
4. Effects of Structure on Mesophase Thermal Stability 18
5. Homologous Series 22
6. Materials for Device Applications 24
7. Summary 27

Chapter 3
Introduction to the Molecular Theory of Nematic Liquid Crystals ● Peter J. Wojtowicz

1. Introduction 31
2. Symmetry and the Order Parameter 32

3. The Molecular Potential 34
4. The Orientational Distribution Function 35
5. Thermodynamics of the Nematic Phase 37
6. Fluctuations at T_c 41

Chapter 4
Generalized Mean Field Theory of Nematic Liquid Crystals ● Peter J. Wojtowicz

1. Introduction 45
2. The Pair Interaction Potential 46
3. The Mean Field Approximation 47
4. Statistical Thermodynamics 51
5. Nature of the Parameters U_L 52
6. The Need for Higher Order Terms in V_1 54

Chapter 5
Hard Rod Model of the Nematic–Isotropic Phase Transition ● Ping Sheng

1. Introduction 59
2. Derivation of Onsager Equations 61
3. Solution of Onsager Equations in a Simplified Case 66

Chapter 6
Nematic Order: The Long Range Orientational Distribution Function ● E. B. Priestley

1. Introduction 71
2. The Orientational Distribution Function 72
3. Macroscopic Definition of Nematic Order 74
4. Relationship Between Microscopic and Macroscopic Order Parameters ... 75
5. Experimental Measurements 77
5.1 Measurements of $\langle P_2(\cos\theta) \rangle$ Based on Macroscopic Anisotropies 77
5.2 Measurements of $\langle P_2(\cos\theta) \rangle$ Based on Microscopic Anisotropies 78
6. Experimental Data 79

Chapter 7
Introduction to the Molecular Theory of Smectic-A Liquid Crystals ● Peter J. Wojtowicz

1. Introduction 83
2. Symmetry, Structure and Order Parameters 84
3. Phase Diagrams 87
4. The Molecular Potential 88
5. Statistical Thermodynamics 91
6. Numerical Results 93
7. Improved Theory 96
8. The Possibility of Second-Order Transitions 99
   Appendix 100

Chapter 8
Introduction to the Elastic Continuum Theory of Liquid Crystals ●
Ping Sheng

1.	Introduction	103
2.	The Fundamental Equation of the Continuum Theory of Liquid Crystals	104
3.	Applications of the Elastic Continuum Theory	110
3.1	Twisted Nematic Cell	110
3.2	Magnetic Coherence Length	112
3.3	Fréedericksz Transition	115
3.4	Field-Induced Cholesteric—Nematic Transition	120
4.	Concluding Remarks	125
	Appendix	126

Chapter 9
Electrohydrodynamic Instabilities in Nematic Liquid Crystals ●
Dietrich Meyerhofer

1.	Introduction	129
2.	Nature of the Instability and the Balance of Forces	131
3.	Dielectric Response	131
4.	Hydrodynamic Effects	132
5.	The Boundary Value Problem in the Conduction Regime	134
6.	The Torque Balance Equation	136
7.	Numerical Results and Comparison with Experiment	139
8.	Range of Applicability	140

Chapter 10
The Landau–de Gennes Theory of Liquid Crystal Phase Transitions ●
Ping Sheng and E. B. Priestley

1.	Introduction	143
2.	Derivation of the Fundamental Equations of the Landau–de Gennes Theory	145
2.1	The Partition Function	145
2.2	The Landau Expansion	150
2.3	Generalization of the Landau Expansion to Liquid Crystals	153
3.	Thermodynamic Properties of Liquid Crystal Phase Transitions	165
4.	Fluctuation Phenomena	168
4.1	Homophase Fluctuations in the Isotropic Phase	169
4.2	Heterophase Fluctuations	182
5.	Observation of Fluctuations Using Light Scattering	189
6.	Magnetic Birefringence and the Paranematic Susceptibility	193
	Appendix A	195
	Appendix B	198

Chapter 11
Introduction to the Optical Properties of Cholesteric and Chiral Nematic Liquid Crystals ● E. B. Priestley

1.	Introduction	203
2.	Maxwell's Equations	205

3.	Discussion	211
4.	Conclusion	215
	Appendix A	216
	Appendix B	216

Chapter 12
Liquid-Crystal Displays—Packaging and Surface Treatments ● L. A. Goodman

1.	Introduction	219
2.	Packaging	219
3.	Electrodes	220
4.	Surface Orientation	222
5.	Influence of Packaging on Surface Orientation	230
6.	Summary	231

Chapter 13
Pressure Effects in Sealed Liquid-Crystal Cells ● Richard Williams

1.	Introduction	235
2.	Effect of Temperature Change	237
3.	Effect of Glass Thickness	238
4.	The Case of a Rigid Container	239

Chapter 14
Liquid-Crystal Displays—Electro-optic Effects and Addressing Techniques ● L. A. Goodman

1.	Introduction	241
2.	Electro-optic Phenomena	242
2.1	Field-Induced Birefringence	242
2.2	Twisted Nematic Effect	245
2.3	Guest–Host Effect	248
2.4	Cholesteric-to-Nematic Transition	249
2.5	Dynamic Scattering	251
2.6	Storage Mode	255
2.7	Transient Response	258
3.	Display-Related Parameters	259
3.1	Display Life	259
3.2	Temperature Dependence	260
4.	Addressing Techniques	261
4.1	Matrix Addressing	261
4.2	Beam Scanning	270
5.	Summary	273

Chapter 15
Liquid-Crystal Optical Waveguides ● D. J. Channin

1.	Introduction	281
2.	Guided Optical Waves	282
3.	Phase Matching and Coupling	286

CONTENTS

4.	Scattering	287
5.	Liquid Crystal Waveguides	288
6.	Conclusions	294

Chapter 16
The Electro-optic Transfer Function in Nematic Liquids ● Alan Sussman

1.	Introduction	297
2.	Geometrical Considerations in Optical Measurements	299
3.	Field Effects—Negative Dielectric Anisotropy	303
4.	Field Effects—Positive Dielectric Anisotropy	305
5.	Hydrodynamic Effects—Diffraction by Domains	308
6.	Dynamic Scattering	312
7.	Photoconductor Control	314

Chapter 17
Electrochemistry in Nematic Liquid-Crystal Solvents ● Alan Sussman

1.	Introduction	319
2.	Equilibrium Properties of Bulk Solutions	320
3.	Electrochemical Reactions	328

Chapter 18
Lyotropic Liquid Crystals and Biological Membranes: The Crucial Role of Water ● Peter J. Wojtowicz

1.	Introduction	333
2.	Lyotropic Liquid Crystals	334
2.1	Constituents of Lyotropics	335
2.2	Micelles	336
2.3	Structure of Lyotropics	338
3.	Biological Membranes	340
3.1	Constituents of Membranes	341
3.2	Structures of Membranes	342
4.	Interaction of Amphiphilic Compounds with Water	344
4.1	Solubility of Hydrocarbons in Water	345
4.2	Solubility of Ionized Species in Water	346
4.3	Aggregation of Amphiphilic Compounds	347
5.	Conclusion	349

Appendix ... 351
Index ... 353

1

Liquid Crystal Mesophases

E. B. Priestley

RCA Laboratories, Princeton, N. J. 08540

1. Mesophases

Everyday experience has led to universal familiarity with substances that undergo a single transition from the solid to the isotropic liquid phase. The melting of ice at 0°C to form liquid water is perhaps the most common such phase transition. There are, however, many organic materials that exhibit more than a single transition in passing from solid to liquid, thereby necessitating the existence of one or more intermediate phases. It is not surprising that the molecular ordering in these intermediate phases, known as "mesophases", lies between that of a solid and that of an isotropic liquid. The partial ordering of the molecules in a given mesophase may be either translational or rotational, or both. Clearly, translational order can be realized regardless of molecular shape, whereas rotational order has meaning only when the constituent molecules are nonspherical (elongated). Thus, there is good reason to expect molecular structure to be an important factor in determining the kind and extent of ordering in any particular mesophase.

Two basically different types of mesophases have been observed. First, there are those that retain a 3-dimensional crystal lattice, but are characterized by substantial rotational disorder (i.e., disordered crystal mesophases), and second, there are those with no lattice, which are therefore fluid, but nevertheless exhibit considerable rotational order (i.e., ordered fluid mesophases). Molecular structure *is* in fact important and, generally speaking, molecules comprising one of these two types of mesophase are distinctly different in shape from molecules comprising the other. Indeed, with the possible exception of some polymorphous smectic materials, there are no known substances that show both disordered crystal and ordered fluid mesophases.

1.1 Disordered Crystal Mesophases

Disordered crystal mesophases are known as "plastic crystals".[1] In most cases plastic crystals are composed of "globular" (i.e. essentially spherical) molecules, for which the barriers to rotation are small relative to the lattice energy. As the temperature of such a material is raised, a point is reached at which the molecules become energetic enough to overcome these rotational energy barriers, but not sufficiently energetic to break up the lattice. The result is a phase in which the molecules are translationally well ordered but rotationally disordered, i.e., a disordered, or plastic crystal. Further increase in the temperature will result eventually in the molecules becoming energetic enough to destroy the lattice, at which point a transition to the isotropic liquid occurs. Perhaps the most striking property of plastic crystal mesophases is the ease with which they may be deformed under stress. It is this softness or "plasticity" that gives these mesophases their name.

A detailed discussion of plastic crystals is outside the scope of this chapter. However, those interested in pursuing the subject further are directed to Ref. [1] which provides an excellent elementary review of the properties of plastic crystals.

1.2 Ordered Fluid Mesophases

Ordered fluid mesophases are commonly called "liquid crystals"[2-6] and are most often composed of elongated molecules. In these mesophases, the molecules show some degree of rotational order (and in some cases partial translational order as well) even though the crystal lattice has been destroyed. Lack of a lattice requires that these mesophases be fluid; they are, however, *ordered* fluid phases. It is this simultaneous possession of liquid-like (fluidity) and solid-like (molecular order) character in a single phase that makes liquid crystals unique and gives rise to so many interesting properties.

In what follows, the various ordered fluid mesophases are described, with particular attention being given to the nature of the molecular ordering in each case. Also, some of the consequences of simultaneous liquid-like and solid-like behavior in a single phase are discussed qualitatively. Later chapters will deal at length with many subjects we can consider only briefly here.

2. Types of Liquid Crystals

Two types of liquid crystal mesophases must be differentiated, viz. thermotropic and lyotropic. Thermotropic liquid crystals are of interest both from the standpoint of basic research and also for applications in electro-optic displays, temperature and pressure sensors, etc. Lyotropic liquid crystals, on the other hand, are of great interest biologically and appear to play an important role in living systems.

2.1 Thermotropic Liquid Crystals

The term "thermotropic" arises because transitions involving these mesophases are most naturally effected by changing temperature. Materials showing thermotropic liquid crystal phases are usually organic substances with molecular structures typified by those of cholesteryl nonanoate and N-(p-methoxybenzylidene)-p'-n-butylaniline (MBBA) shown in Fig. 1. Axial ratios of 4-8 and molecular weights of 200-500 gm/mol are typical for thermotropic liquid crystal mesogens. In this type of liquid crystal, every molecule participates on an equal basis in the long range ordering.

2.2 Lyotropic Liquid Crystals

Solutions of rod-like entities in a normally isotropic solvent often form liquid-crystal phases for sufficiently high solute concentration. These anisotropic solution mesophases are called "lyotropic liquid crystals".[7-9] Although the rod-like entities are usually quite large compared with typical thermotropic liquid-crystal mesogens, their axial ratios are seldom greater than ~15. Deoxyribonucleic acid (DNA), certain viruses (e.g., tobacco mosaic virus (TMV)), and many synthetic polypeptides all form lyotropic mesophases when dissolved in an appropriate solvent (usually water) in suitable concentration. The conformation of most of these materials is quite temperature sensitive, i.e. the rods themselves are rather unstable with respect to temperature changes. This essentially eliminates the possibility of thermally inducing phase transitions involving lyotropic mesophases. A more natural parameter which can be varied to produce such transitions is the solute concentration. The principal interaction producing long range order in lyotropic liquid crystals is the solute–solvent interaction; solute–solute interactions are of secondary importance. To a good approximation, then, only the rod-like entities (solute) participate in the long range ordering.

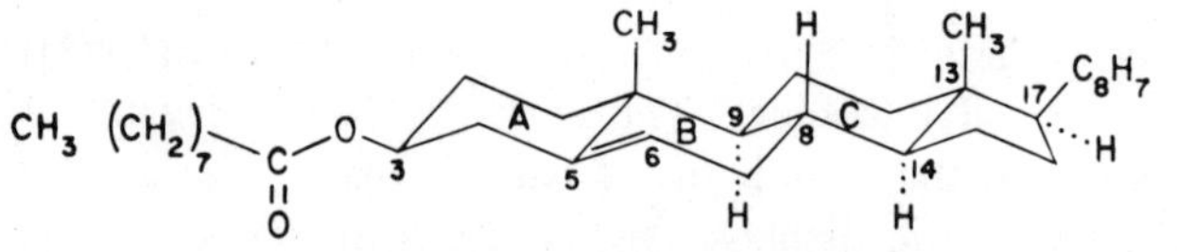

CHOLESTERYL NONANOATE

(a)

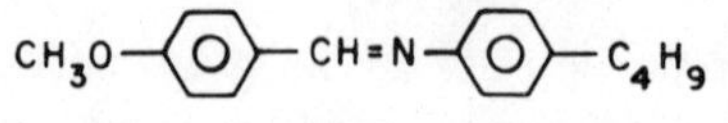

N-(P-METHOXY BENZYLIDENE)-P'-BUTYLANILINE (MBBA)

(b)

Fig. 1—Examples of molecular structures that give rise to thermotropic mesophases.

3. Classification According to Molecular Order

With the distinction between thermotropic and lyotropic mesophases in mind, we proceed to the classification of these mesophases using a scheme based primarily upon their symmetry. This scheme, first proposed by Friedel in 1922,[10] distinguishes three major classes—the nematic, the cholesteric, and the smectic.

3.1 Nematic Order

The molecular order characteristic of nematic liquid crystals is shown schematically in Fig. 2. Two features are immediately apparent from the figure:

(1) There is long range orientational order, i.e., the molecules tend to align parallel to each other.

(2) The nematic phase is fluid, i.e., there is no long range correlation of the molecular center of mass positions.

In the state of thermal equilibrium the nematic phase has symmetry ∞/mm and is therefore uniaxial. The direction of the principal axis n̂ (the director) is arbitrary in space.

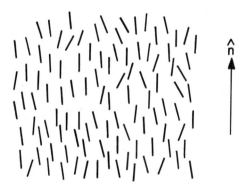

Fig. 2—Schematic representation of nematic order.

3.2 Cholesteric Order

Fig. 3 shows the equilibrium structure of the cholesteric phase. As in the nematic phase, lack of long range translational order imparts fluidity to the cholesteric phase. On a local scale, it is evident that cholesteric and nematic ordering are very similar. However, on a larger scale the cholesteric director n̂ follows a helix of the form

$$n_x = \cos(q_o z + \varphi)$$
$$n_y = \sin(q_o z + \varphi)$$
$$n_z = 0$$

where both the direction of the helix axis z in space and the magnitude of the phase angle ϕ are arbitrary. Thus the structure of a cholesteric liquid crystal is periodic with a spatial period given by

$$L \rightrightarrows \frac{\pi}{|q_o|}.$$

The sign of q_o distinguishes between left and right helicies and its magnitude determines the spatial period. When L is comparable to optical wavelengths, the periodicity results in strong Bragg scattering of light. If the wavelength of the scattered light happens to be in the visible region of the spectrum, the cholesteric phase will appear brightly colored.

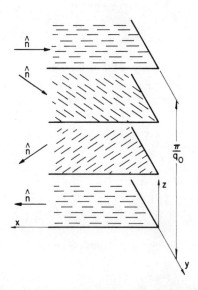

Fig. 3—Schematic representation of cholesteric order.

It is interesting to note that a nematic liquid crystal is really nothing more than a cholesteric with $q_o = 0$ (infinite pitch). In fact, the two are subclasses of the same family, the distinction being whether the equilibrium value of q_o is identically zero, or finite. If the constituent molecules are *optically inactive*, i.e., are superimposable on their mirror image, then the mesophase will be nematic. If, on the other hand, the constituent molecules are *optically active*, i.e., are not superimposable on their mirror image, then the mesophase will be cholesteric (except if the molecule and its mirror image are present in precisely equal amounts, i.e. a "racemic" mixture, in which case the mesophase will again be nematic).

Finally, a comment on nomenclature is in order. Cholesteric liquid crystals get their name historically from the fact that the first materials that were observed to exhibit the characteristic helical structure were esters of cholesterol. It would be useful to continue identifying as "cholesteric" those mesophases whose constituent molecules are derivatives of cholesterol, and to use the term "chiral nematic" to identify mesophases formed by optically active, non-steroidal molecules.

3.3 Smectic Order

As many as eight smectic phases have been tentatively identified; however, except for three that have been reasonably well characterized, considerable uncertainty still exists about the exact nature of the molecular ordering in these phases.

We will discuss only the three best understood smectic phases, the smectic A, C, and B phases. All three appear to have one common feature, viz. one degree of translational ordering, resulting in a layered structure. As a consequence of this partial translational ordering, the smectic phases are much more viscous than either the nematic or cholesteric phase.

Smectic A Order

Within the layers of a smectic A mesophase the molecules are aligned parallel to the layer normal and are uncorrelated with respect to center of mass position, except over very short distances. Thus, the layers are individually fluid, with a substantial probability for inter-layer diffusion as well. The layer thickness, determined from x-ray scattering data, is essentially identical to the full molecular length. At thermal equilibrium the smectic A phase is optically uniaxial due to the infinite-fold rotational symmetry about an axis parallel to the layer normal. A schematic representation of smectic A order is shown in Fig. 4(a).

Smectic C Order

Smectic C order is depicted in Fig. 4(b). X-ray scattering data from several smectic C phases indicates a layer thickness significantly less than the molecular length. This has been interpreted as evidence for a uniform tilting of the molecular axes with respect to the layer normal. The fact that the smectic C phase is optically biaxial is further evidence in support of a tilt angle. Tilt angles of up to 45° have been observed and in some materials the tilt angle has been found to be temperature dependent. As in the smectic A phase, the layers are individually fluid and inter-layer diffusion can occur, although most likely with somewhat lower probability.

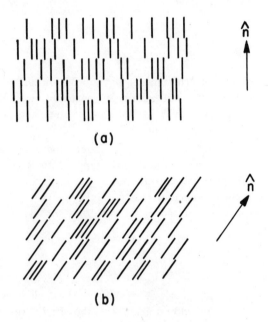

Fig. 4—Schematic representation of two types of smectic order: (a) smectic A order; (b) smectic C order.

Smectic B Order

In addition to the layered structure, x-ray scattering data indicate ordering of the constituent molecules *within* the layers of the smectic B phase. Hence, the layers are no longer fluid, in contrast to the smectic A and C phases. However, the mechanical properties of the smectic B phase are quite different from those that would be expected for a material having full 3-dimensional order; thus, we are forced to conclude that the ordering, whatever its detailed nature, cannot be of the sort familiar in solids. It has been suggested that the smectic B phase may in fact be a plastic crystal.[11] If so, this would be very interesting as it would provide the first opportunity to investigate both disordered-crystal and ordered-fluid mesophases in a single material (for any polymorphous smectic substance having a smectic B phase). This model is not the only one possible for the smectic B phase, however. It could also be that this phase is composed of a collection of 2-dimensional solid layers coupled by very weak forces, such that the layers could slip over one another quite easily.[11] Although, in principle, these two models are experimentally distinguishable, there are no data that differentiate between them at present. Smectic B phases can be either biaxial or uniaxial depending upon whether or not there is a finite tilt angle of the sort discussed above.

4. Polymorphism in Thermotropic Liquid Crystals

Many thermotropic materials have been observed to pass through more than one mesophase between the solid and isotropic liquid phases. Such materials are said to be "polymorphous". One can predict the order of stability of these mesophases on a scale of increasing temperature simply by utilizing the fact that raising the temperature of any material results in progressive destruction of molecular order. Thus, the more ordered the mesophase, the closer in temperature it lies to the solid phase. From the description of the various types of order given in Sec. 3, we can immediately draw the following conclusions:

(1) For a material having nematic and smectic trimorphous phases, the order of mesophase stability with increasing temperature will be
solid → smectic B → smectic C → smectic A → nematic → isotropic
This order of stability is in complete agreement with experimental observation.

(2) For a material having nematic and/or smectic phases, but not all those listed in (1), the order of stability can be obtained from that shown in (1) by simply deleting those phases not present. This is also confirmed experimentally.

(3) For materials having both cholesteric and smectic mesophases, the order is identical to that shown in (1) except the word "nematic" is replaced by "cholesteric". The following have been experimentally observed:

(a) solid → cholesteric → isotropic

(b) solid → smectic A → cholesteric → isotropic.

Finally, we note that there are no known examples of polymorphism involving both nematic and cholesteric mesophases (recall the discussion of molecular symmetry in Sec. 3.2). Of course, in the presence of an external electric or magnetic field, a cholesteric liquid crystal can be forced into a nematic structure. Such field induced distortions have been studied extensively.

5. Molecular Structure of Thermotropic Mesogens

At present there is no way of predicting *with certainty* whether or not a given molecule will exhibit liquid-crystal mesophases. However, the presence of common structural features in the majority of thermotropic liquid-crystal mesogens makes possible certain generalizations regarding the types of molecules most likely to show liquid-crystalline behavior. The two structural features that appear essential are (1) the constituent molecules must be elongated and (2) they must be rigid. It should be borne in mind that these are *only* generalizations, and exceptions do exist.

We have already encountered one group of structurally related compounds that often form liquid-crystal phases, viz. the substituted cholesterols. It is evident from Fig. 1(a) that these molecules are elongated and quite rigid, thereby satisfying the two criteria above. As mentioned earlier, all mesophases produced by molecules of this type have finite pitch due to the optical activity of the parent cholesterol molecule.

As a second group, we consider molecules whose structures can be represented schematically as

with $n = 0$, 1, or 2. Many of these molecules exhibit nematic and smectic mesophases and, if optically active, may give rise to chiral nematic mesophases. It is apparent even from this crude representation that the requirement of elongated molecules is satisfied by this structure. The need for rigidity is satisfied, in general, by restricting the linkage groups to those containing multiple bonds. Fig. 5 shows three nematogenic molecules that serve to illustrate the structure for $n = 0$, 1, and 2, respectively. In the p-pentyl-p'-cyanobiphenyl (PCB) molecule, the biphenyl is considered as a single aromatic group. Thus $n = 0$, there are no linkage groups, and the substituents are $-C_5H_{11}$ and $-CN$. The p-azoxyanisole (PAA) molecule illustrates the case for n = 1. Its two phenyl aromatic groups are linked by an azoxy group

LIQUID CRYSTAL MESOPHASES

and both substituents are CH_3O-. Notice the double bond between the nitrogens of the linkage group. An example of the $n = 2$ case is provided by the 2,6-di-(p-methoxybenzylideneamino)-naphthalene molecule. The aromatic groups are phenyl, naphthyl and phenyl respectively, with two $-CH=N-$ linkage groups and two CH_3O- substituents. Again the linkage groups contain double bonds to provide the requisite rigidity. Brown et al[2] list many other aromatic linkage, and substituent groups that can be combined in the manner indicated to form potentially mesomorphic molecules.

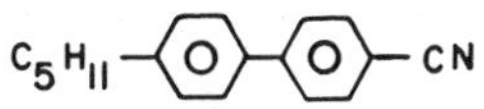

C_5H_{11}—⟨O⟩—⟨O⟩—CN

P–PENTYL–P′–CYANOBIPHENYL (PCB)

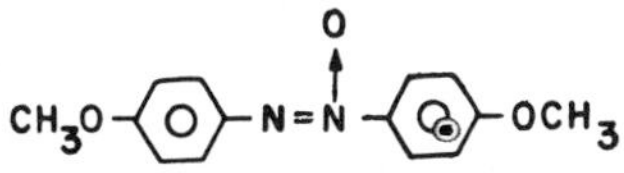

CH_3O—⟨O⟩—N=N—⟨O⟩—OCH_3
           ↑
           O

P–AZOXYANISOLE (PAA)

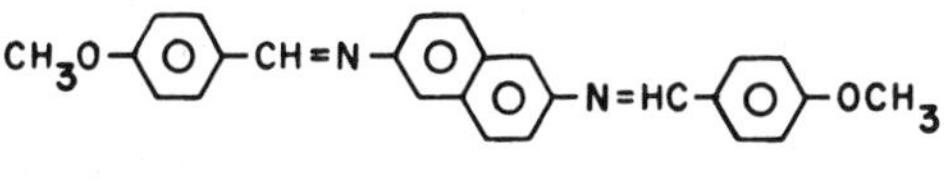

CH_3O—⟨O⟩—CH=N—⟨OO⟩—N=HC—⟨O⟩—OCH_3

2,6–DI–(P–METHOXY BENZYLIDENEAMINO)–NAPHTHALENE

Fig. 5—Examples of themotropic nematogens.

A word of caution is in order at this point. This purely "mechanical" approach to "organic synthesis" is not to be taken too seriously. It has many obvious shortcomings, not the least of which is its total disregard for the laws of chemical combination. In spite of this, one can very quickly write down a great many structures that do not violate any chemical principles and that have a good chance of showing liquid crystal mesophases. However, one should not be at all surprised by the appearance of the occasional "dud".

Finally, there are other classes of materials, such as the substituted monocarboxylic acids, that also form liquid-crystal mesophases. However, they represent relatively few of the known thermotropic mesogens. For more details concerning the relationship of molecular structure to liquid crystallinity and for a discussion of these other classes of mesomorphic materials, the reader is referred to the following chapter[12] and references therein.

6. Properties of Ordered Fluid Mesophases

The combination of molecular order and fluidity in a single phase results in several remarkable properties unique to liquid crystals. By now, it is quite evident that the constituent molecules of liquid crystal mesophases are structurally very anisotropic. Because of this shape anisotropy, all the molecular response functions, such as the electronic polarizability, are anisotropic. The long range order in the liquid-crystal phases prevents this molecular anisotropy from being completely averaged to zero, so that all the *macroscopic* response functions of the bulk material, such as the dielectric constant, are anisotropic as well. We have, therefore, a flexible fluid medium whose response to external perturbations is anisotropic.

Consider the dielectric tensor $\epsilon_{\alpha\beta}$. In the uniaxial nematic phase (chosing the z-axis parallel to the nematic axis), $\epsilon_{\alpha\beta}$ has the form

$$\epsilon_{\alpha\beta} = \begin{pmatrix} \epsilon_\perp & 0 & 0 \\ 0 & \epsilon_\perp & 0 \\ 0 & 0 & \epsilon_\parallel \end{pmatrix}$$

with an anisotropy defined by $\Delta\epsilon = \epsilon_\parallel - \epsilon_\perp$. Here $\epsilon_\parallel$ and $\epsilon_\perp$ refer to the dielectric constant parallel and perpendicular to the nematic axis, respectively. The response of a nematic liquid crystal to an external electric field depends on both the sign and magnitude of $\Delta\epsilon$. If $\Delta\epsilon$ is positive, the lowest energy state in the presence of the field will be that in which the nematic axis lies parallel to the field. For a negative $\Delta\epsilon$, the lowest energy state will be that in which the nematic axis is perpendicular to the field. Due to the fluid nature of the phase, the field strength necessary to cause such realignment of the nematic axis is not very large. This ability to control the orientation of the nematic axis by means of weak external fields is the basis for several applications of nematic liquid crystals in optical display devices.

Cholesteric liquid crystals have also been put to practical use. In this case, it is primarily the sensitivity of the pitch to changes in temperature, pressure, etc. that are of interest. Recall that when the pitch of a cholesteric is equal to an optical wavelength, Bragg scattering occurs. It is evident that by choosing a cholesteric of appropriate pitch, changes in temperature and pressure can be monitored by means of the accompanying color change of the material. Cholesteric liquid crystals are especially useful in applications where large-area temperature or pressure profiles must be determined.

Largely because of their high viscosity, smectic liquid crystals have not come into widespread use. While it seems unlikely they will be useful in any application where speed of response is important, they may be valuable as storage media.[13]

References

[1] J. G. Aston, "Plastic Crystals," in **Physics and Chemistry of the Organic Solid State**, pp. 543-583, ed. by D. Fox, M. M. Labes, and A. Weissberger, Interscience Pub., N.Y., N.Y. (1963).
[2] G. H. Brown, J. W. Doane, and V. D. Neff, **A Review of the Structure and Physical Properties of Liquid Crystals**, CRC Press, Cleveland, Ohio (1971).
[3] G. Durand and J. D. Litster, "Recent Advances in Liquid Crystals," in **Annual Reviews of Materials Science**, Vol. 3, pp. 269-292, ed. by R. A. Huggins, Annual Reviews, Inc., Palo Alto, Calif. (1973).
[4] I. G. Chistyakov, "Liquid Crystals," **Sov. Phys. Usp.**, Vol. 9, p. 551 (1967).
[5] A. Saupe, "Recent Results in the Field of Liquid Crystals," **Angew. Chem. Int. Ed.** (English), Vol. 7, p. 97 (1968).
[6] G. W. Gray, **Molecular Structure and the Properties of Liquid Crystals**, Academic Press, N.Y., N.Y. (1962).
[7] A. S. C. Lawrence, "Lyotropic Mesomorphism in Lipid-Water Systems," **Mol. Cryst. Liquid Cryst.**, Vol. 7, p. 1 (1969).
[8] P. A. Winsor, "Binary and Multicomponent Solutions of Amphiphilic Compounds," **Chem. Rev.**, Vol. 68, p. 1 (1968).
[9] P. Ekwall, L. Mandell, and K. Fontell, "Solubilization in Micelles and Mesophases and the Transition from Normal to Reversed Structures," **Mol. Cryst. Liquid Cryst.**, Vol. 8, p. 157 (1969).
[10] G. Friedel, "Les états Mésomorphes de la Matière," **Ann. de Physique**, Vol. 18, p. 273 (1922).
[11] P. G. de Gennes, "Some Remarks on the Polymorphism of Smectics," **Mol. Cryst. Liquid Cryst.**, Vol. 21, p. 49 (1973).
[12] A. W. Levine, "Structure-Property Relationships in Thermotropic Liquid Crystals," Chapter 2.
[13] F. J. Kahn, "IR-Laser-Addressed Thermo-Optic Smectic Liquid-Crystal Storage Displays," **Appl. Phys. Lett.**, Vol. 22, p. 111 (1973).

2

Structure-Property Relationships in Thermotropic Organic Liquid Crystals

Aaron W. Levine
RCA Laboratories, Princeton, N. J. 08540

1. Introduction

The phenomenon of thermotropic liquid crystallinity has been known[1] at least since 1888. Since the early observations by Reinitzer[1,2] and Lehmann[3] of unusual melting behavior in certain organic compounds, mesogenic compounds have been both actively sought and incidentally discovered. Several relatively recent discoveries[4-9] have shown the technological utility of organic mesophases. Substantial impetus has thus been provided for the systematic investigation of the relationships between molecular structure and liquid crystallinity.

The literature of organic liquid crystals has been reviewed several times.[10-13] Reviews of those properties and compounds usable in some electro-optical devices have also appeared.[14] The intention of this chapter is to present a brief summary of thermotropic mesophasic behavior in organic compounds as related to molecular structure. Recent activity in this field will be indicated in an attempt to show current trends in theory and experimentation. The cited reviews and the more recent symposia reprints and abstracts[15] are commended to the interested readers.

2. Organic Mesophases

Organic liquid crystals may be broadly classified as smectic, nematic. or cholesteric. These states are similar in that they exist between the fully crystallized solid state and the isotropic melt and differ in the extent of lattice order preserved in the mesophase. In the smectic phase, molecules are constrained to be parallel with their neighbors in layers, and translational motion between these layers is of low probability. In the nematic phase, molecules are able to translate in any direction with respect to their immediate neighbors but are still constrained to be parallel with them along the "nematic director". When an otherwise nematogenic molecule also possesses chirality, a twisting of the nematic director occurs and the resulting structure is termed cholesteric. This phase, so named because it is frequently encountered in derivatives of cholesterol, possesses some unique properties as a result of its twist but can, within limits, be considered as a nematic phase for purposes of structural arguments. Fig. 1 illustrates these three types of ordering.

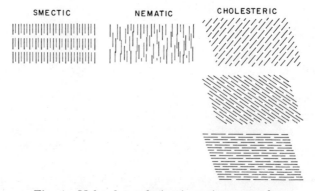

Fig. 1—Molecular ordering in various mesophases.

The molecules in Fig. 1 are illustrated as being rods. When one considers that the various mesophases are differentiated from each other, and from the isotropic liquid, by the degree of freedom of molecular motion, it is reasonable to hypothesize that intermolecular attractive forces play an important role in mesophase stability. For a given level of molecular attractions, long range ordering of molecules would clearly be encountered more often as the molecules become less spherical in shape. Thus, one would anticipate that molecular associations capable of withstanding temperatures higher than the crystalline melting point, i.e., mesophases, would be more frequently observed for molecules of rod-like shape. In addition to a molecular shape that is considerably longer than it is wide (linear molecules), mesophasic thermal stability is favored by molecular rigidity, permanent dipoles within the molecules, and a high level of molecular polarizability. Nearly all mesogenic compounds contain multiple bonds along their long axes, aromatic nuclei, and either polar or long-chain terminal groups.

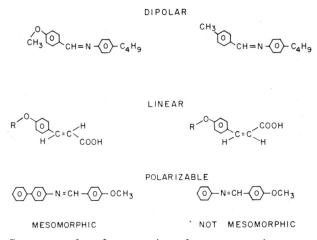

Fig. 2—Some examples of mesogenic and non-mesogenic compounds illustrating some general structural requirements.

Fig. 2 illustrates these points with some well-known mesogens and closely related compounds whose melting behavior is normal. Note that, in this chapter, the term melting point is used to refer to the solid-to-mesomorphic (or isotropic) transition while the abbreviations C, S, N, Ch, and I will be used when referring to transitions involving the crystalline, smectic, nematic, cholesteric, and isotropic states, respectively. Thus, the melting point of a nematogen refers to the C-N transition while N-I will describe the nematic clearing transition.

3. General Structural Features of Mesogens

The vast majority of compounds exhibiting liquid crystalline properties may be regarded as possessing a central linkage and end groups (structure 1). When the central linkage is small, such as —C=N— or —C≡C—, aromatic nuclei are nearly always present. Mesogens containing both carbocyclic and heterocyclic aromatic residues are known

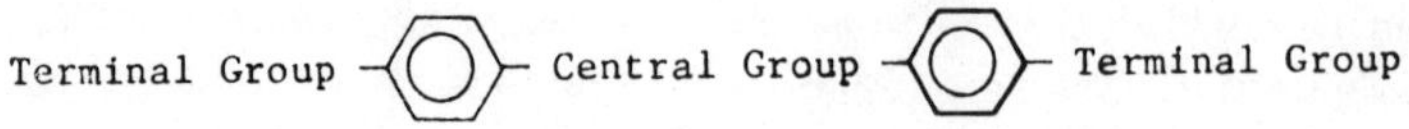

Terminal Group —⌬— Central Group —⌬— Terminal Group

Structure 1

and, in a sufficiently long molecule, substitution other than at *para* positions is permissible. Central groups that require no highly polarizable aromatic residues are exemplified by the Δ^5-steroid ring system and the carboxylic acid dimers. Fig. 3 shows the more common central linkages.

acid dimer	$-CH=CH-$ olefin	$-C\equiv C-$ acetylene
Δ^5-steroid	$-CH=N-$ azomethine (Schiff's base)	$-CH=\overset{\downarrow}{N}-$ O nitrone
$-N=N-$ azo	$-N\overset{\downarrow}{=}N-$ O azoxy	$-\underset{\parallel}{C}-O-$ O ester

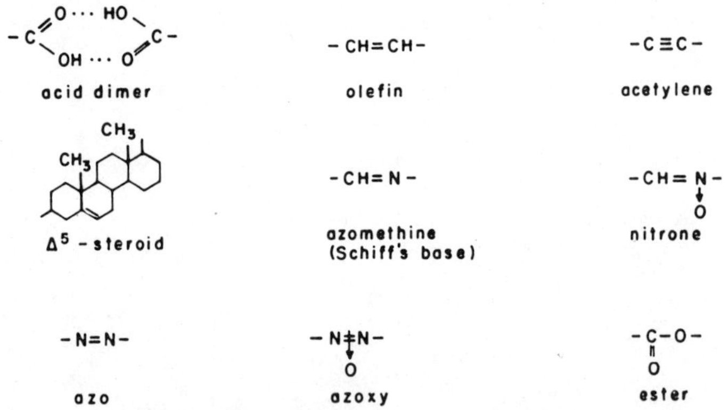

Fig. 3—Common central linkages of mesogenic compounds of Structure 1.

Terminal groups vary widely in chemical nature. Fig. 4 presents those most frequently encountered. The terminal group frequently determines which mesophase will be observed. This will be discussed in detail in the section on homologous series.

4. Effect of Structure on Mesophase Thermal Stability

Of the transitions observed upon melting a liquid crystalline material, the melting point is least correlated with molecular structure. This is so because the thermal stability of the solid state is principally influenced by attractive forces in the crystal lattice. Since these short-range associations are only secondarily related to molecular structure, the melting point is not usually influenced in an obvious manner by molecular structure.

The forces responsible for the mesophase, however, are primarily dipole-dipole (both permanent and induced) and dispersion forces, which, since the molecules in the mesophase or melt are free to rotate, should be correlated with molecular structure. If the mesophase is to survive a vibrational excitation that has already exceeded the lattice energy, intermolecular polar attractions are required. In the following considerations, the mesomorphic thermal stability will be used to compare the relative efficiency of various structures in stabilizing a mesophase. It is important to understand that we are refering here to the maximum temperature at which a particular liquid crystalline

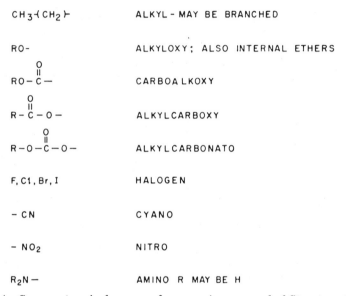

Fig. 4—Common terminal groups of mesogenic compound of Structure 1.

behavior is observed (i.e., the S-N, S-I, N-I, S-Ch, and Ch-I transitions). Thermal stability is not to be confused with the mesomorphic range, which is the difference between the isotropic transition and melting temperatures.

For a mesophase to obtain, it is generally necessary that the terminal groups of the molecules contain permanent dipoles. For example, the N-I transition of 4-octyloxybenzoic acid[16] is some 40° higher than that of 4-nonylbenzoic acid[17] despite the essentially equivalent size of the two molecules. Similarly, the 4-alkoxy-4'-cyanobiphenyls form substantially more stable mesophases than 4-alkyl-4'-cyanobiphenyls.[18] Polarity of the termini, however, frequently gives rise to very strong intermolecular attractions. In such cases, the melting points of the compound may be raised so high that the mesophase cannot survive.

For example, the end-group polarity of 6-hydroxy-2-naphthoic acid (C−I=250°) is certainly greater than that of 6-methoxy-2-naphthoic acid (C−N=206°; N−I=219°).[19] In this case, hydrogen bonding is thought to be responsible for the nonmesogenic property of the hydroxy acid.

Frequently, more than one mesophase occurs upon heating a compound. Under these conditions, not just the presence of polar interactions but their direction is important. Thus, smectic states are stabilized by multiple dipoles acting transverse to the molecular axis (lateral attractions), while terminal attractions appear to be more important in determining nematic thermal stability. It is often found that the lower members of homologous series are purely or predominantly nematic, while in the higher members, the smectic state encroaches on, and eventually prevents observation of, the nematic. The ratio of lateral to terminal attractions increases for an homologous series due to increased dispersion forces and to increased shielding of terminal dipolar attractions of, for example, a normal alkoxy or acyloxy terminal group.

An additional difficulty in assessing the effects of end-group polarity arises when compounds with very differently sized termini are compared. Since it is usually difficult to change polarity substantially without also changing the size of the polar substituent, great care is necessary to predict mesophase stability. Thus, Brown and Shaw[11] point out that N-I temperatures generally fall if the polarity of the termini increase for a given molecular structure other than carboxylic acid dimers. Aaron, Byron, and Gray[20] found, however, that for anils of general structure 2, any substituent X provides a more thermally stable nematic state than does X=H.

$$X-\bigcirc-CH=N-\bigcirc-\bigcirc-OCH_3$$

Structure 2

The polarity and polarizability of the central groups of mesogenic compounds are more clearly correlated with thermal stability. Thus, 4-alkoxy-4′-biphenyl carboxylic acids[21] form more stable mesophases than corresponding 4-alkoxybenzoic acids.[16] Similarly, esters of cholesterol form generally more thermally stable mesophases than esters of cholestanol.[22] These differ only by the presence of a double bond, as shown in Fig. 5. Again, certain conjugated unsaturated aliphatic carboxylic acids show liquid crystalline phases,[23] while normal aliphatic acids are not mesogenic. For compounds based on central linkages that allow approximate linearity, such as azo or azomethine, mesophase thermal stability generally increases with increasing polarity or polarizability of the central linkage. Thus, N,N'-diarylazo nematogens clear some 9° higher than corresponding Schiff bases.[24] Oxidation of both classes of compounds (to azoxy and nitrone, respectively) raises the N-I points by substantial amounts.

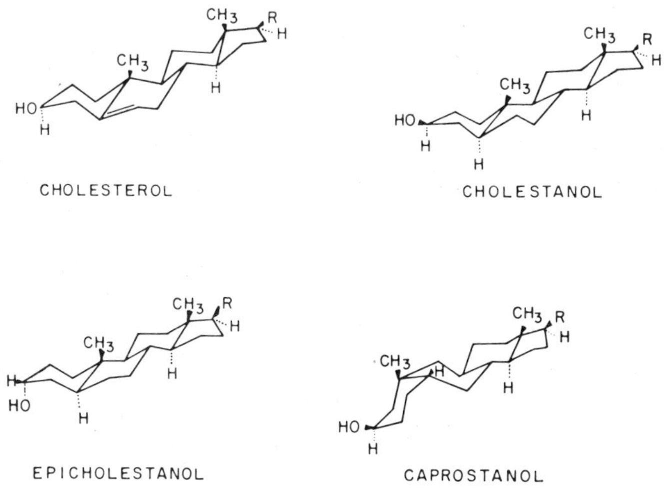

Fig. 5—Ring structures for steroid derivatives discussed in text.

When the central linkage does not provide a linear molecule, however, mesophase stability suffers. Mesophases exhibited by derivatives of phenyl benzoate, for example, are generally some 35° less stable than those of correspondingly substituted azobenzenes.[24] The effect of linearity, or at least planarity, of molecules on mesophase stability is further evidenced on comparison of *trans*-4-alkoxycinnamic acids[25] and *trans*-stilbenes[26] with their corresponding *cis* isomers, which demonstrates that severely nonlinear molecules show no mesomorphic properties. Again, derivatives of epicholestanol[22] and caprostanol (Fig. 5) are not known to be mesogenic.

The geometrical anisotropy of mesogenic compounds is certainly their most obvious common feature. Since lateral attractions play an important role in mesophase formation and stability, the effect of molecular breadth has been studied rather extensively. Generally, smectic states suffer more from molecular broadening than do nematics.

Of course, when a substituent is added that changes both molecular breadth and dipole attractions, a balancing of forces is observed, as in the 5-substituted-6-alkoxy-2-naphthoic acids[19] where broadening is partially attenuated due to molecular geometry. Even more unusual is the effect of addition of a 3-nitro group to 4-amino-4′ nitro-p-terphenyl which causes liquid crystallinity by changing terminal-group interactions.[27] A similar effect is known for stilbene derivatives[28] and several examples of Schiff bases exhibiting increased nematic thermal stability as a result of 2-hydroxylation have been reported.[29] By comparing the thermal stabilities of the cholesteric phase of 2-, 4-, and 2, 4-substituted benzoate esters of cholesterol, a marked effect of broadening is seen.[22, 30] This is perhaps surprising considering the size and conformation of these compounds.

As the previous discussion has indicated, despite the intuitive feeling that mesophase thermal stability should be correlated with chemical structure, the correlations are less than crystal clear. When large changes are made in the structure of a molecule in order to study polar effects, for example, other changes frequently occur that reinforce or counterbalance the particular effect under study. Even so small a change as reversing the termini of 4,4′-disubstituted benzylideneanilines causes some drastic changes in melting behavior.[24, 31]

5. Homologous Series

One of the more common types of recent investigation into structure-property relationships has been the synthesis of groups of molecules differing from one another only by the number of methylene groups in a terminal substituent. There appear to be two principal motivations for this type of study. First, the search for technologically useful liquid crystals is usually directed toward materials of specific nematic temperature ranges. The present lack of ability to predict such ranges, coupled with the usual availability of general syntheses for homologous series, often makes this type of investigation fruitful. To date, few, if any, pure liquid crystals have been found that are usable in display applications. Thus, all device manufacturing utilizes mixtures of mesogenic compounds. In many cases, these mixtures are based on homologous series in order to obtain eutectic mixtures or to guard against deleterious effects of material instabilities such as Schiff base trans-substitution.[32]

The second reason for studying homologous series is more fundamental. Most structure-property studies involving changes in polarity of groups, polarizability, and molecular geometry, require substantial differences to be present among a collection of compounds. It is true, for example, that a methyl, a cyano, and a nitro terminal group impart very different polarity to a given molecule. The fact that these groups are also quite different in size and shape, however, makes interpretation of structural effects difficult. In proceeding along an homologous series, only very small changes are made at each step. It is thus hoped that more specific conclusions would be possible when the effects on mesophase behavior are correlated with molecular size and shape.

It was Gray who first recognized[10] that if the temperatures of a mesophase-to-isotropic transition were plotted against the length of the hydrocarbon chain being homologated, a smooth curve could be drawn through the points. The same seemed true for mesophase-to-mesophase transition temperatures. Of some seventy series available at that time, seven general types of curves were found that could be used to correlate all of the series. The curves differ from one another in significant ways but some general conclusions are possible as enumerated below. It should be noted that subsequently reported homologous series[33] also can be correlated by one or another of the general curves of Gray, although, again, substantial differences in detail of the fit among series that fit the same curve type are seen.

(1) The mesophase-to-isotropic transition temperature usually falls with increasing chain length. A few exceptions to this rule are known among compounds containing a biphenyl moiety which, due to substitution, is restricted from adopting a coplanar conformation. When the two aromatic nuclei are not coplanar, it is possible for the termini of one molecule to interact strongly with one of the rings of another molecule, causing additional attractions and a more thermally stable nematic state as the terminus grows in length. A similar argument has been proposed for the appearance of mesomorphism in certain Schiff bases.[34]

(2) When more than one mesophase is possible, the smectic state increases in thermal stability, at the expense of the nematic or cholesteric phase, as chain length increases. It is often seen that, while the lower homologues are purely nematic, the higher are purely smectic and the intermediate homologues exhibit both mesophases.

(3) Single smooth curves are usually not available to fit all of the points. It is most often found that transition temperatures for compounds with an odd number of carbon atoms in the chain are correlated by one smooth curve, while homologues with even numbers of carbon atoms fit another smooth curve. The transition temperature difference between adjacent odd and even members of the series may be only a few degrees or several tens of degrees depending on the

particular series and the length of the chain. The odd and even curves frequently converge at chain lengths of 8-10 carbon atoms but this is not a necessary condition. This so-called "odd-even effect" is usually justified in terms of a variation in conformation and, therefore, attractive forces as the methylene chain grows.

6. Materials for Device Applications

With the exception of cholesteric liquid crystal temperature indicators and a few devices using smectic fluids,[9,35] all liquid crystal displays rely on the anisotropic properties of nematic liquids or mixtures of nematics and cholesterics. In general, the material in a device is uniformly aligned in such a way that, upon application of an electric field, the dielectric anisotropy of the liquid crystal material causes it to become reoriented. Since light propagating in the direction of the long molecular axes of nematic liquid crystals is unaffected while light propagating in the transverse directions may be, the reorientation of molecules will be detectable if the original alignment was properly chosen. This is covered in detail in a number of papers[4,6-8,36-40] and is summarized in Table 1. For the present purpose, it is only necessary to appreciate the role of dielectric anisotropy in determining the suitability of various liquid crystals for particular types of displays.

Dielectric anisotropy is defined as the difference of the dielectric constants in the directions parallel and perpendicular to the long molecular axis, i.e., $\Delta\epsilon = \epsilon_\parallel - \epsilon_\perp$. A linear molecule containing an on-axis polar group would have positive dielectric anisotropy ($\Delta\epsilon > 0$). Since few mesogens are entirely linear, most have a small perpendicular dielectric constant. Thus, in the absence of a strong dipole along the molecular axis, Schiff bases, esters, stilbenes, acid dimers, etc., exhibit small negative dielectric anisotropy, in the range 0 to -1. Molecules with substantial off-axis dipolar groups, such as α-substituted stilbenes,[41] exhibit correspondingly larger negative values of $\Delta\epsilon$. Diaryl azo nematogens are apparently more linear than corresponding diaryl Schiff bases and frequently exhibit positive dielectric anisotropy.[42] Even azoxy compounds which would appear to have a large off-axis dipole ($N\overset{\overset{O}{\uparrow}}{=}N$) can occasionally have[42] $\Delta\epsilon > 0$ although the more frequently exhibited property for such compounds is negative dielectric anisotropy.

If a nematogen is substituted with a strongly dipolar moiety along its major axis, positive anisotropy is ordinarily obtained. Thus, while 4-ethoxybenzylidene-4'-n-butyl aniline (EBBA) has weak negative anisotropy, 4-ethoxybenzylidene-4'-cyano aniline (PEBAB) has a strongly positive dielectric anisotropy. The cyano group ($-C\equiv N:$) is the most commonly employed substituent for this purpose, and most molecules bearing an on-axis CN substituent have $\Delta\epsilon \sim +10$ or more. Other substituents can be used, and halogen,

Table 1—Summary of Important Nematic Liquid Devices

Device Type	Phenomenon	Original Alignment	Observable Effect Field off → on	Polarizers Required	Materials Considerations	Material Example
Dynamic scattering	Hydrodynamic turbulence	Perpendicular or parallel	Clear → scattering (white)	0	Negative Δε conductivity required	p-Azoxyanisole
Deformation of aligned phases	Reorientation	Perpendicular	Change in polarization	2	Negative Δε	MBBA[a]
Reflective storage mode	Production of stable scattering centers	Perpendicular or parallel	Clear ↔ scattering either storable with field = 0	0	Negative Δε + cholesteric	MBBA + cholesteryl oleate
Twisted nematic	Reorientation	Parallel	Change in polarization	2	Positive Δε	MBBA + PEBAB
Guest–host	Reorientation	Parallel	Color → colorless	1	Positive Δε pleochroic dye added	MBBA + PEBAB + indophenol blue
Field-induced phase change	Cholesteric ↔ nematic	—	Scattering ↔ clear	0	Positive Δε cholesteric added	MBBA + PEBAB + cholesteryl oleyl carbonate

[a] 4-Methoxybenzylidene-4'-n-butyl aniline.

nitro, and dialkylamino have been tried. There is one report of stable nematogens with on-axis hydroxyl groups,[43] but measurements of the anisotropy were not made.

Dielectric anisotropy is an additive molar property. Thus, a small amount of PEBAB [$\Delta\epsilon \sim 10$] (about 10–15 mol %) dissolved in MBBA [$\Delta\epsilon \sim -0.2$] will provide a material suitable for twisted nematic devices. The threshold will, of course, be higher for this mixture than for a pure positive one such as 4-pentyl-4'-cyanobiphenyl,[44,45] where the dielectric anisotropy is much higher. There are other influences on the threshold voltage for liquid crystal cells, principally the materials' elastic constants and, in the case of dynamic scattering, material viscosity. The response times also are dependent upon elastic constants, viscosity, and dielectric anisotropy. These factors are discussed at length in a review by Goodman.[46]

Probably the single most important consideration for a useful liquid crystal material is its nematic (or cholesteric) temperature range. Since the devices under consideration are operative only when a mesophase is present, a reasonably wide range is required if a commercial product is to be made. Most devices currently sold have nematic ranges of about 60–70°C and NI transition temperatures of 60–70°C. While this is satisfactory for now, devices of the future, particularly those intended for outdoor applications or for use by the military, will require broader ranges. Since no single material yet discovered has a satisfactory nematic range for device applications, all currently marketed devices contain mixtures of liquid crystals. Probably the most common mixture is the eutectic of 4-methoxybenzylidene-4'-butylaniline (MBBA) and 4-ethoxybenzylidene-4'-butyl aniline (EBBA).[47] This eutectic has been variously observed to have CN of $-10°C$,[48] $0°C$,[47] and higher.[49] It does have a strong tendency to supercool, and is certainly usable over the range 5–60°C. Many commercial dynamic scattering devices contain this mixture, and modifications of it with both positive and negative dielectric anisotropy additives are also in common use.

The problem of preparing new useful mixtures is a complex one, even if consideration is given only to the nematic range. Trial-and-error attempts to find eutectics in other homologous series or using compounds of differing molecular structure have met with reasonable success.[14,33,44] With the growing number of liquid crystals available for making mixtures, combined with the need for mixtures designed for specific purposes, an improved method of obtaining eutectics has been sought. Some indication of the need for a method to predict the nematic ranges of mixtures is seen when one considers that, from 100 individual compounds, 79,375,395 unique eutectic mixtures are possible, if only 2-, 3-, 4-, and 5-component mixtures are considered. Each of these has an unending number of noneutectic compositions and it is thus easy to see how the probability of finding the few required mixtures by trial and error rapidly becomes infinitesimal, even for the most prolific laboratory chemist.

The relationship between clearing temperature of a nematic mixture and the NI temperatures of its components is closely represented by Eq. [1],

$$\text{NI}_{\text{mixture}} = \sum_i \chi_i \text{NI}_i \qquad [1]$$

where NI is in degrees centigrade and χ represents mole fraction. There have been occasional observations of significance deviations from this relationship, but they are relatively few. Predicting the CN temperature of a mixture is quite a bit less straightforward. To date, the best approximation reported utilizes the Schroeder–Van Laar equation

$$\ln \chi_i = [-\Delta H_i/R][(1/T_E) - (1/T_i)] \qquad [2]$$

in which the eutectic melting point, T_E, is related to the mole fractions, χ_i, CN temperatures, T_i, and molar enthalpies of fusion, ΔH_i, of the components of the mixture, and the gas constant, R. For an N-component mixture, N equations of the form of Eq. [2] may be written and solved simultaneously. Several authors have reported[50-53] moderate to very good success in using this method to predict CN temperatures of mixtures of up to five components, provided the components are structurally similar.

An alternative method of solution[54] applies the fact that the sum of the mole fractions in any mixture must be unity. Making use of this relationship and solving Eq. [2] for mole fraction, Eq. [3] is obtained:

$$\sum_i \{[\Delta H_i/R] \exp[(1/T_i) - (1/T_E)]\} - 1 = 0 \qquad [3]$$

The solution of Eq. [3] by the Newton–Raphson method[55] is readily programed and runs very quickly on a digital computer, provided the initial guess for T_E is chosen with care. This method allows rapid cursory evaluation of the potential mixture utility of new compounds.

While this type of prediction is quite useful, it can fail badly when complex mixtures of structurally different molecules are considered. Even in such cases, however, the success rate in finding useful mixtures is much better than that resulting from undirected experimentation. In addition, the method also provides data useful in improving the model upon which the approximation is based. It may be expected that, in the near future, a more exact prediction of mixture properties will be possible.

7. Summary

Organic compounds that form mesophases upon melting are characterized by being long, narrow, linear molecules. Both permanent dipoles and polarizable moieties are required. The thermal stability of the mesophase or mesophases formed depends in large measure on subtle structural, steric, and electronic effects in the central and terminal groups. Structure-property relationships have been investigated by systematic variation in polarity, polarizability, molecular size, and nature of the central linkage and termini. Some general principles have evolved but the field must not yet be considered fully understood.

Liquid crystal materials for device applications are mixtures, usually of eutectic composition. Preparation of such mixtures requires consideration of dielectric anisotropy, viscosity, and nematic range as principal properties. A method of estimating nematic ranges of mixtures is available, but requires modification.

References

[1] F. Reinitzer, **Monatsh.**, Vol. 9, p. 421 (1888).
[2] F. Reinitzer, "History of Liquid Crystals," **Ann. Physik.**, Vol. 27, p. 213 (1908).
[3] O. Lehmann, "Liquid Crystals," **Ber.**, Vol. 41, p. 3774 (1908).
[4] G. H. Heilmeier, L. A. Zanoni, and L. A. Barton, "Dynamic Scattering: A New Electro-optic Effect in Certain Classes of Nematic Liquid Crystals," **Proc. IEEE**, Vol. 56, p. 1162 (1968).
[5] G. H. Heilmeier, J. A. Castellano, and L. A. Zanoni, "Guest–Host Interactions in Nematic Liquid Crystals," **Mol. Cryst. Liq. Cryst.**, Vol. 8, p. 293 (1969).
[6] G. H. Heilmeier and J. E. Goldmacher, "A New Electric Field Controlled Reflective Optical Storage Effect in Mixed Liquid Crystal Systems," **Proc. IEEE**, Vol. 57, p. 34 (1969).
[7] G. H. Heilmeier and J. E. Goldmacher, "Electric-Field-Induced Cholesteric-Nematic Phase Change in Liquid Crystals," **J. Chem. Phys.**, Vol. 51. p. 1258 (1969).
[8] M. Schadt and W. Helfrich, "Voltage Dependent Optical Activity of a Twisted Nematic Liquid Crystal," **Appl. Phys. Lett.**, Vol. 18, p. 127 (1971).
[9] F. J. Kahn, "IR-Laser-Addressed Thermo-Optic Smectic Liquid Crystal Storage Displays," **Appl. Phys. Lett.**, Vol. 22, p. 111 (1973).
[10] G. W. Gray, **Molecular Structure and the Properties of Liquid Crystals**, Academic Press, N. Y., N. Y. (1962).
[11] G. H. Brown and W. G. Shaw, "The Mesomorphic State Liquid Crystals," **Chem. Rev.**, Vol. 57, p. 1049 (1957).
[12] A. Saupe, "Recent Results in the Field of Liquid Crystals," **Angew. Chem.** (International Ed. in English), Vol. 7, p. 97 (1968).
[13] G. H. Brown, J. W. Doane, and V. D. Neff, **A Review of the Structure and Physical Properties of Liquid Crystals**, CRC Press, Cleveland, Ohio (1971).
[14] J. A. Castellano, "Liquid Crystals for Electro-Optical Application," **RCA Rev.**, Vol. 33, p. 296 (1972); L. T. Creagh, "Nematic Liquid Crystal Materials for Displays," **Proc. IEEE**, Vol. 61, p. 814 (1973).
[15] See, e.g., R. S. Porter and J. F. Johnson, Eds., **Ordered Fluids and Liquid Crystals**, Amer. Chem. Soc., Wash., D.C. (1967); G. H. Brown, G. J. Diene, and M. M. Labes, Eds., **Liquid Crystals**, Gordon and Breach, N. Y., N. Y. (1966); J. F. Johnson and R. S. Porter, Eds., **Liquid Crystals and Ordered Fluids**, Plenum Press, N. Y., N. Y. (1970).
[16] G. W. Gray and B. Jones, "The Mesomorphic Transition Points of the p-n-Alkoxybenzoic Acids," **J. Chem. Soc.**, p. 4179, Part IV 1953.
[17] C. Weygand and R. Gabler, **Z. Phys. Chem.** (Leipzig), Vol. 46B, p. 270 (1940).
[18] G. W. Gray, K. J. Harrison, and T. W. Nash, Abstracts of the 166th Natl. Meeting of the A.C.S., Chicago, 1973, (Abstract Coll-142).
[19] G. W. Gray and B. Jones, "Mesomorphism and Chemical Constitution. Part IV," **J. Chem. Soc.**, p. 236, Part I 1955.
[20] See Ref. (10), p. 183.

[21] G. W. Gray, B. Jones, and F. Marson, "Mesomorphism and Chemical Constitution. Part VIII." **J. Chem. Soc.**, p. 393, Part I 1957.
[22] C. Wiegand, **Z. Naturforsch**, Vol. 4b, p. 249 (1949).
[23] W. Maier and K. Markau, **Z. Phys. Chem.** (Frankfort), Vol. 28, p. 190 (1961).
[24] L. K. Knaak, H. M. Rosenberg, M. P. Serve, "Estimation of Nematic-Isotropic Points of Nematic Liquid Crystals," **Mol. Cryst. Liq. Cryst.**, Vol. 17, p. 171 (1972).
[25] G. W. Gray and B. Jones, "Mesomorphism and Chemical Constitution. Part II," **J. Chem. Soc.**, p. 1467, Part II 1954.
[26] W. R. Young, A. Aviram, and R. J. Cox, **Angew. Chem.** (International Ed. in English), Vol. 10, p. 410 (1971) and references therein.
[27] P. Culling, G. W. Gray, and D. Lewis, "Mesomorphism and Polymorphism in Simple Derivatives of p-Terphenyl," **J. Chem. Soc.**, p. 2699 (1960).
[28] R. J. Cox, "Liquid Crystal Properties of Methyl Substituted Stilbenes," **Mol. Cryst. Liq. Cryst.**, Vol. 19, p. 111 (1972).
[29] I. Teucher, C. M. Paleos, and M. M. Labes, "Properties of Structurally Stabilized Anil-Type Nematic Liquid Crystals," **Mol. Cryst. Liq. Cryst.**, Vol. 11, p. 187 (1970).
[30] See Ref. (10), p. 194.
[31] H. M. Rosenberg and R. A. Champa, "The Effect on Thermal Nematic Stability of Schiff's Bases Upon Reversal of Terminal Substituents," **Mol. Cryst. Liq. Cryst.**, Vol. 11, p. 191 (1970).
[32] H. Sorkin and A. Denny, "Equilibrium Properties of Schiff-Base Liquid-Crystal Mixtures," **RCA Rev.**, Vol. 34, p. 308 (1973).
[33] See, e.g., the following papers in **Mol. Cryst. Liq. Cryst.**: J. A. Castellano, M. T. McCaffrey, and J. E. Goldmacher, "Nematic Materials Derived from p-Alkylcarbonato-p-Alkoxyphenyl Benzoates," Vol. 12, p. 345 (1971); R. D. Ennulat and A. J. Brown, "Mesomorphism of Homologous Series," Vol. 12, p. 367 (1971); G. W. Gray and K. J. Harrison, Vol. 13, p. 37 (1971); M. J. Rafuse and R. A. Soref, "Carbonate Schiff Base Nematic Liquid Crystals: Synthesis and Electro-optic Properties," Vol. 18, p. 95 (1972); M. T. McCaffrey and J. A. Castellano, "The Mesomorphic Behavior of Homologous p-Alkoxy-p'-Acyloxyazoxybenzenes," Vol. 18, p. 209 (1972); and C. S. Oh, "The Effect of Heterocyclic Nitrogen on the Mesomorphic Behavior of 4-Alkoxybenzylidene-2'-alkoxy-5'-aminopyridenes," Vol. 19, p. 95 (1972).
[34] V. I. Minkin, Y. A. Zhdanov, E. A. Madyantzeva, and Y. A. Ostroumov, "The Problem of Acoplanarity of Aromatic Azomethines," **Tetrahedron**. Vol. 23, p. 3651 (1967); also more recent calculations by H. B. Burgi and J. D. Dunitz, "Molecular Conformity of Benzylideneanilines," **Helv. Chim. Acta**, Vol. 54, p. 1255 (1971).
[35] C. Tani, "Novel Electro-Optical Storage Effect in a Certain Smectic Liquid Crystal," **Appl. Phys. Lett.**, Vol. 19, p. 241 (1971).
[36] J. A. Castellano, E. F. Pasierb, G. H. Heilmeier, W. Helfrich, and M. T. McCaffrey, "Electronically Tuned Optical Filter," Final Report to NASA, April 1970. Contract No. NAS12-638By.
[37] M. F. Schiekel and F. Fahrenschon, "Deformation of Nematic Liquid Crystals with Vertical Orientation in Electrical Fields," **Appl. Phys. Lett.**, Vol. 19, p. 391 (1971).
[38] L. T. Creagh, A. R. Kmetz, and R. A. Reynolds, "Performance Characteristics of Nematic Liquid Crystal Display Devices," **IEEE Trans. Electron Devices**, p. 672 (1971).
[39] F. J. Kahn, "Electric Field-Induced Orientational Deformation of Nematic Liquid Crystals: Tunable Birefringence," **Appl. Phys. Lett.**, Vol. 20, p. 199 (1972).
[40] J. A. Castellano, E. F. Pasierb, C. S. Oh, and M. T. McCaffrey, "Electronically Tuned Optical Filters," Final Report to NASA, January 1972. Contract No. NAS1-10490, NTIS No. N72-20612.
[41] W. R. Young, A. Aviram, and R. J. Cox, "New Non-Planar trans-Stilbenes Exhibiting Nematic Phases at Room Temperature," **Angew. Chem. Int. Ed.** in English, Vol. 10, p. 410 (1971); W. H. DeJeu and J. Van der Veen, "Instabilities in Electric Fields of a Nematic Liquid Crystal with Large Negative Dielectric Anisotropy," **Phys. Lett.**, Vol. 44A, p. 277 (1973).
[42] W. H. DeJeu and T. W. Lathouwers, "Dielectric Properties of Some Nematic Liquid Crystals for Dynamic Scattering Displays," **Chem. Phys. Lett.**, Vol. 28, p. 239 (1974) and Ref. 5 therein.
[43] J. E. Goldmacher and M. T. McCaffrey, "Nematic Mesomorphism in Benzylidene Anils Containing a Terminal Alcohol Group," in J. F. Johnson and R. S. Porter, eds, **Liquid Crystals and Ordered Fluids**, Plenum Press, N.Y., N.Y., 1970, p. 375.
[44] G. W. Gray, K. J. Harrison, and J. A. Nash, "New Family of Nematic Liquid Crystals for Displays," **Electron. Lett.**, Vol. 9, p. 130 (1973).
[45] A. Ashford, J. Constant, J. Kirton, and E. P. Raynes, "Electro-Optic Performance of a New Room-Temperature Nematic Liquid Crystal," **Electron. Lett.**, Vol. 9, p. 118 (1973).
[46] L. A. Goodman, "Liquid Crystal Displays," **J. Vac. Sci. Tech.**, Vol. 10, p. 804 (1973); Also see Chapter 14.
[47] E. L. Strebel, "Eutectic Mixture of Para-Alkoxybenzylidene-Para-n-Alkyl Anilines," U.S. Patent 3,809,656, May 7, 1974.
[48] L. T. Creagh, "Nematic Liquid Crystal Materials for Displays," **Proc. IEEE**, Vol. 61, p. 814 (1973).

[49] A. W. Levine, unpublished results.
[50] J. A. Castellano, E. F. Pasierb, A. Sussman, C. S. Oh, D. Meyerhofer, M. T. McCaffrey, and R. N. Friel, "Liquid Crystal Systems for Electro-Optical Storage Effects," Final Report to the AFML, Wright Patterson AFB, December 1971. Contract No. F33615-70-C-1590. NTIS No. AD760-173.
[51] U. Bonne, J. P. Cummings, J. Hicks, A. E. Johnson, B. L. Person, D. Soathoff, S. Scholdt, and J. L. Stevens, "Properties and Limitations of Liquid Crystals for Aircraft Displays," Final Report to the ONR, Department of the Navy, October 1972. Contract No. N00014-71-C-0262. NTIS No. AD751-667.
[52] E. C.-H. Hsu and J. F. Johnson, "Phase Diagrams of Binary Nematic Mesophase Systems," **Mol. Cryst. Liq. Cryst.**, Vol. 20, p. 177 (1973).
[53] D. S. Holme, E. P. Raynes, and K. J. Harrison, "Eutectic Mixtures of Nematic 4'-Substituted 4-Cyanobiphenyls," **J. Chem. Soc. Chem. Commun.**, p. 98 (1974).
[54] A. W. Levine and R. J. Beshinske, unpublished results.
[55] J. B. Scarborough, "Numerical Mathematical Analysis," Fifth Edition, The Johns Hopkins Press, Baltimore, 1962, p. 199ff.

3

Introduction to the Molecular Theory of Nematic Liquid Crystals

Peter J. Wojtowicz
RCA Laboratories, Princeton, N.J. 08540

1. Introduction

In this chapter we consider the very simplest approach to the molecular theory of liquid crystals. We shall approach the theory phenomenologically, treating the problem of the existence of the nematic phase as an order-disorder phenomenon. Using the observed symmetry of the nematic phase we shall identify an order parameter and then attempt to find an expression for the orientational potential energy of a molecule in the nematic liquid in terms of this order parameter. Such an expression is easily found in the mean field approximation. Once this is accomplished, expressions for the orientational molecular distribution function are derived and the thermodynamic functions simply calculated. The character of the transformation from nematic liquid crystal to isotropic fluid is then revealed by the theory, and the nature of the fluctuations near the transition temperature can be explored.

The results of this simple theory will be compared with experiment and will be found to account qualitatively for many of the observed features of nematic liquids. Though derived in a simple phenomenological way, this theory turns out to be equivalent to the well-known pioneering theory of Maier and Saupe.[1] The relation of this theory to more rigorous or complete approaches, as well as the required generalizations of this theory will be considered in a subsequent chapter.[2]

2. Symmetry and the Order Parameter

The identification of the appropriate order parameter for nematic liquid crystals is aided by a consideration of the observed structure and symmetry of the phase. As in any liquid, the molecules in the nematic phase have no translational order; i.e., the centers of mass of the molecules are distributed at random throughout the volume of the liquid. Experiments of many varieties, however, do demonstrate that the nematic phase differs from ordinary liquids in that it is anisotropic. The symmetry, in fact, is cylindrical; that is, there exists a unique axis along which the properties of the phase display one set of values, while another set of values is exhibited in all directions perpendicular to this axis. The symmetry axis is traditionally referred to as the "director". The optical properties of nematics provide an example of how the cylindrical symmetry is manifest. For light passing parallel to the director, optical isotropy is observed, while for all directions perpendicular to the director, optical birefringence is observed. Rays polarized parallel to the director have a different index of refraction from those polarized perpendicular to the director.

Many experiments demonstrate that the anisotropy of nematics arises because of the tendency of the rod-like molecules in the fluid to align their long axis parallel to the director. This is shown schematically in Fig. 1(a). The director is denoted by the symbol $\hat{n}$; the rod-like molecules are represented by the short lines. Note that at finite temperatures, the thermal motion of the molecules prevents perfect alignment with $\hat{n}$; the orientations of the molecules are in fact distributed in angle, but with the director as the most probable, or the most populated, direction.

If we look more closely at the orientation of a single molecule with respect to the director, we find that the cylindrical symmetry of the phase requires just a single order parameter to describe the structure. In Fig. 1(b), we let the director lie along the z-axis of a fixed rectangular coordinate system. The orientation of the rod-like molecule can then be described using the three Eulerian angles shown. Because of the cylindrical symmetry, no order in the angles ψ (rotation about the long molecular axis) or φ (rotation in the azimuthal direction) is permitted. If any angle ψ or φ was in some way preferred, a symmetry lower than cylindrical would necessarily result. Thus, we are left with the remaining angle θ as the only one for which any degree of order can exist. If any angle θ is preferred, then the resultant symmetry is indeed cylindrical; the most probable, or the prefered, angle in nematics has been shown experimentally to be $\theta = 0$, the long molecular axis parallel to n̂. If there is no preference for a particular θ, then all such angles become equally probable and complete isotropy obtains; this is the isotropic normal liquid phase.

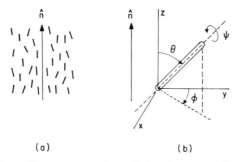

Fig. 1—(a) Schematic representation of the structure of a nematic liquid crystal. (b) The Euler angles required to describe the orientation of a molecule in a nematic liquid.

The observed symmetry and structure of the nematic liquid has enabled us to establish that a single order parameter will suffice to describe the structure of the phase. Clearly, ordering in the polar angle θ distinguishes the nematic structure from the isotropic liquid. However, θ itself is not a convenient order parameter. By analogy with ferromagnetism, one might expect that the projection of the molecules along n̂, $\cos\theta$, would be a natural order parameter. This is not quite correct, however, since the electron spins in magnetism have a definite polarity, the heads (north poles) being distinctly different from the tails (south poles). In nematic liquids, the experiments on most materials (there are apparently some exceptions[3]) demonstrate that the heads or tails of the molecules are not distinguished in the nematic structure. That is, as many head ends point up as tail ends; these, furthermore, are randomly distributed throughout the fluid. Thus, to prevent the order parameter from describing situations in which the nematic phase becomes polarized (as in a ferromagnet polarized with north poles up) we will use $\cos^2\theta$ rather than $\cos\theta$ to describe the structure. But now we must realize that it is not the instantaneous value of $\cos^2\theta$ for a given molecule that we need. Rather, we desire the average value of $\cos^2\theta$, $<\cos^2\theta>$, averaged over all molecules in the liquid. When all the molecules are fully aligned with $\hat{n}$, all $\theta = 0$ and $<\cos^2\theta> = 1$. On the other hand, if the molecules are randomly distributed in direction, all values of θ are equally likely and $<\cos^2\theta> = \frac{1}{3}$.

By tradition, the order parameter in any order–disorder problem is always taken such that it is unity in the perfectly ordered phase and vanishes for the completely disordered phase. Examination of the average values described above shows that the proper order parameter for the nematic liquid crystal is

$$<P_2> = \frac{1}{2}(3<\cos^2\theta> - 1). \qquad [1]$$

Clearly, $<P_2> = 1$ for the completely ordered nematic phase and $<P_2> = 0$ for the disordered isotropic phase. The symbol $<P_2>$ is used for the order parameter because we recognize the particular combination in Eq. [1] to be the second-order Legendre polynominal, $P_2(\cos\theta) = (3\cos^2\theta - 1)/2$. We will continue to use this symbol in this chapter and in the next[2] for its clarity when considering generalizations to the present theory. In the original theory[1] as well as in much of the literature, the symbol S is used to represent $<P_2>$. Values of $<P^2>$ between 0 and 1 describe degrees of ordering intermediate between completely isotropic and completely ordered. It is the task of order–disorder theory to (a) determine the temperature dependence of $<P_2>$, (b) calculate the thermodynamic and other properties in terms of $<P^2>$, and (c) demonstrate the precise way in which the transformation from finite $<P_2>$ to zero order occurs. We will now examine these questions.

3. The Molecular Potential

The stability of the nematic liquid crystal results from interactions between the constituent molecules. Without going into their nature at this time, it is clear that there must exist interactions that cause the molecules to prefer to align parallel to each other (and to the mean direction of alignment, the director). In the spirit of the mean field aproximation, we can attempt to mimic these intermolecular interactions with an effective single-molecule potential function V.

The potental V must have the correct orientation dependence; that is, it should be a minimum when the molecule is parallel to the director (parallel to the average direction of all the other molecules it interacts with) and a maximum when the molecule is perpendicular to this preferred condition. As we have seen above, the angular dependence of $-P_2(\cos\theta) = -(3\cos^2\theta - 1)/2$ is sufficient for this purpose. The potential V representing the field of intermolecular interaction forces should, furthermore, be a minimum when the phase is highly ordered and should vanish when the phase becomes disordered. Thus, V should be proportional to the degree of order, $<P_2>$. Finally, V should contain a factor v to describe the overall strength of the intermolecular interactions; the differences between various materials will be accounted for by allowing the strength v to vary from one substance to another. Putting all the above together we arrive at mean field approximation to the orientational potential energy function of a single molecule:

$$V(\cos\theta) = -vP_2(\cos\theta)<P_2>. \qquad [2]$$

By mean field approximation we mean that the interactions between individual molecules are represented by a potential of average force, ignoring the fact that the individual behaviors and interactions of molecules can be widely distributed about the average. This theory therefore ignores fluctuations in the short-range order (mutual alignment of two neighboring molecules); the results of the theory will then only be approximate in so far as near-neighbor fluctuation effects have been neglected in the derivation of V, as well as in subsequent calculations. The strengths and weaknesses of the mean field approximation are particularly well documented in the case of magnetism;[4] similar considerations will apply to the case of nematic liquid crystals.

4. The Orientational Distribution Function

Having derived an approximate expression for the potential energy of a single molecule in the nematic phase, we can now obtain the orientational distribution function. This function, which shall be denoted by $\rho(\cos\theta)$, describes how the molecules are distributed among the possible directions about the director; it gives the probability of finding a molecule at some prescribed angle θ from n̂. With this function we can compute the average values of various quantities of interest pertaining to the nematic phase.

The rules of classical statistical mechanics give the orientational distribution function in terms of the potential function V as:

$$\rho(\cos\theta) = Z^{-1}\exp[-\beta V(\cos\theta)],\qquad [3]$$

$$Z = \int_0^1 \exp[-\beta V(\cos\theta)]d(\cos\theta),$$

where Z is the single-molecule partition function and $\beta = 1/kT$ (k is Boltzmann's constant and T is the temperature). The integration over all possible orientations of the molecule can be restricted to $0 \leqslant \cos\theta \leqslant 1$, since both V and ρ are even functions of $\cos\theta$. As it stands, however, Eq. [3] is still useless for the calculation of average values. The reason is that the function V (and hence ρ) contains the order parameter $<P_2>$, and this parameter is an as yet undetermined function of the temperature. Its temperature dependence, however, can be determined as follows. $<P_2>$ is, afterall, just the average value of the second Legendre function for a given molecule. Therefore, Eq. [3] can be used formally to express this fact:

$$<P_2> = \int_0^1 P_2(\cos\theta)\,\rho(\cos\theta)\,d(\cos\theta).\qquad [4]$$

Expanding this equation to display its pertinent factors,

$$<P_2> = \frac{\int_0^1 P_2(\cos\theta)\exp[\beta v P_2(\cos\theta)<P_2>]d(\cos\theta)}{\int_0^1 \exp[\beta v P_2(\cos\theta)<P_2>]d(\cos\theta)}.\qquad [5]$$

We now see in Eq. [5] a self-consistent equation for the determination of the temperature dependence of $<P_2>$. The order parameter $<P_2>$ appears on both the left and right hand sides of the equation. For every temperature T (or β) we can use a computer to obtain the value (or values) of $<P_2>$ that satisfies the self-consistency equation. This process has been accomplished and the results are depicted in Fig. 2. $<P_2> = 0$ is a solution at all temperatures; this is the disordered phase, the normal isotropic liquid. For temperatures T below 0.22284 v/k, two other solutions to Eq. [5] appear. The upper branch tends to unity at absolute zero and represents the nematic phase, i.e., all molecules trying to align with the director. The lower branch tends to $-\frac{1}{2}$ at absolute zero and represents a phase in which the molecules attempt to line up perpendicular to the director without azimuthal order. This phase also has cylindrical symmetry but has not yet been seen experimentally. That it ever will is rather unlikely, since we will see later on that it is unstable with respect to the parallel-aligned nematic phase. In the temperature range where Eq. [5] provides three solutions, we require a criterion to determine which one of the three possibilities actually exists. The laws of thermodynamics provide that the observed solution, the stable phase, will be that one having the minimum free energy.

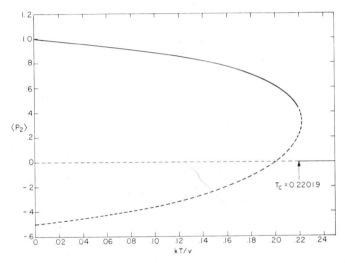

Fig. 2—Temperature dependence of the order parameter obtained from solving the self-consistency Equation, Eq. [5]. The stable equilibrium solutions are shown as the solid lines.

5. Thermodynamics of the Nematic Phase

The derivation of the free energy from the orientational distribution function is straightforward. The free energy F is given by $E-TS$, where E is the internal energy and S the entropy. The energy is computed by taking the average value of the potential,

$$E = \frac{1}{2} N <V> = \frac{1}{2} N \int_0^1 V(\cos\theta)\, \rho(\cos\theta)\, d(\cos\theta). \qquad [6]$$

The factor N comes in because we are dealing with a system of N molecules. The factor ½ is required to avoid counting intermolecular interactions twice (pair interactions have been approximated by a single molecule potential V). The entropy is computed by taking the average value of the logarithm of the distribution function:

$$S = -Nk <\ln\rho> = \frac{N}{T}<V> + Nk\ln Z, \qquad [7]$$

where the quantities on the extereme right hand side have been defined in Eqs. [6] and [3]. Combining Eqs. [6] and [7],

$$F = -NkT\ln Z - \frac{1}{2}N<V>. \qquad [8]$$

At first sight this equation seems unusual with the presence of the second term. The necessity for its existence can be verified immediately in two ways. If we take the derivative $(\partial F/\partial <P_2>)_T$ and set it equal to zero we regain Eq. [5], the self consistency equation for $<P_2>$. Thus, as required by thermodynamics, the self-consistent solutions to our problem must be those that represent the extrema of the free energy. Another verification of the correctness of Eq. [8] comes from forming the derivative $[d\beta F/d\beta]$. Again, as required by thermodynamics, Eq. [6] for the internal energy results. The reason for the appearance of the second term in Eq. [8] is the replacement of pair interactions by temperature-dependent single molecule potentials.[5]

The free energy for each of the three branches of the order parameter, Fig. 2, is computed from Eq. [8] by successive substitution of the three different values of $<P_2>$ for each T. The free energy for the $<P_2>=0$ branch is constant with T and equal to zero. The free energy of the negative $<P_2>$ branch is negative but small in magnitude. The free energy for the positive $<P_2>$ branch is negative (with magnitude larger than that for the negative $<P_2>$ branch) up to a temperature of 0.22019 v/k. The remainder of the positive $<P_2>$ branch has positive values of F. Thus the stable phases to be observed are as follows: from $T=0$ to $T=0.22019$ v/k the nematic phase is stable. The order parameter decreases from unity to a minimum value of 0.4289 at $T=0.22019$ v/k. For temperatures above 0.22019 v/k the isotropic phase with vanishing order parameter is stable (the stable phases are shown as the solid lines in Fig. 2). At $T_c=0.22019$ v/k we

MOLECULAR THEORY OF NEMATIC CRYSTALS

see that we have a first-order phase transition with the order parameter discontinuously changing from $<P_2> = 0.4289$ to $<P_2> = 0$. T_c is usually called a critical temperature, although in much of the liquid-crystal literature it is referred to as the clearing temperature.

Numerical values of the equilibrium order parameter $<P_2>$ for various temperatures between zero and $T_c = 0.22019 \ v/k$ have been found on the computer and are presented in Table 1 as an aid to anyone wishing to perform numerical calculations based on this or the Maier-Saupe[1] mean field theory.

Table 1—Equilibrium values of the order parameter $<P_2>$ as a function of temperature in the nematic range

kT/v	$<P_2>$	kT/v	$<P_2>$
0	1.0000	.201	.6091
.01	.9899	.202	.6032
.02	.9794	.203	.5971
.03	.9687	.204	.5908
.04	.9576	.205	.5843
.05	.9461	.206	.5776
.06	.9342	.207	.5706
.07	.9218	.208	.5634
.08	.9089	.209	.5558
.09	.8953	.210	.5479
.10	.8810	.211	.5396
.11	.8657	.212	.5309
.12	.8493	.213	.5217
.13	.8315	.214	.5120
.14	.8210	.215	.5016
.15	.7910	.216	.4904
.16	.7655	.217	.4783
.17	.7372	.218	.4649
.18	.7041	.219	.4500
.19	.6644	.220	.4327
.20	.6148	.22019	.4289

We can get a better feel for Fig. 2 if we discuss it in terms of some real temperatures. A typical nematic liquid crystal used in device applications will have its clearing point T_c at about 50°C. Then room temperature will correspond to about $T = 0.2 \ v/k$. Thus, the degree of order at room temperature will be about $<P_2> = 0.615$. From room temperature upwards, $<P_2>$ will gradually decrease to about 0.429 at the clearing point. Degrees of order larger than 0.75 to 0.8 are rarely if ever seen. Long before such order can be accomplished by lowering the temperature, the smectic phase and/or the solid crystalline phase become more stable and spontaneously appear. The general trend of the temperature dependence of $<P_2>$ displayed in Fig. 2 is in qualitative agreement with experimental determinations (the methods by which $<P_2>$ is measured are discussed by E. B. Priestley in a separate lecture in this series). A more detailed comparison for several real materials is deferred to the next chapter.[2]

The first-order phase transition at T is very much like the phase transition corresponding to the melting of a solid into its liquid phase. That is, discontinuities in the volume, the internal energy (latent heat), and the entropy of the system are observed. The present theory cannot account for the small volume changes seen at T_c, since no distance (volume) dependence of the potential V was assumed. We have neglected this aspect in the present exercise, since the volume changes actually observed over the short nematic ranges of most materials are not very significant; theories that include the volume dependence are available in the literature.[1,6,7]

The latent heat of the transition from nematic to isotropic liquid can be calculated from Eq. [6] and the fact that $<P_2>$ changes from 0.4289 to zero at T_c:

$$\Delta E(T_c) = 0.830 \, T_c \text{ cal/mol}. \qquad [9]$$

The entropy of the transition can be computed from the formula $\Delta S = \Delta E / T_c$ so that

$$\Delta S(T_c) = 0.830 \text{ cal/mol}°\text{K}. \qquad [10]$$

The entropy change is a measure of the change in order at T_c and conveniently comes out of the theory as a number without any parameters to be found or adjusted. First, we note that this is a very small entropy change. Typical entropy changes for solid to liquid transitions of similar organic materials run around 25 cal/mol°K. Thus, the nematic phase at T_c contains only a very slight amount of order, certainly a lot less than a crystalline solid. The smallness of $\Delta S(T_c)$ is, of course, in keeping with the fact that the nematic phase possesses order in only a single degree of freedom, the θ angle of the individual molecules. Secondly, we can compare Eq. [10] with some experimental results; the agreement is quite satisfactory for such a simple theory. As an example,[8] the series of homologous 4,4'-d-n-alkoxyazoxybenzenes from methyl to decyl displays entropy changes in the range from 0.3 to 1.9 cal/mol°K, with most values being around 0.6 to 0.8.

The results obtained here and in the previous several sections are completely equivalent to the mean field theory derived by Maier and Saupe;[1] Eqs. [2], [3], [5], and [8] are identical to those presented in this classic series of papers. Their approach is, of course, more systematic than presented here, and the volume dependence of the potential is treated explicitly. In the next chapter[2] we will examine a more systematic development of the theory and touch upon the volume dependence of the potential.

6. Fluctuations at T_c

All nematic to isotropic liquid phase transitions have been observed to be first order with a small volume change and a latent heat. The present simple theory, as well as more sophisticated versions in the literature, also display first-order changes in agreement with the experiments. Many articles in the literature, however, refer to this phase change as "nearly second order." Such remarks are not only misleading but are entirely incorrect. The phase transition in nematics cannot be anything but first order. The symmetry of the problem demands it. Fig. 2 shows one way of looking at this: note that the curves of $<P_2>$ vs T for positive and negative $<P_2>$ are not symmetrical about $<P_2> = 0$. Because of this, the curves of positive $<P_2>$ must intercept the abscissa with finite positive slope. In the temperature range just above where this occurs, moreover, (positive) $<P_2>$ is double valued. Thus, there is no way that a state of finite $<P_2>$ can transform into a state of zero $<P_2>$ except discontinuously. In the case of ferromagnetism or antiferromagnetism, on the other hand, the curves of order parameter vs T are symmetrical about the T-axis; the positive branch is the mirror image of the negative branch. In addition, the free energies of these two branches are identical. The order parameter curves then approach the abscissa symmetrically and with infinite slope. In these cases transformations from states of finite order to ones of vanishing order can occur continuously, and second-order phase changes are allowed.

Why then did the phrase "nearly second order" arise in the literature? One reason is that the latent heat and volume changes at the nematic-isotropic transition are very small and closely approximate the vanishing changes of second-order phase transitions. The most prominent reason, however, is that many properties of the isotropic liquid display critical behavior on cooling to temperatures close to T_c (this phenomenon has been termed "pretransitional behavior" in the liquid-crystal literature). That is, certain response functions (such as the magnetically induced birefringence[9]) are seen to diverge as T_c is approached from above. The divergence of response functions at T_c is one of the hallmarks of second-order phase changes.[10] In ferromagnetism, as an example, the magnetic susceptibility diverges as $(T-T_c)^{-1.35}$ as the Curie point is approached from above. In the case of the isotropic liquid above its clearing point, the divergence never actually occurs; the first-order phase change to the nematic phase takes place just a few degrees before. Thus the origin of the remark "nearly second order".

Now, critical behavior and the divergence of response functions are not usually seen at first-order phase transitions. No pretransitional behavior is observed in any of the properties of a liquid about to freeze into its crystalline solid. Why then is there pretransitional behavior at the nematic-isotropic transition? My own feeling is that the responsible factor is the small entropy of transition, ΔS, Eq. [10]. The divergence of response functions occur because of the presence of fluctuations. If in a disordered phase a fluctuation occurs such that a very small region of the material rearranges itself into a locally highly ordered structure, that region will have an enhanced response to some external stimulus; the corresponding response function will therefore increase in magnitude. If the fluctuations occur to a very great extent (as they do when second-order phase transitions are approached) then the response functions diverge. It is the small ΔS for nematics (compared to the ΔS of the melting/freezing transition) that permits a substantial amount of fluctuations, giving rise to pretransitional behavior.

We can make this somewhat more quantitative. According to Landau and Lifshitz,[11] the probability of a fluctuation whose change in entropy (order) is Δs is proportional to $\exp(\Delta s/k)$; here Δs is not the molar entropy of the transition, the ΔS of Eq. [10], but the entropy of ordering of a small region of the material. If this small region contains n molecules (which order to a degree normally displayed by a nematic just at T_c), then Δs can be related to ΔS, and the probability of such a fluctuation becomes proportional to

$$\exp(-n\Delta S/R), \qquad [11]$$

where R is the universal gas constant. The negative sign appears because Δs describes fluctuations from disordered to ordered states while ΔS is concerned with the transition from ordered to disordered phases. Consider now a fluctuation in which ten molecules in the isotropic phase get together and become as ordered as the nematic phase at or near its T_c. We find from Eqs. [10] and [11] that the probability of occurrence of such a fluctuation is proportional to about e^{-4}. For a fluctuation in liquid in which ten molecules get together and become as ordered as the crystalline solid, on the other hand, Eq. [11] plus the observation that $\Delta S \sim 25$ cal/mol°K give the probability as $\sim e^{-125}$. Thus we see that a significant fluctuation in the ordering of ten molecules is $\sim e^{120}$ more probable just above the isotropic-nematic transition than it is just above a typical freezing point. This, then, is why pretransitional behavior is observed in nematics but not in the case of crystalline solids.

References

[1] W. Maier and A. Saupe, **Z. Naturforschg.**, Vol. 14a, p. 882 (1959) and Vol. 15a, p. 287 (1960).

[2] P. J. Wojtowicz, "Generalized Mean Field Theory of Nematic Liquid Crystals," Chapter 4.

[3] R. Williams, "Optical-Rotary Power and Linear Electro-Optic Effect in Nematic Liquid Crystals of p-Azoxyanisole," **J. Chem. Phys.**, Vol. 50, p. 1324 (1969) and D. Meyerhofer, A. Sussman, and R. Williams, "Electro-Optic and Hydrodynamic Properties of Nematic Liquid Films with Free Surfaces," **J. Appl. Phys.**, Vol. 43, p. 3685 (1972).

[4] J. S. Smart, **Effective Field Theories of Magnetism**, W. B. Saunders Co., Philadelphia, Pa. (1966).

[5] E. R. Callen and H. B. Callen, "Anisotropic Magnetization," **J. Phys. Chem. Solids**, Vol. 16, p. 310 (1960).

[6] S. Chandrasekhar and N. V. Madhusudana, "Molecular Statical Theory of Nematic Liquid Crystals," **Acta Cryst.**, Vol. A27, p. 303 (1971).

[7] R. L. Humphries, P. G. James, and G. R. Luckhurst, **J. Chem. Soc., Faraday Trans. II**, Vol. 68, p. 1031 (1972).

[8] G. H. Brown, J. W. Doane and D. D. Neff, **A Review of the Structure and Physical Properties of Liquid Crystals**, p. 43, CRC Press, Cleveland, Ohio (1971).

[9] T. W. Stinson and J. D. Litster, "Pretransitional Phenomena in the Isotropic Phase of a Nematic Liquid Crystal," **Phys. Rev. Lett.**, Vol. 25, p. 503 (1970).

[10] H. E. Stanley, **Introduction to Phase Transitions and Critical Phenomena**, Oxford University Press, N.Y., N.Y. (1971).

[11] L. D. Landau and E. M. Lifshitz, **Statistical Physics**, p. 344, Pergamon Press Ltd., London (1958).

Generalized Mean Field Theory of Nematic Liquid Crystals

Peter J. Wojtowicz
RCA Laboratories, Princeton, N. J. 08540

1. Introduction

In the previous chapter[1] we examined a simple version of the molecular theory of nematic liquid crystals. The problem was treated as an order-disorder phenomenon with the solution based on a phenomenologically derived single-molecule orientational potential. The results of this development were found to be equivalent to the well known mean field theory of Maier and Saupe.[2]

In this chapter we will consider a more systematic approach to the molecular theory of the nematic state. In deriving the theory we will follow the development of Humphries, James, and Luckhurst.[3] We start with a completely general pairwise intermolecular interaction potential. After expanding in a series of appropriate spherical harmonics we will systematically average the pair-interaction potential to obtain a generalized version of the single-molecule potential function in the mean field approximation. The Maier-Saupe theory[2] will result from the retention of only the first term in the generalized potential. More general versions of the theory result from the inclusion of higher-order terms. In the final sections of the chapter we will discuss the volume dependence of the interaction potential, examine the reasons why higher-order terms in the potential appear to be required, and consider the agreement between the theory and various experiments.

2. The Pair Interaction Potential

The stability of the nematic liquid-crystal phase arises from the existence of strong interactions between pairs of the constituent molecules. As in normal liquids, the potential of interactions will have attractive contributions to provide the cohesion of the fluid, and repulsive contributions which prevent the interpenetration of the molecules. In the case of the rod-like molecules of nematics, however, these interactions are highly anisotropic. That is, the forces acting between such molecules depend not only on their separation but also, and most importantly, on their mutual orientations. From the symmetry and structure of the nematic phase, we see that the rod-like molecules in fact interact in a manner that favors the parallel alignment of neighboring molecules.

A complete understanding of the nature of the intermolecular interactions is not available. The precise mathematical form of the pair potential is not known. In the absence of such detailed information, we proceed by assuming a perfectly general form for the pair potential and then deriving the theory systematically from it.

Many coordinates are required to describe the orientation dependent interaction between a pair of asymmetric molecules. Fig. 1(a) depicts the situation; in addition to the parameter r which gives the separation of the centers of gravity, we require the three Eulerian angles of each of the two molecules. Since in nematics there appears to be no ordering of the molecules about their long axes, we can eliminate the angles ψ_i by assuming the molecules to be axially symmetric. The angles θ_i and ϕ_i can then be considered as polar angles with the intermolecular vector r as the common polar axis. The intermolecular pair potential V_{12} is then a function of five coordinates:

$$V_{12} = V_{12}(r, \theta_1, \phi_1, \theta_2, \phi_2). \qquad [1]$$

Pople[4] has demonstrated a very powerful expansion for the pair potential between axially symmetric molecules:

$$V_{12} = 4\pi \sum_{L_1 L_2 m} U_{L_1 L_2 m}(r) Y_{L_1 m}(\theta_1, \phi_1) Y_{L_2 m}^*(\theta_2, \phi_2), \qquad [2]$$

where the $Y_{Lm}(\theta, \phi)$ are the usual spherical harmonics. For axially symmetric molecules and for the coordinates defined in Fig. 1(a), the potential V_{12} really depends only on r, θ_1, θ_2, and the combination $\phi_1 - \phi_2$; thus, the appearance of only a single m index in Eq. [2].

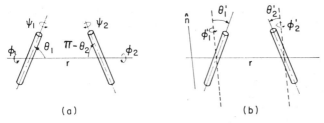

Fig. 1—The coordinate systems required to describe the interaction between two asymmetric molecules: (a) the intermolecular vector r is the mutual polar axis and (b) the director $\hat{n}$ is the polar axis for each molecule.

This expression is particularly convenient, because (a) it separates the distance and orientation dependencies of the potential and (b) the coefficients of the expansion, the $U_{L_1 L_2 m}(r)$, are found[5] to decrease rapidly with increasing L_1 and L_2. If the molecules have reflection symmetry (or if we assume that we can neglect the ordering of heads and tails of the molecules), then only terms with even L_1 and L_2 are required in Eq. [2].

3. The Mean Field Approximation

A rigorous molecular theory of a fluid system based on a pairwise interaction potential as complicated as Eq. [2] is impossibly difficult. A simple but adequate approach is to derive a theory in the mean field approximation. That is, we derive a single molecule potential that serves to orient the molecule along the symmetry axis (the director) of the nematic phase. The single-molecule potential represents (approximately) the mean field of intermolecular forces acting on a given molecule. Mean field theories have been found capable of describing the qualitative behavior of many different cooperative phenomena. The theories are not quantitatively correct, however, since intermolecular short-range order and fluctuation effects are neglected. The properties of mean field theories in the case of magnetism have been described by Smart.[6]

In order to derive a mean field approximation to the potential, we first have to express V_{12} in terms of a polar coordinate system based on the director, n̂, as the polar axis. The coordinate axes for the molecules 1 and 2 must be rotated from that shown in Fig. 1(a) to that shown in Fig. 1(b). The primed angles now describe the orientations of the molecules with respect to the new rotated coordinate system. Mathematically, the rotation of the coordinate axes transforms the spherical harmonics into the form

$$Y_{Lm}(\theta, \phi) = \sum_p D_{pm}^L Y_{Lp}(\theta', \phi'), \qquad [3]$$

where the D_{pm}^L are the elements of the Wigner rotation matrices.[7] In the new coordinate system V_{12} assumes the form

$$V_{12} = 4\pi \sum_{L_1 L_2 m} \sum_{pq} U_{L_1 L_2 m}(r) Y_{L_1 p}(\theta_1', \phi_1') Y_{L_2 q}^*(\theta_2', \phi_2') (D_{pm}^{L_1})(D_{qm}^{L_2})^*.$$

$$[4]$$

To obtain the single molecule potential V_1 in the mean field approximation, it is necessary to take three successive averages of the function V_{12}. First, we average over all orientations of the intermolecular vector r. Next, we average V_{12} over all orientations of molecule 2. Finally, we average V_{12} over all values of the intermolecular separation r. The combination of these three averaging processes provides us with the potential energy of a given single molecule as a function of its orientation with respect to the director. This potential energy, moreover, is that experienced by the molecule when subjected to the average force fields of neighboring molecules, each averaged over all its possible positions and orientations—thus the descriptive phrase, "mean field approximation".

The averaging of V_{12} over all orientations of the intermolecular vector r has an influence only on the Wigner rotation matrices:

$$<(D_{pm}^{L_1})(D_{qm}^{L_2})^*> = \delta_{L_1 L_2} \delta_{pq} (2L+1)^{-1}, \qquad [5]$$

where $<>$ denotes the average value and the δ_{kl} are the usual Kronicker delta functions. The simple result expressed in Eq. [5] is valid only if the distribution function for the intermolecular vectors is spherically symmetric. Since the symmetry of the nematic phase is cylindrical, this result will only be an approximation. A somewhat different expression obtains if the cylindrical symmetry is taken into account;[3] we will return to this point later. Use of Eq. [5] in taking the first average of V_{12} gives

$$<V_{12}> = 4\pi \sum_{Lpm} U_{LLm}(r)(2L+1)^{-1} Y_{Lp}(\theta_1', \phi_1') Y_{Lp}^*(\theta_2' \phi_2'). \quad [6]$$

The averaging of $<V_{12}>$ over all orientations of molecule 2 has influence only on the spherical harmonics in θ_2', ϕ_2':

$$<Y_{Lp}^*(\theta_2', \phi_2')> = \int_0^{2\pi} \int_0^{\pi} Y_{Lp}^*(\theta_2', \phi_2') \rho_1(\theta_2') \sin\theta_2' d\theta_2' d\phi_2', \quad [7]$$

where $\rho_1(\theta_2')$ is the orientational molecular distribution function that describes how molecule 2 is distributed among all possible directions about the director (this function has been described in the previous chapter[1] and will be considered further below). Because of the cylindrical symmetry of the nematic phase, ρ_1 can be a function only of the polar angle θ; no ϕ dependence is permitted. Then, since ρ_1 is not a function of ϕ_2', the ϕ_2' integrals vanish except for $p=0$ and

$$<Y_{Lp}^*(\theta_2', \phi_2')> = \delta_{p0} \left(\frac{2L+1}{4\pi}\right)^{1/2} <P_L>, \quad [8]$$

$$<P_L> = \int_0^1 P_L(\cos\theta) \rho_1(\cos\theta) d(\cos\theta), \quad [9]$$

where the P_L are the Lth order Legendre polynomials and where Eq. [9] is the Lth generalization of $<P_2>$, the order parameter for the nematic phase (as described in Ref. [1]). Use of Eqs. [8] and [9] in taking the average of $<V_{12}>$ yields

$$\ll V_{12} \gg = \sum_{Lm} U_{LLm}(r) P_L(\cos\theta_1') <P_L>. \quad [10]$$

The averaging of $\ll V_{12} \gg$ over all values of the intermolecular separation r has influence only on the $U(r)$:

$$< U_{LLm}(r) > = \frac{1}{n} \int U_{LLm}(r) n_2(r) \, dr. \qquad [11]$$

Here $n_2(r)$ is the molecular distribution function for the separation of pairs of molecules and n is the number density of molecules in the fluid. Use of Eq. [11] in taking the average of $\lll V_{12} \ggg$ gives

$$\lll V_{12} \ggg = \sum_{Lm} < U_{LLm}(r) > P_L(\cos \theta_1') < P_L >. \qquad [12]$$

This is just the desired single-molecule potential in the mean field approximation; we will write it as

$$V_1(\cos \theta) = \sum_L U_L < P_L > P_L(\cos \theta), \qquad [13]$$

$$U_L = \sum_m < U_{LLm}(r) >,$$

where we have dropped the subscript and prime on the θ. Going back to the discussion following Eq. [2], we assume no preference between the heads and tails of the molecules and restrict the L sum in Eq. [13] to even values of L. The $L = 0$ term, moreover can be discarded, since it is merely an additive constant. The first few terms of Eq. [13] are then

$$V_1(\cos \theta) = U_2 < P_2 > P_2(\cos \theta) + U_4 < P_4 > P_4(\cos \theta)$$
$$+ U_6 < P_6 > P_6(\cos \theta) + \cdots \qquad [14]$$

As shown in Ref. [3], if the true cylindrical symmetry of the distribution function for the orientation of the intermolecular vector were taken into account, terms proportional to

$$< U_{L_1 L_2 m}(r) > (P_{L_1} < P_{L_2} > + P_{L_2} < P_{L_1} >) \qquad [15]$$

would also appear in the single molecule potential V_1. If only the first term in Eq. [14] is retained, we obtain the mean field theory of Maier and Saupe[2] and the equivalent theory of the previous chapter[1] ($v = -U_2$). Since the original expansion on which V_1 is based (Eq. [2]) converges rapidly, retention of only the first term in V_1 should provide a good approximation to the theory. Comparison with experiment

MEAN FIELD THEORY OF NEMATIC CRYSTALS

shows that this is indeed the case. The necessity for including higher-order terms is quite apparent in some of the experiments, however, and we will examine this point shortly.

4. Statistical Thermodynamics

As described in the previous chapter,[1] the orientational distribution function corresponding to the single-molecule potential has the form

$$\rho_1(\cos\theta) = Z_1^{-1} \exp[-\beta V_1(\cos\theta)], \quad [16]$$

$$Z_1 = \int_0^1 \exp[-\beta V_1(\cos\theta)] d(\cos\theta),$$

where Z_1 is the single molecule partition function and $\beta = 1/kT$ (k is Boltzmann's constant and T the temperature). The integration over all possible orientations of the molecule can be restricted to $0 \leqslant \cos\theta \leqslant 1$, since V_1 and ρ_1 are even functions of $\cos\theta$. Again, as it stands, Eq. [16] is not complete and not yet useful in computing average values. V_1 does after all contain the as yet unknown averages, $<P_L>$. The self-consistent determination of the temperature dependence of the $<P_L>$ is realized by combining their definition (Eq. [9]) with Eq. [16]:

$$<P_L> = \frac{\int_0^1 P_L(\cos\theta) \exp[-\beta V_1(\cos\theta)] d(\cos\theta)}{\int_0^1 \exp[-\beta V_1(\cos\theta)] d(\cos\theta)}. \quad [17]$$

There is one such self-consistency equation for each term L included in the potential V_1, Eq. [13]. In each of these equations one of the $<P_L>$ appears on the left-hand side, and all the included $<P_L>$ appear in the integrals on the right-hand side. The simultaneous solution of these equations yields the temperature dependence of all the $<P_L>$ originally included in the potential V_1. In particular, the solution for $<P_2>$ gives the temperature dependence of the traditional order parameter of the nematic phase. One of the solutions provided by the set of equations, Eq. [17], is $<P_L> = 0$, all L; this is the isotropic normal liquid. To find which of all the possible solutions is the physically observed stable solution, we must compute the free energy and determine the solution giving the minimum free energy.

The internal energy, entropy, and free energy are obtained in exactly the same way as computed in the simpler theory of the previous chapter:

$$E = -\frac{1}{2}N<V_1> = -\frac{1}{2}N\sum_L U_L <P_L>^2, \qquad [18]$$

$$S = -Nk<\ln\rho_1> = \frac{N}{T}\sum_L U_L <P_L>^2 + Nk\ln Z_1, \qquad [19]$$

$$F = -NkT\ln Z_1 - \frac{1}{2}N\sum_L U_L <P_L>^2. \qquad [20]$$

Just as in the case of the simpler theory,[1] the free energy is found to include additional terms beyond the usually expected $\ln Z_1$ term. The form is correct, however, and arises because we have approximated the action of the pair potential V_{12} by the temperature-dependent single-molecule potential V_1. Note that setting the partial derivatives $(\partial F/\partial <P_L>)_{T, P_{L'}}$ to zero regains the required self-consistency equations, Eq. [17]. Furthermore, testing Eqs. [18] and [20], we see that they do satisfy the required thermodynamic identity, $E = (d\beta F/d\beta)$.

5. Nature of the Parameters U_L

The main physics of the problem, the character of the intermolecular interactions, is contained in the parameters U_L of the one molecule potential V_1. In spite of their central importance in the theory, however, not all that much is known about their magnitude or volume dependence. Several attempts to estimate their properties have been made, but very little in the way of definitive results has been obtained. What is usually done in practice is to treat the U_L as simple volume-dependent functions with constants determined by fitting the theory to experimental results.

The first attempt to elucidate the nature of the intermolecular forces in nematics was performed by Maier and Saupe.[2] They did a calculation of the London dispersion forces (induced dipole–induced dipole) expected between asymmetric molecules. Since the polarizabilities of elongated molecules are anisotropic, the interactions involving these polarizabilities must likewise be anisotropic. Indeed, Maier and Saupe found (within their approximations) an attractive interaction between pairs of molecules that led to the form

$$V_1 \approx -\frac{A}{V^2} <P_2> P_2(\cos\theta). \quad [21]$$

Thus, U_2 is determined from these calculations to be of the form $-A/V^2$, where A is a constant to be determined from experiment and V is the mean molecular volume. The particular volume dependence should not be surprising; the potential of dispersion forces (whether between symmetric or asymmetric molecules) is expected to depend on the intermolecular separation as r^{-6}. On the average, then, this translates into the above V^{-2} dependence.

Chandrasekhar and Madhusudana[8] have also considered the calculation of the coefficients U_L required in V_1. The first contribution that these authors examined was the permanent dipole–permanent dipole forces. These were shown to vary as r^{-3} and provided a V^{-1} dependence to U_1. It was shown however, that this term vanished when the pair potential V_{12} is averaged over a spherical molecular distribution function. The authors thus discard this term and provide further arguments for its neglect based on the empirical result that permanent dipoles apparently play a minor role in providing the stability of the nematic phase. The second contribution considered was the dispersion forces based on induced dipole–induced dipole interactions and induced dipole–induced quadrupole interactions. As mentioned above, the first of these gives a V^{-2} dependent contribution, while the second provides a contribution depending on $V^{-8/3}$. The final contribution considered was the repulsive forces that arise from the mutual impenetrability of the molecules. Assuming (by analogy with the behavior of the better understood, more simple molecules) that the repulsive potential varies as r^{-12}, this mechanism provides a contribution to the U_L that varies as V^{-4}. All these contributions were then lumped together and an average volume dependence was assumed; the authors chose to allow the U_L to have a V^{-3} dependence.

Humphries, James, and Luckhurst[3] in their development of the mean field theory decided to allow each of the U_L to have the same volume dependence, $V^{-\gamma}$, where γ is a parameter to be determined by fitting theory to experiment. The use of an arbitrary parameter γ gives great latitude in reproducing experimental results and much success can be expected. The procedure is, however, dangerous. In the first place, the use of γ in the theory provides still another adjustable parameter; with enough adjustable parameters, of course, any experimental observation can be reproduced (but without learning any physics). Second, the introduction of an arbitrary $V^{-\gamma}$ dependence can lead to abuses. There is one instance in the literature where experimental data was force fit using a value of γ equal to 10. Certainly, we know of no substantial attractive intermolecular potential that varies as strongly as r^{-30}.

Great care should be used in choosing and using various forms for the U_L in Eq. [13]. The simplest (and possibly least controversial) procedure is to assume that the U_L are simply constants independent of the volume and to restrict the fitting of theory to experiment to a narrow range of temperature near T_c. Since the volume changes are small over a short temperature interval, the approximation of constant U_L will be quite adequate and will keep the use of the theory simple and straightforward. The region near T_c, moreover, is the most interesting; it is here that $<P_2>$ has its strongest temperature dependence.[1]

6. The Need for Higher Order Terms in V_1

As stated in Sec. 3, the retention of only the first term in Eq. [14] leads to the mean field theory of Maier and Saupe[2] and the equivalent theory of the previous chapter.[1] This version of the theory has been shown to provide a good qualitative picture of the nematic phase and its transition to the isotropic liquid. What, then, is it about the experimental facts that indicate the necessity of higher order terms in V_1?

In Fig. 2 we display (schematically) the comparison of several experiments with the mean field theory, including only the term in U_2. We use $T - T_c$ as the abscissa, so that the two materials depicted can be conveniently compared on the same graph. The dashed line is the theoretical curve computed from the mean field theory using only the term in U_2. We note that there is a striking difference in the properties of the two materials, PAA (para-azoxyanisole) and PAP (para-azoxyphenetol). We note further that the simple theory cannot account for the order parameter of either material. Fig. 2 is, in fact, clear evidence for the necessity of higher-order terms in the potential. That the two substances have different values of T_c must mean that the interaction potentials differ by at least a multiplicative constant. However, the fact that the order parameter of PAP is everywhere larger than that of PAA cannot be explained by allowing the interactions to differ only by a multiplicative constant. Such a large difference in the values of $<P_2>$ can only be accounted for by assuming that the potentials are in some essential way different in the two materials. This difference must be a difference in both the sign and magnitude of the higher-order terms in V_1.

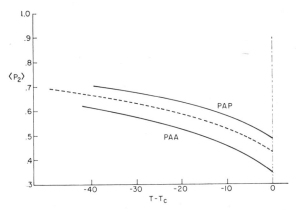

Fig. 2—Schematic representation of the temperature dependence of the order parameter for PAA and PAP. The dotted line is the result of the simple mean field theory.

The mean field theory with higher-order terms included is capable of reproducing the experimental data of Fig. 2. The addition of the term in U_4, Eq. [14], is found to be sufficient.[9] The data on PAP can be accounted for by having $U_4 \approx +0.12\ U_2$, while the data on PAA are fit by assuming $U_4 \approx -0.19\ U_2$. In the case of PAP, U_4 adds to the ordering effects of U_2 and $<P_2>$ is raised above that predicted by the simple theory. In the case of PAA, on the other hand, U_4 decreases the ordering effects of the U_2 term and a lowered $<P_2>$ results. Similar results on these two materials have also been reported by Chandrasekhar and Madhusudana.[8]

The most dramatic evidence that the interaction potential must be quite complicated comes from recent experiments by Jen, Clark, Pershan, and Priestley[10] on MBBA (N-(p'-methoxybenzylidene)-p-n butylaniline). Using the Raman scattering technique these authors were able to determine the temperature dependences of both $<P_2>$ and $<P_4>$. The experimental data are shown (with their error bars) as the dots in Fig. 2. As is readily apparent, the value of $<P_4>$ becomes quite small and then turns negative (!) just before the transition temperature is reached. The simple version of the mean field theory is just not capable of handling this situation; the mean field theory with U_2 only is represented by the solid lines. Note in particular that $<P_4>$ is predicted to remain positive right up to the transition point. The addition of the term in U_4 does not help very much; the dashed lines represent the results[10] of the theory with $U_4 \approx -0.55\ U_2$. While $<P_2>$ has been brought into agreement with experiment, $<P_4>$, though reduced in magnitude, still remains quite positive.

The physical meaning of a negative $<P_4>$ has been discussed by Jen et al.[10] These authors have demonstrated that a negative $<P_4>$ implies the existence of a broad orientational distribution function $\rho_1(\cos\theta)$, Eq. [16]. In the simple mean field theory (U_2 term only), the distribution function is sharply peaked at $\theta = 0$; that is, the probability of locating a molecule whose long axis is at an angle θ with respect to the director is extremely high for angles near zero and quite small when the angle θ becomes appreciable. In the case of MBBA, however, the probability for finite angles is almost as high as for $\theta = 0$. It therefore follows that the interaction potential V_1 required to describe the behavior of MBBA must be of a form which (at high temperatures at least) provides a very shallow minimum at $\theta = 0$ and/or subsidiary minimum at some finite angle. In attempting to reproduce the experimental results shown in Fig. 3, P. Sheng and I have tried numerous potential functions that have this feature and that are compatible with the systematic development, Eqs. [14] and [15]. In addition to terms in U_2 and U_4, Eq. [14], we have also included terms in $(<P_2>P_4 + <P_4>P_2)$, Eq. [15]. Though it has been possible to reduce the magnitude of $<P_4>$ drastically, the desired behavior as depicted in Fig. 3 has not been obtained.

Our most promising attempt at describing the experiments on MBBA was realized by using the following single molecule potential:

$$V_1(\cos\theta) = U<P_2>P_2(\cos\theta) + U'<f>f(\cos\theta),$$
$$f(\cos\theta) = P_2(\cos\theta) - 2P_4(\cos\theta) + P_6(\cos\theta).$$

[22]

This potential is obtained by taking very special combinations of terms from both Eqs. [14] and [15.] This combination was purposely chosen so that the part in U' gave a strong minimum in orientational potential energy at an angle far from the director ($\theta = 52°$, in fact). Even with this potential function (specifically engineered to give the desired result) we were unsuccessful in explaining the data on MBBA. For certain values of U'/U we did get negative values of $<P_4>$ of the right magnitude allright; it just happened, however, that the nematic phase became unstable with respect to other more complex structures. The molecular origin of the negative $<P_4>$ thus remains a mystery.

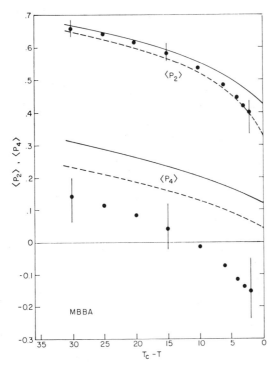

Fig. 3—Temperature dependence of the parameters $<P_2>$ and $<P_4>$ for the compound MBBA (data of Ref. [10]). The lines are theoretical curves described in the text.

A final comment. It may be that the source of the negative $<P_4>$ is not to be found in a complicated form of the potential V_1 but in improving the mean field theory itself. As is well known from studies of other cooperative phenomena, the mean field theory is only a first approximation and theories that attempt to include short-range order and fluctuation effects always show great improvements in reproducing the details of experiments. Accordingly P. Sheng and I[11] have derived a theory of the nematic phase based on the well known "constant coupling" theory of ferromagnetism (see Ref. [6]). Though short-range order is explicitely included and pair-potential functions including terms through P_4 (cos θ_{12}) have been used, the temperature dependence of $<P_4>$ for MBBA has not been realized. The explanation of the negative $<P_4>$ in MBBA, therefore, constitutes one of the outstanding theoretical problems of the nematic liquid-crystal state.

References

[1] P. J. Wojtowicz, "Introduction to the Molecular Theory of Nematic Liquid Crystals," Chapter 3.
[2] W. Maier and A. Saupe, **Z. Naturforschg.**, Vol. 14a, p 882 (1959), and Vol 15a, p. 287 (1960).
[3] R. L. Humphries, P. G. James and G. R. Luckhurst, "Molecular Field Treatment of Nematic Liquid Crystals," **J. Chem. Soc., Faraday Trans. II**, Vol. 68, p. 1031 (1972).
[4] J. A. Pople, "The Statistical Mechanics of Assemblies of Axially Symmetric Molecules. I. General Theory," **Proc. Roy. Soc.**, Vol. A221, p. 498 (1954).
[5] J. R. Sweet and W. A. Steele, "Statistical Mechanics of Linear Molecules. I. Potential Energy Functions," **J. Chem. Phys.**, Vol. 47, p. 3022 (1967).
[6] J. S. Smart, **Effective Field Theories of Magnetism**, W. B. Saunders Co., Phila., Pa. (1966).
[7] M. E. Rose, **Elementary Theory of Angular Momentum**, J. Wiley and Sons, N. Y., N. Y. (1957).
[8] S. Chandrasekhar and N. V. Madhusudana, "Molecular Statistical Theory of Nematic Liquid Crystals," **Acta Cryst.**, Vol. A27, p. 303 (1971).
[9] G. R. Luckhurst, unpublished notes.
[10] S. Jen, N. A. Clark, P. S. Pershan, and E. B. Priestley, "Raman Scattering from a Nematic Liquid Crystal: Orientational Statistics," **Phys. Rev. Lett.**, Vol. 31, No. 26, p. 1552 (1973).
[11] P. Sheng and P. J. Wojtowicz, to be published.

5

Hard Rod Model of the Nematic-Isotropic Phase Transition

Ping Sheng

RCA Laboratories, Princeton, N. J. 08540

1. Introduction

In the previous chapters we have seen how an anisotropic, attractive interaction between the molecules of the form $P_2(\cos\theta_{12})$ can give rise to a first-order nematic-isotropic phase transition. The origin of the anisotropy lies in the fact that almost all the liquid-crystal molecules are elongated, rod-like, and fairly rigid (at least in the central portion of the molecule). It is clear, however, that besides the anisotropic attractive interaction there must also be an anisotropic steric interaction that is due to the impenetrability of the molecules.

It is natural to ask what effect, if any, the steric interaction might have on the nematic-isotropic phase transition. Onsager recognized that a system of hard rods, without any attractive interaction, can have a first-order transition from the isotropic phase to the anisotropic phase as the density is increased.[1-5] To see how this can come about, we note that in a gas of hard rods there are two kinds of entropy. One is the entropy due to the translational degrees of freedom, and the other is the orientational entropy. In addition, there is a coupling between these two kinds of entropy that can be described as follows. When two hard rods lie at an angle with respect to each other, the excluded volume (the volume into which the center of mass of one molecule cannot move due to the impenetrability of the other molecule) is always larger than that when the two hard rods are parallel. This fact can easily be illustrated in two dimensions. For simplicity let us take two lines of length L. When they are perpendicular to each other, the excluded area is L^2. When they are parallel, the excluded area vanishes. This is shown in Fig. 1. It is only a short step to

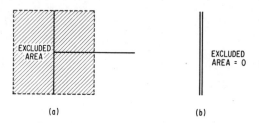

Fig. 1—Excluded area of two lines. In (a), the excluded area between two perpendicular lines is indicated by shade. In (b), the lines are parallel, and the excluded area is zero.

visualize the same situation in three dimensions where the excluded area is now replaced by excluded volume. The importance of excluded volume is that the translational entropy favors parallel alignment of the hard rods because this arrangement gives less excluded volume and, therefore, more free space for the molecules to jostle around. However, parallel alignment represents a state of low orientational entropy. Therefore, a competition exists between the tendency to maximize the translational entropy and the tendency to maximize the orientational entropy.

In the limit of zero density the tendency to maximize the orientational entropy always wins because each molecule rarely collides with another molecule, and the gain in excluded volume due to parallel alignment would only be a minimal addition to the already large volume of space within which each molecule can move about. When the density is increased, however, the excluded volume effect becomes more and more important. We know that in the limit of tight-packing density, the hard rods must be parallel. A transition between the isotropic and the anisotropic states therefore must occur at some intermediate density. But this kind of general argument cannot tell us whether the transition is smooth or abrupt. To be more precise, we will derive the Onsager equations and actually solve them in a simplified case.

2. Derivation of Onsager Equations

Consider a classical system of N hard rods in a volume V. The partition function Z, including the angular part, is[6]

$$Z(N,V,kT) = \frac{1}{N!h^{6N}} \int \cdots \int d^N p_\theta d^N p_\phi d^N p_\psi d^N\theta d^N\phi d^N\psi$$

$$\times \int \cdots \int d^{3N}x d^{3N}p \exp\left\{-\left[\sum_{i=1}^N \left(\frac{p_{\theta i}^2}{2I_1} + \frac{(p_{\phi i} - p_{\psi i}\cos\theta_i)^2}{2I_1 \sin^2\theta_i}\right.\right.\right.$$

$$\left.\left.\left. + \frac{p_{\psi i}^2}{2I_2} + \frac{p_{xi}^2 + p_{yi}^2 + p_{zi}^2}{2m}\right) + \sum_{i<j}^N v_{ij}(x,\theta,\phi)\right]/kT\right\} \quad [1]$$

Here k is the Boltzmann constant, T is the temperature, h is Plank's constant, θ, ϕ, ψ are the usual Euler angles (following the notation used in Goldstein, *Classical Mechanics*), p_θ, p_ϕ, p_ψ are the canonical conjugate angular momenta, I_1, I_2 are the two principal moments of inertia of a rigid rod, m is the mass of the rod, v_{ij} is the interparticle steric interaction, and p_x, p_y, p_z are the linear momenta of the hard rod. The kinetic energy part of the integral can be immediately integrated. Integration over the translational momenta yields a term $(2\pi mkT)^{3N/2}$. Integration over the angular momenta gives a term

$$(2\pi I_1\sqrt{2\pi I_2})^N \prod_{i=1}^N \sin\theta_i$$

Since nothing in the integrand depends on ψ, this angle can also be integrated to yield $(2\pi)^N$. Grouping the constants and combining the

$$\prod_{i=1}^N \sin\theta_i$$

factor with $d^N\theta$ yields

$$Z(N,V,kT) = \frac{(2\pi)^N (2\pi I_1^{2/3} I_2^{1/3} kT)^{3N/2} (2\pi mkT)^{3N/2}}{N!h^{6N}}$$

$$\times \int \cdots \int d^N\cos\theta d^N\phi \int \cdots \int d^{3N}x \exp\left\{-\sum_{i<j}^N v_{ij}/kT\right\}$$

$$= \frac{1}{N! \eta^N \lambda^{3N}} \int \cdots \int d^N\Omega d^{3N}x \exp\left\{-\sum_{i<j}^{N} \frac{v_{ij}}{kT}\right\}, \quad [2]$$

where $\eta = \hbar^3/[kTI_1^{2/3}(2\pi I_2)^{1/3}]^{3/2}$ is a dimensionless constant ($\hbar = h/2\pi$), $\lambda = h/\sqrt{2\pi mkT}$ is the thermal wavelength, and Ω is the solid angle. The angular integrals in Eq. [2] can be approximated to arbitrary accuracy by summations in the following manner. Divide the unit sphere into K cells, each containing a solid angle $\Delta\omega$. The orientational distribution of the rods is specified by the integers N_a, which give the number of rods oriented in the direction of ath angular cell. The angular integrals can then be replaced by sum over all the possible partitions $\{N_1, N_2, \ldots N_a, \ldots N_K\}$ with $N_1 + N_2 + \ldots + N_K = N$, multiplied by a factor $[N!/(N_1!N_2!\ldots N_K!)]$ for the number of times the partition $\{N_1, \ldots N_K\}$ has been counted in the angular integrals[5];

$$Z(N, V, kT) = \frac{1}{N! \eta^N \lambda^{3N}} \sum_{\{N_a\}} \frac{N!(\Delta\omega)^N}{N_1! \cdots N_a! \cdots N_K!}$$

$$\times \int \cdots \int d^{3N}x \exp\left[-\sum_{i<j}^{N} \frac{v_{ij}}{kT}\right] \quad (3a)$$

$$= \sum_{\{N_a\}} Z(\{N_a\}, N, V, kT), \quad [3b]$$

where $Z(\{N_a\}, N, V, kT) = \dfrac{(\Delta\omega)^N}{\eta^N \lambda^{3N} \prod_{a=1}^{K}(N_a!)} \int \cdots \int d^{3N}x \, \text{emp}\left\{-\sum_{i<j}^{N}\frac{v_{ij}}{kT}\right\}$

Let us pick the maximum term in the summation

$\sum_{\{N_a\}} Z(\{N_a\}, N, V, kT)$ and label it $Z(\{N_a\})_{\max}$.

Since the summation can have at most N^K terms, we have the following inequality:

$$Z(\{N_a\})_{\max} < Z(N, V, kT) < N^K Z(\{N_a\})_{\max}. \quad [4]$$

Taking the logarithm and dividing through by N, we have

$$\frac{1}{N}\ln Z\left(\{N_a\}\right)_{max} < \frac{1}{N}\ln Z < \frac{1}{N}\left(\ln Z\left(\{N_a\}\right)_{max} + K\ln N\right). \quad [5]$$

As $N \to \infty$, the term $(\ln N)/N$ approaches zero and $(\ln Z)/N$ can be exactly replaced by $[\ln Z(\{N_a\})_{max}]/N$. Therefore, our task can be reduced to the calculation of $Z\left(\{N_a\}, N, V, kT\right)$ as long as we remember to maximize the result with respect to $\{N_a\}$ afterwards.

Now $Z\left(\{N_a\}, N, V, kT\right)$ can be put in the following form:

$$Z\left(\{N_a\}, N, V, kT\right) = \frac{1}{\eta^N \lambda^{3N}} \frac{(\Delta\omega)^N}{\prod_{a=1}^{K}(N_a!)} \int \cdots \int d^{3N}x \exp\left\{-\sum_{i<j}^{N}\frac{v_{ij}}{kT}\right\}$$

$$= \frac{1}{\eta^N \lambda^{3N}} \frac{(\Delta\omega)^N}{\prod_{a=1}^{K}(N_a!)} \int \cdots \int d^{3N}x \prod_{i<j}^{N} \exp\left\{-\frac{v_{ij}}{kT}\right\}. \quad [6]$$

Define $\Phi_{ij} \equiv \exp\{-v_{ij}/(kT)\} - 1$. With v_{ij} denoting the steric repulsion, Φ_{ij} has the value zero when the two molecules are not in contact but takes the value -1 when the center of mass of one molecule enters the "excluded volume" of two hard rods. With this definition of Φ_{ij} we can rewrite the integrand as

$$\prod_{i<j}^{N}\exp\left\{-\frac{v_{ij}}{kT}\right\} = \prod_{i<j}^{N}(1+\Phi_{ij}) = 1 + \sum_{i<j}^{N}\Phi_{ij} + \sum_{i<j,l<m}^{N}\Phi_{ij}\Phi_{lm} + \ldots \quad [7]$$

The first term represents the behavior of the ideal gas, and the second term is the first-order correction to this behavior due to steric repulsion between the molecules. If we limit ourselves to densities that are not too high, we can neglect higher-order terms, which represent the effects of excluded volume between more than two molecules. With this crucial approximation, we have

$$Z\left(\{N_a\}, N, V, kT\right) \simeq \frac{(\Delta\omega)^N}{\eta^N\lambda^{3N}} \frac{1}{\prod_{a=1}^{K} N_a!} [V^N + V^{N-2}\sum_{i<j}^{N}\iint \Phi_{ij}\, d^3x_i\, d^3x_j]$$

$$= \frac{(\Delta\omega)^N V^N}{\eta^N \lambda^{3N} \prod_{\alpha=1}^{K}(N_\alpha!)} [1 + \frac{1}{V}\frac{1}{2}\sum_{i \neq j}^{N} \int \Phi_{ij}\, d^3 x_{ij}]. \qquad [8]$$

We note that Φ_{ij} depends on the relative separation x_{ij} and the relative orientation between two molecules. Therefore, a better way to label the indices should be $\Phi_{\alpha\beta}(x_{ij})$, where α and β denote the angular orientations of the two molecules and x_{ij} the relative separation between their centers of mass. The summation over i and j can be rewritten as

$$\sum_{i \neq j}^{N} \int \Phi_{\alpha\beta}(x_{ij})\, d^3 x_{ij} = \sum_{\alpha,\beta=1}^{K} N_\alpha (N_\beta - \delta_{\alpha\beta}) \int \Phi_{\alpha\beta}(x)\, d^3 x. \qquad [9]$$

As $N \to \infty$, the $\delta_{\alpha\beta}$ term can be neglected. The integral of $\Phi_{\alpha\beta}(x)$ is just the negative of excluded volume $V_{\alpha\beta}^{\text{ex.}}$. $Z(\{N_\alpha\}, N, V, kT)$ is now in the form

$$Z(\{N_\alpha\}, N, V, kT) = \frac{(\Delta\omega)^N V^N}{\eta^N \lambda^{3N} \prod_{\alpha=1}^{K} N_\alpha!}\left[1 - \frac{1}{2V}\sum_{\alpha,\beta=1}^{K} N_\alpha N_\beta V_{\alpha\beta}^{\text{ex.}}\right]. \qquad [10]$$

Taking the logarithm, dividing by N, and using Sterling's approximation $\ln N_\alpha! \simeq N_\alpha \ln N_\alpha$, we obtain

$$\frac{\ln[Z(\{N_\alpha\}, N, V, kT)]}{N} = \ln\frac{(\Delta\omega)V}{\eta\lambda^3} - \sum_{\alpha=1}^{K}\left(\frac{N_\alpha}{N}\ln\frac{N_\alpha}{N} + \frac{N_\alpha}{N}\ln N\right)$$

$$+ \frac{1}{N}\ln\left[1 - \frac{N^2}{2V}\sum_{\alpha,\beta}^{K}\frac{N_\alpha N_\beta}{N\ N}V_{\alpha\beta}^{\text{ex.}}\right]$$

$$= \ln\left[\frac{(\Delta\omega)V}{\eta\lambda^3 N}\right] - \sum_{\alpha=1}^{K}\frac{N_\alpha}{N}\ln\frac{N_\alpha}{N} + \ln\left[1 - \frac{N^2}{2V}\sum_{\alpha,\beta}^{K}\frac{N_\alpha N_\beta}{N\ N}V_{\alpha\beta}^{\text{ex.}}\right]^{1/N}$$

$$\simeq \ln\left[\frac{(\Delta\omega)V}{\eta\lambda^3 N}\right] - \sum_{\alpha=1}^{K}\frac{N_\alpha}{N}\ln\frac{N_\alpha}{N} + \ln\left[1 - \frac{N}{2V}\sum_{\alpha,\beta}^{K}\frac{N_\alpha N_\beta}{N\ N}V_{\alpha\beta}^{\text{ex.}}\right]$$

$$\simeq \ln\frac{(\Delta\omega)}{\eta\lambda^3 \rho} - \sum_{\alpha=1}^{K}\frac{N_\alpha}{N}\ln\frac{N_\alpha}{N} - \frac{\rho}{2}\sum_{\alpha,\beta}^{K}\frac{N_\alpha N_\beta}{N\ N}V_{\alpha\beta}^{\text{ex.}}. \qquad [11]$$

Here $\rho \equiv N/V$ is assumed to be small.

Let us define an angular distribution function f such that $N_a = N \cdot f(\Omega_a) \cdot (\Delta\omega)$. Substituting N_a/N by $f(\Omega_a)$ and making the replacement

$$\sum_\alpha (\Delta\omega) \to \int d\Omega,$$

we can put Eq. [11] in the form

$$\frac{1}{N}\ln Z(\{N_a\}, N, V, kT) = -\ln \eta\lambda^3\rho - \int d\Omega f(\Omega) \ln f(\Omega)$$

$$- \frac{\rho}{2}\int\int d\Omega d\Omega' f(\Omega)f(\Omega')V^{\text{ex.}}(\Omega - \Omega'). \qquad [12]$$

The second term on the right-hand side of Eq. [12] represents the orientational entropy, and the third term represents the effect of excluded volume. In order to maximize the right hand side of Eq. [12] with respect to $f(\Omega)$, we will use the Euler–Lagrange equation. This equation says if one wants to maximize the integral

$$I[y(x)] = \int_a^b F\left(y, \frac{dy}{dx}, x\right) dx$$

by varying the function $y(x)$, the correct $y(x)$ must satisfy the equation

$$\frac{\partial F}{\partial y} - \frac{d}{dx}\frac{\partial F}{\partial\left(\dfrac{dy}{dx}\right)} = 0.$$

If in addition $y(x)$ must satisfy the normalization condition

$$\int_a^b y(x)\,dx = 1, \text{ then the equation for } y(x) \text{ becomes}$$

$$\frac{\partial F}{\partial y} - \frac{d}{dx}\frac{\partial F}{\partial\left(\dfrac{dy}{dx}\right)} - \nu = 0,$$

where ν is the Lagrange multiplier. Applying the Euler–Lagrange equation to Eq. [12] gives

$$\ln f(\Omega) + 1 + \nu + \rho\int f(\Omega')\, V^{\text{ex.}}(\Omega - \Omega')\, d\Omega' = 0. \qquad [13]$$

Eqs. [12] and [13] are the Onsager equations. In principle, Eq. [13] can be solved to get $f(\Omega)$ as a function of ρ and ν. The quantity ν is then determined by the normalization condition $\int f(\Omega)d\Omega = 1$. When there is more than one solution to Eq. [13], the results must be put into Eq. [12] in order to select the one solution that maximizes the right-hand side of Eq. [12]. Label that solution $f^o(\Omega)$. The free energy is then

$$\frac{F}{NkT} = \ln \eta\lambda^3\rho + \int d\Omega f^o(\Omega)\ln f^o(\Omega) + \frac{\rho}{2}\int\int d\Omega d\Omega' f^o(\Omega) f^o(\Omega')\, V^{\text{ex.}}(\Omega - \Omega'). \qquad [14]$$

Other thermodynamic quantities can be obtained directly from Eq. [14].

3. Solution of Onsager Equations in a Simplified Case

In practice, Eq. [13] is a nonlinear integral equation and its exact solution in the general case is difficult. Therefore, we will solve the Onsager equations in a very simplified case. Let us constrain the hard rods to point in only three orthogonal directions, say x, y, and z. Choose z as the preferred direction. If the fraction of hard rods pointing in the x-direction is r, the same fraction should point in the y-direction, since x and y directions are equivalent by uniaxial symmetry of the nematic phase. The fraction pointing in the z-direction is then $1 - 2r$. The order parameter

$$S \equiv \int^1 P_2(\cos\theta)\, f(\cos\theta)\, d(\cos\theta)$$

in this case is simply $(1 - 3r)$. Eq. [12] can now be written

$$\frac{Z_r(N, V, kT)}{N} = -\ln \eta\lambda^3\rho - 2r \ln r - (1 - 2r) \ln (1 - 2r)$$

$$- \rho r (2 - 3r) V_\perp^{ex} - \frac{\rho}{2} (1 - 4r + 6r^2) V_\parallel^{ex}$$

$$= -\ln \frac{\eta\lambda^3\rho}{3} - \frac{1}{3} (1 + 2S) \ln (1 + 2S) - \frac{2}{3} (1 - S) \ln (1 - S)$$

$$- \frac{\rho(1 - S^2)}{3} V_\perp^{ex} - \frac{\rho}{2} \frac{1 + 2S^2}{3} V_\parallel^{ex} . \qquad [15]$$

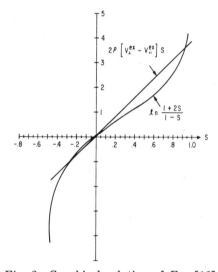

Fig. 2—Graphical solution of Eq. [16].

Here $V_\perp^{ex}$ and $V_\parallel^{ex}$ are the excluded volumes of two hard rods when they are respectively, perpendicular and parallel to each other.

To maximize $Z_r (N, V, kT)/N$, we merely differentiate with respect to S and set the result equal to zero. This gives

$$\ln \frac{1 + 2S}{1 - S} = 2\rho [V_\perp^{ex} - V_\parallel^{ex}] S , \qquad [16]$$

which is the equivalent of Eq. [13]. In order to calculate $V^{\text{ex.}}$, we will take the rods to have the shape of rectangular parallelepiped with length L and width D. Then

$$V_\perp^{\text{ex}} = \left(4 + 2\frac{L}{D} + 2\frac{D}{L}\right) D^2 L = \left(4 + 2l + \frac{2}{l}\right) V_o, \qquad [17a]$$

$$V_\parallel^{\text{ex}} = 8D^2L = 8V_o, \qquad [17b]$$

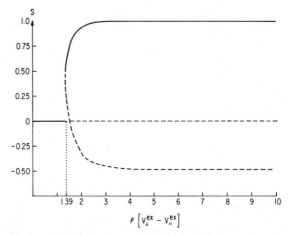

Fig. 3—Solution of Eq. [16] as a function of $\rho[V_\perp^{\text{ex}} - V_\parallel^{\text{ex}}]$. Dark lines indicate the branch that maximizes the right hand side of Eq. [15].

where $V_o = D^2L$ is the volume of a parallelepiped and $l = L/D$ is the ratio of length to breadth. Eq. [16] can be solved graphically by plotting the right- and left-hand sides on the same graph and locating the points of intersection. This is shown in Fig. 2. The results are displayed in Fig. 3 as a function of $\rho[V_\perp^{\text{ex}} - V_\parallel^{\text{ex}}]$. Dark lines indicate the stable solution. Label these values of S by $S^o(\rho)$. The free energy is

$$\frac{F}{NkT} = \ln\left[\frac{N\eta\lambda^3}{3V}\right] + \frac{1}{3}(1 + 2S^o(\rho)) \ln(1 + 2S^o(\rho))$$

$$+ \frac{2}{3}(1 - S^o(\rho)) \ln(1 - S^o(\rho))$$

$$+ 2\frac{NV_o}{V}\left[\frac{1 - (S^o(\rho))^2}{3}\left(l + \frac{1}{l} - 2\right) + 2\right]. \qquad [18]$$

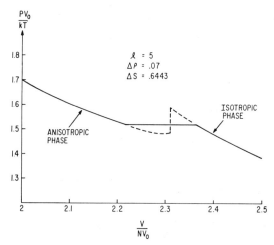

Fig. 4—Reduced pressure PV_o/kT plotted as a function of reduced volume V/V_oN for $l=5$.

The pressure is obtained by differentiating F with respect to V:

$$\frac{PV_o}{kT} = -\frac{V_o}{kT}\frac{\partial F}{\partial V} = \frac{V_oN}{V} + 2\left(\frac{V_oN}{V}\right)^2\left[\frac{1-(S^o(\rho))^2}{3}\left(l+\frac{1}{l}-2\right)+2\right],$$

[19]

where we remember that $\partial F/\partial S^o(\rho) = 0$ by definition of $S^o(\rho)$.

In Figs. 4 and 5, reduced pressure PV_o/kT is plotted versus reduced volume V/V_oN for two values of l. It should be noted that there is a

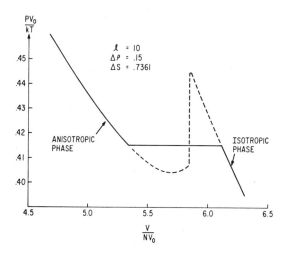

Fig. 5—Reduced pressure PV_o/kT plotted as a function of reduced volume V/V_oN for $l=10$.

van der Waals loop in each curve. Using the standard Maxwell equal-area construction, we obtain the magnitudes of density and order parameter discontinuities ($\Delta\rho$ and ΔS) across the transition. Their values are typically $\Delta\rho \sim 0.1$ and $\Delta S \sim 0.7$, which are close to the results obtained by the variational solution of the exact Onsager equations. How well do these values agree with the experimental results? Typical experimental values for $\Delta\rho$ and ΔS are $\sim .01$ and .4, respectively. Therefore, there exists a large discrepancy between the theoretical predictions and the experimental results.

What went wrong? The answer probably lies in the neglect of short-range order. In real liquid crystals there are short-range attractive interactions that favor parallel alignment of the molecules. This means that parallel molecules prefer to lie close to each other and form "bundles". In this kind of configuration, a molecule would most probably interact with those molecules pointing in similar orientations, thus reducing the excluded-volume effect. A lattice-gas model calculation[7] that includes the short-range anisotropic attractive interaction as well as the steric repulsion has shown that "bundling" would indeed soften the isotropic-anisotropic phase transition and decrease the values of $\Delta\rho$ and ΔS. Therefore, the joint consideration of short-range order and steric repulsion seems to be the direction for improving the hard-rod model.

References

[1] L. Onsager, "The Effects of Shapes on the Interaction of Colloidal Particles," **Ann. N.Y. Acad. Sci.**, Vol. 51, p. 627 (1949).

[2] R. Zwanzig, "First-Order Phase Transitions in a Gas of Long Thin Rods," **J. Chem. Phys.**, Vol. 39, p. 1714 (1963).

[3] P. J. Flory, "Phase Equilibriums in Solutions of Rodlike Particles," **Proc. Roy. Soc.** (London), Vol. A234, p. 73 (1956).

[4] E. A. DiMarzio, "Statistics of Orientation Effects in Linear Polymer Molecules," **J. Chem. Phys.**, Vol. 35, p. 658 (1961).

[5] G. Lasher, "Nematic Ordering of Hard Rods Derived from a Scaled Particle Treatment," **J. Chem. Phys.**, Vol. 53, p. 4141 (1970).

[6] R. H. Fowler, **Statistical Mechanics**, p. 62, Cambridge University Press, Cambridge, England (1966).

[7] P. Sheng, "Effects of Bundling in a Lattice Gas Model of Liquid Crystals," **J. Chem. Phys.**, Vol. 59, p.1942 (1973).

Nematic Order: The Long Range Orientational Distribution Function

E. B. Priestley
RCA Laboratories, Princeton, N. J. 08540

1. Introduction

Earlier,[1] in Chapter 3, the existence of the nematic phase was presented as an example of an order–disorder phenomenon. The symmetry and structure of the nematic phase were used to identify the natural order parameter $<P_2(\cos\theta)>$, where $P_2(\cos\theta)$ is the second Legendre polynomial and the angular brackets denote a statistical average over the orientational distribution function $f(\cos\theta)$. The orientational potential energy of a single molecule was shown, in the mean field approximation, to be

$$V(\cos\theta) = -v <P_2(\cos\theta)> P_2(\cos\theta) \qquad [1]$$

where v is a number that scales with the strength of the intermolecular interaction. Using the rules of classical statistical mechanics, the theoretical orientational distribution function $\rho(\cos\theta)$ was given in terms of the mean field potential $V(\cos\theta)$ as

$$\rho(\cos\theta) = Z^{-1} \exp[-\beta V(\cos\theta)] \qquad [2]$$

with

$$Z = \int_0^1 \exp[-\beta V(\cos\theta)] \, d(\cos\theta).$$

In Eq. [2] Z is the single-particle partition function and $\beta = 1/kT$, where k is Boltamann's constant and T is the absolute temperature.

It is important at this point to distinguish between $f(\cos\theta)$ and $\rho(\cos\theta)$. Given that the nematic phase is truly uniaxial, $f(\cos\theta)$ completely describes the long range orientational molecular order and therefore can be regarded as the exact or true single-particle distribution function. On the other hand, $\rho(\cos\theta)$ is a theoretical approximation to $f(\cos\theta)$. The extent to which it represents a *good* approximation depends upon how closely $V(\cos\theta)$ approximates the true potential of mean torque.

There is an obvious motivation for determining as much as possible experimentally about $f(\cos\theta)$; it describes the long range orientational ordering of the molecules which, of course, is the single feature that distinguishes the nematic phase from the isotropic phase. The average values of various quantities of interest pertaining to the nematic phase could be computed if $f(\cos\theta)$ were known with precision.

2. The Orientational Distribution Function

The main question we want to answer in this chapter is "What do experiments tell us about the orientational distribution function?". To answer this question it is instructive to expand $f(\cos\theta)$ in a formal mathematical sense as[2]

$$f(\cos\theta) = \sum_{\substack{l \\ even}} \frac{2l+1}{2} <P_l(\cos\theta)> P_l(\cos\theta), \qquad [3]$$

where the $P_l(\cos\theta)$ are the lth even order Legendre polynomials. Notice that the expansion has the correct symmetry and that $f(\cos\theta)$ is normalized. The coefficients $<P_l(\cos\theta)>$ are defined by

$$\langle P_l(\cos\theta)\rangle = \int_{-1}^{1} P_l(\cos\theta) f(\cos\theta) d(\cos\theta). \qquad [4]$$

Explicitly, the first three coefficients are

$$\langle P_o(\cos\theta)\rangle = 1,$$

$$\langle P_2(\cos\theta)\rangle = \frac{1}{2}(3\langle\cos^2\theta\rangle - 1), \qquad [5]$$

$$\langle P_4(\cos\theta)\rangle = \frac{1}{8}(35\langle\cos^4\theta\rangle - 30\langle\cos^2\theta\rangle + 3).$$

There is no practical experimental means of measuring $f(\cos\theta)$ directly. Rather, individual experiments each detect some property of the nematic phase averaged over $f(\cos\theta)$. Thus, experiments are sensitive to $f(\cos\theta)$ only through its influence on these averaged properties. The most commonly measured such property is the temperature dependence of $\langle\cos^2\theta\rangle$ and, hence, the temperature dependence of the order parameter $\langle P_2(\cos\theta)\rangle$. Recent measurements of $\langle\cos^4\theta\rangle$ as a function of temperature for two nematic materials[3] have permitted the evaluation of $\langle P_4(\cos\theta)\rangle$ over the entire nematic temperature range as well. Therefore, the answer to the question posed at the beginning of this section is that experiments measure the coefficients in the expansion of $f(\cos\theta)$, (Eq. [3]). In other words, experiments measure different moments of the orientational distribution function. These experimentally determined moments can be used to evaluate truncated series approximations to $f(\cos\theta)$. It is convenient to label these approximate distribution functions as $f^N(\cos\theta)$ where N is the Roman numeral corresponding to the number of terms retained in the truncated series.

In the discussion so far, the constituent molecules of the nematic phase have been assumed to be representable by simple rigid rods so as to make the definition of a molecular axis unambiguous. The order parameter $\langle P_2(\cos\theta)\rangle$ has been defined from a microscopic point of view by treating it as a statistical average of individual molecular behavior. In most cases this is an adequate description since real nematogens most often do behave like simple rigid rods. However, if the molecules do not behave like simple rigid rods, this microscopic description of order is no longer adequate and we must find some other means for specifying the degree of order. For this, we turn to a consideration of macroscopic response functions of the nematic phase and show how a suitable order parameter can be defined using these response functions.

3. Macroscopic Definition of Nematic Order

In deriving a macroscopic order parameter, we will use the diamagnetic susceptibility as an example. However, any other macroscopic property, e.g., refractive index or dielectric response, could be used as well.

Consider the relationship between magnetic moment $\vec{M}$ and magnetic field $\vec{H}$

$$M_\alpha = \chi_{\alpha\beta} H_\beta; \qquad \alpha,\beta = x,y,z. \tag{6}$$

$\chi_{\alpha\beta}$ represents the $\alpha\beta$-component of the diamagnetic susceptibility tensor $\overleftrightarrow{\chi}$ and, for static fields, $\chi_{\alpha\beta} = \chi_{\beta\alpha}$. In Eq. [6], the summation convention over repeated indices is implied. For the uniaxial nematic phase, we can write $\overleftrightarrow{\chi}$ in the diagonal, but completely general form

$$\overleftrightarrow{\chi} = \begin{pmatrix} \chi_\perp & 0 & 0 \\ 0 & \chi_\perp & 0 \\ 0 & 0 & \chi_\parallel \end{pmatrix}, \tag{7}$$

where $\chi_\parallel$ and $\chi_\perp$ refer to the susceptibility parallel and perpendicular to the symmetry axis, respectively. $\overleftrightarrow{\chi}$ itself is not a useful order parameter because it does not vanish in the isotropic phase, but has the value

$$\begin{aligned}\chi &= \frac{1}{3}(2\chi_\perp + \chi_\parallel) \\ &= \frac{1}{3}\chi_{\gamma\gamma}.\end{aligned} \tag{8}$$

However, if we extract the anisotropic part of $\overleftrightarrow{\chi}$, viz $\overleftrightarrow{\chi}^a$ defined by

$$\chi^a{}_{\alpha\beta} = \chi_{\alpha\beta} - \frac{1}{3}\chi_{\gamma\gamma}\delta_{\alpha\beta}, \tag{9}$$

which does vanish in the isotropic phase, we can use it to define a suitable order parameter.

Combining Eqs. [7] and [9] we find for the nematic phase

$$\overleftrightarrow{\chi^a} = \frac{2}{3} \Delta\chi \begin{pmatrix} -1/2 & 0 & 0 \\ 0 & -1/2 & 0 \\ 0 & 0 & 1 \end{pmatrix}, \qquad [10]$$

where $\Delta\chi = \chi_\parallel - \chi_\perp$. It should be noted that $\overleftrightarrow{\chi^a}$ is temperature dependent because the magnitude of $\Delta\chi$ is temperature dependent. If we denote the maximum possible diamagnetic anisotropy (i.e., the anisotropy for a perfectly ordered nematic phase) as $(\Delta\chi)_{max}$, we can define the desired order parameter as

$$Q_{\alpha\beta} = \frac{3}{2} \frac{\chi^a{}_{\alpha\beta}}{(\Delta\chi)_{max}}. \qquad [11]$$

Notice that $Q_{\alpha\beta}$ has values between 0 (isotropic phase) and 1 (perfectly ordered nematic phase) and that it has been derived without any assumption about molecular rigidity. $Q_{\alpha\beta}$ is a valid measure of the order for any nematic liquid crystal. As mentioned earlier, $Q_{\alpha\beta}$ could have been defined in terms of the anisotropy in any other macroscopic response function.

One might wonder if there is any relationship between $Q_{\alpha\beta}$ and $<P_2 (\cos \theta)>$. Indeed, *in nematic phases whose constituent molecules can be approximated by simple rigid rods*, these two measures of the long range order are related. The next section deals explicitly with this relationship.

4. Relationship Between Microscopic and Macroscopic Order Parameters

The diamagnetic susceptibility $\overleftrightarrow{\zeta}$ of a single rod-like molecule can be described by two parameters $\zeta_\parallel$ and $\zeta_\perp$, representing the molecular response to a magnetic field applied parallel and perpendicular to the molecular axis, respectively. In tensor notation the magnetic susceptibility of the molecule is given by

$$\overleftrightarrow{\zeta} = \begin{pmatrix} \zeta_\perp & 0 & 0 \\ 0 & \zeta_\perp & 0 \\ 0 & 0 & \zeta_\parallel \end{pmatrix}. \qquad [12]$$

To find the macroscopic diamagnetic anisotropy we first rotate $\overleftrightarrow{\zeta}$ by an arbitrary rotation $R(\phi, \theta, \psi)$, where ϕ, θ, and ψ are the Euler angles of the rotation, and then perform a statistical average. In a uniaxial nematic, the angles ϕ and ψ are isotropically distributed so averages over them can be carried out explicitly. The orientational distribution function $f(\cos\theta)$ describes the distribution of the angle θ, and averages over θ are designated by angular brackets.

From the general formalism for rotation of second rank tensors[4] it can be shown that

$$\zeta_{xx}^{ROT} = \frac{1}{3}(2\zeta_\perp + \zeta_\parallel) + \frac{1}{6}(\zeta_\parallel - \zeta_\perp)[3(1-\cos^2\theta)\cos 2\psi - (3\cos^2\theta - 1)].$$

and [13]

$$\zeta_{zz}^{ROT} = \frac{1}{3}(2\zeta_\perp + \zeta_\parallel) + \frac{1}{3}(\zeta_\parallel - \zeta_\perp)(3\cos^2\theta - 1),$$

where ζ_{ij}^{ROT} are components of the rotated molecular diamagnetic susceptibility tensor. The macroscopic diamagnetic anisotropy is then

$$\Delta\chi = N[<\zeta_{zz}^{ROT}> - <\zeta_{xx}^{ROT}>],\quad [14]$$

where N is the number of molecules per cubic centimeter. Taking the averages of the ζ_{ij}^{ROT} and substituting into Eq. [14] gives

$$\Delta\chi = N(\zeta_\parallel - \zeta_\perp)<P_2(\cos\theta)>.\quad [15]$$

But $N(\zeta_\parallel - \zeta_\perp)$ is simply the macroscopic diamagnetic anistropy of the perfectly ordered nematic phase, i.e.,

$$(\Delta\chi)_{max} = N(\zeta_\parallel - \zeta_\perp).\quad [16]$$

Thus,

$$\Delta\chi = (\Delta\chi)_{max}<P_2(\cos\theta)>.\quad [17]$$

Combining Eqs. [10], [11], and [17] yields

$$Q_{\alpha\beta} = <P_2(\cos\theta)>\begin{pmatrix} -1/2 & 0 & 0 \\ 0 & -1/2 & 0 \\ 0 & 0 & 1 \end{pmatrix}\quad [18]$$

for a nematic liquid crystal composed of rod-like molecules.

NEMATIC ORDER

As noted at the beginning of Section 3, $Q_{\alpha\beta}$ can be defined equally well in terms of other macroscopic response functions such as the dielectric tensor $\overleftrightarrow{\epsilon}$. However, a theoretical relationship between $\Delta\epsilon$ (macroscopic measure of the order) and $<P_2(\cos\theta)>$ (microscopic measure of the order) cannot be derived in the absence of certain questionable assumptions. This inability to establish a rigorous relationship analogous to Eq. [14] between the dielectric anisotropy $\Delta\epsilon$ and the polarizability α_{ij} results from complicated depolarization effects caused by the relatively large near-neighbor electrostatic interaction. Nevertheless, it has been established empirically[5] that the analog of Eq. [17] does hold, i.e., the macroscopic anisotropy does scale directly with the microscopic order parameter $<P_2(\cos\theta)>$. Consequently, electrical and optical anisotropy measurements continue to be used to measure $<P_2(\cos\theta)>$ in nematics composed of rod-like molecules, even though theoretical justification for this is presently lacking.

5. Experimental Measurements

Various experimental methods have been used to examine molecular ordering in nematic liquid crystals[6]. All the methods fall into one of two groups—those that measure the anisotropy in some macroscopic response function and those that measure $<P_2(\cos\theta)>$ directly. As we have already seen, those in the first group also permit a determination of $<P_2(\cos\theta)>$ because of equations analogous to Eg. [17] that relate the macroscopic anisotropy to the microscopic order. In general, there is quite good agreement among the various measures of $<P_2(\cos\theta)>$ for materials that have been studied by one or more methods from each group, indicating that nematogenic molecules can be approximated rather well by hard rods. We turn now to a brief description of the various experimental techniques and refer the interested reader to more detailed discussions in each instance.

5.1 Measurements of $<P_2(\cos\theta)>$ Based on Macroscopic Anisotropies

Little need be said concerning this type of order parameter determination. It is evident from Eq. [17] that the only requirement in evaluating $<P_2(\cos\theta)>$ from macroscopic anisotropy measurements is a knowledge of the maximum possible magnitude of the anisotropy being measured. Clearly, this is the anisotropy that would be measured in a perfectly ordered nematic phase and can be obtained by means of a little algebra from the crystalline anisotropy. Diamagnetic, dielectric, and optical anisotropy measurements have been used extensively to determine the temperature dependence of $<P_2(\cos\theta)>$ for a variety of nematic liquid crystals. Further details about the conduct of these experiments and analysis of the data can be found in References [6] and [7].

5.2 Measurements of $<P_2(\cos\theta)>$ Based on Microscopic Anisotropies

In this section we group techniques that sense the anisotropy in various properties of *individual molecules* as opposed to those discussed above, which sense bulk anisotropies. It is impossible to measure the anisotropy of a single molecule in a nematic phase; rather, these techniques measure the statistical average (both temporal and spatial) of the molecular anisotropy. The averaging process results in the measured anisotropy being proportional to $<P_2(\cos\theta)>$ for most of these techniques. For the technique based on Raman scattering measurements, the anisotropy is related to both $<P_2(\cos\theta)>$ and $<P_4(\cos\theta)>$.

(a) *Magnetic Resonance Techniques*

An excellent review of magnetic resonance methods has been given recently by Brown et al.[8] We simply point out that, for molecules having nuclei with spin $I = 1/2$ (such as H), the dipole-dipole splitting in the NMR spectrum of the nematic phase, due to a proximal pair of such nuclei in the same molecule, is proportional to $<P_2(\cos\theta)>$. In addition, for molecules having nuclei with spin $I \geq 1$ (such as D and ^{14}N), the quadrupole splitting can be used to determine $<P_2(\cos\theta)>$. Finally, anisotropy in the Zeeman and hyperfine splitting observed in the EPR spectrum of free radicals dissolved in nematic liquid crystals also allows one to determine the order parameter, assuming the paramagnetic guest aligns sympathetically with the host nematic material.

(b) *Raman Scattering Technique*

Details of the Raman scattering technique have been published elsewhere.[3] Successful utilization of the technique requires the existence of a (preferably) strong, narrow, anisotropic Raman line associated with an identifiable vibration of the nematogen. It is straightforward to show that the anisotropy in the Raman scattering from such a vibration observed in the laboratory frame of reference contains information about the statistical averages $<P_2(\cos\theta)>$ and $<P_4(\cos\theta)>$. This unique feature of the Raman technique, viz. the ability to determine $<P_4>$ as well as $<P_2>$, stems from the fourth rank tensor nature of the scattering interaction.

NEMATIC ORDER

Having three moments $<P_o>$, $<P_2>$ and $<P_4>$ of the orientational distribution function allows us to evaluate $f^{III}(\cos\theta)$ (See Sec. 2). In the next section we present NMR, optical and magnetic anisotropy, and Raman measurements of $<P_2(\cos\theta)>$, Raman measurements of $<P_4(\cos\theta)>$, and representative plots of $f^{III}(\cos\theta)$.

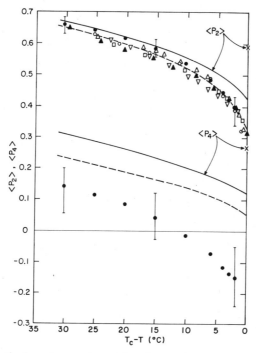

Fig. 1—Theoretical and experimental values of the nematic order parameters $<P_2(\cos\theta)>$ and $<P_4(\cos\theta)>$: solid line, theoretical results of simple mean field theory[9]; dashed line, HJL theory[10]; crosses, Onsager-Lakatos theory[13]; filled circles, Raman measurements; open circles, NMR data on partially deuterated MBBA; squares, relative values obtained from measurements of the optical anisotropy; and triangles, relative values obtained from measurements of the diamagnetic anisotropy.[7] (Reprinted from Ref. [3].)

6. Experimental Data

In Fig. 1 we display the data of Reference [3]. Notice that the $<P_2>$ results from NMR, optical and diamagnetic anisotropy, and Raman measurements all agree very well. This is reasonably compelling evidence that all of these techniques, and the Raman technique in particular, are indeed measuring $<P_2(\cos\theta)>$. The $<P_4>$ data, which come only from the Raman measurements, are seen to become negative near the nematic-isotropic phase transition. The physical meaning and implications of negative $<P_4>$ have been discussed by Jen et al[3] and by P. Wojtowicz[9] in Chapter 4.

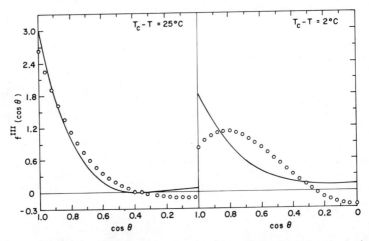

Fig. 2—Plot of the theoretical and experimental truncated orientational distribution function f^{III} ($\cos \theta$); solid line, HJL theory;[10] circles, Raman measurements. (Reprinted from Ref. [3].)

Fig. 2 shows a plot of f^{III} ($\cos \theta$) for two temperatures using the experimental Raman values of $<P_2>$ and $<P_4>$ in Eq. [3]. For comparison, we also plot f^{III} ($\cos \theta$) for the same two temperatures using values of $<P_2>$ and $<P_4>$ calculated from the Humphries-James-Luckhurst model[9,10] after the parameters were adjusted to obtain good agreement with the $<P_2>$ data. Note that f^{III} need not be positive definite because of truncation errors. The principal result to see in Fig. 2 is that the molecules have a stronger tendency to be tipped away from the nematic axis than is predicted by mean field theory. *This tendency is strongest near the nematic-isotropic phase transition.* In addition to mean field theories, other statistical mechanical models of nematic ordering have been developed. In particular there is the Onsager model[11,12] for the ordering of a system of hard rods as a function of density. Using this model, Lakatos[13] has recently calculated values of $<P_l (\cos \theta)>$ for all l. In Fig. 1 we also show these results for $<P_2>$ and $<P_4>$ at the transition density. The ratio of $<P_4>/<P_2>$ as a function of $<P_2>$ is essentially the same as that calculated from mean field theory and it also disagrees with the experimental results.

There has been considerable speculation[3,9] regarding the origin of the discrepancy between experiment and mean field theory. As of this writing, that discrepancy remains unexplained, though extensive work is underway on both the experimental and theoretical fronts in an attempt to better understand the exact nature of nematic ordering and the single-particle orientational distribution function $f(\cos\theta)$.

References

[1] P. J. Wojtowicz, "Introduction to the Molecular Theory of Nematic Liquid Crystals," Chapter 3.

[2] E. B. Priestley, P. S. Pershan, R. B. Meyer, and D. H. Dolphin, "Raman Scattering from Nematic Liquid Crystals. A Determination of the Degree of Ordering," **Vijnana Parishad Anusandhan Patrika**, Vol. 14, p. 93 (1971).

[3] Shen Jen, N. A. Clark, P. S. Pershan, and E. B. Priestley, "Raman Scattering from a Nematic Liquid Crystal: Orientational Statistics," **Phys. Rev. Lett.**, Vol. 31, p. 1552 (1973); E. B. Priestley and P. S. Pershan, "Investigation of Nematic Ordering Using Raman Scattering," **Mol. Cryst. Liquid Cryst.**, Vol. 23, p. 369 (1973); and E. B. Priestley and A. E. Bell (to be published).

[4] E. B. Priestley and P. J. Wojtowicz, unpublished results.

[5] N. V. Madhusudana, R. Shashidhar, and S. Chandrasekhar, "Orientational Order in Anisaldazine in the Nematic Phase," **Mol. Cryst. Liquid Cryst.**, Vol. 13, p. 61 (1971).

[6] A. Saupe and W. Maier, "Methods for the Determination of the Degree of Order in Nematic Liquid-Crystal Layers," **Z. Naturforschg.**, Vol. 16a, p. 816 (1961).

[7] G. Sigaud and H. Gasparoux, **J. Chem. Phys. Physicochim. Biol.**, Vol. 70, p. 669 (1973); I. Haller, "Elastic Constants of the Nematic Liquid Crystalline Phase of p-Methoxybenzylidene-p-n-Butylaniline (MBBA)," **J. Chem. Phys.**, Vol. 57, p. 1400 (1972); P. I. Rose, Fourth International Liquid Crystal Conf., Kent State University, Kent, Ohio, 1972 (to be published) see also Ref. (5).

[8] G. H. Brown, J. W. Doane, and V. D. Neff, **A Review of the Structure and Physical Properties of Liquid Crystals**, The CRC Press, Cleveland, Ohio (1971).

[9] P. J. Wojtowicz, "Generalized Mean Field Theory of Nematic Liquid Crystals," Chapter 4.

[10] R. L. Humphries, P. G. James, and G. R. Luckhurst, "Molecular Field Treatment of Nematic Liquid Crystals," **J. Chem. Soc. Faraday Trans. II**, Vol. 68, p 1031 (1972).

[11] L. Onsager, "The Effects of Shapes on the Interaction of Colloidal Particles," **Ann. N.Y. Acad. Sci.**, Vol. 51, p. 627 (1949).

[12] P. Sheng, "Hard Rod Model of the Nematic-Isotropic Phase Transition," Chapter 5.

[13] K. Lakatos, **J. Status Phys.**, Vol. 2, p. 121 (1970).

Introduction to the Molecular Theory of Smectic-A Liquid Crystals

Peter J. Wojtowicz

RCA Laboratories, Princeton, N. J. 08540

1. Introduction

Contributions to the theory of smectic-A liquid crystals have been made by a number of investigators.[1-5] In all cases the treatments are an extension of the Maier–Saupe[6] mean-field model of nematics examined in a previous chapter.[7] Here we essentially follow the development of McMillan.[3,4]

The symmetry and structure of the smectic-A liquid crystals are reviewed; the natural order parameters are identified. The relationship of the smectic-A phase to the nematic (or cholesteric) and isotropic phases in homologous series is also examined. The McMillan form of the single molecule potential function is then deduced starting from the Kabayashi form of the potential[1,2] and using the formal development presented earlier.[8] The derivation of the statistical thermodynamics then follows, along with a presentation of McMillan's numerical results and a comparison with experiment. Improvements in the theory introduced by Lee et al[5] are also considered. In the last section, the important question of whether the smectic-A to nematic (cholesteric) phase transition can ever be second order is examined.

2. Symmetry, Structure and Order Parameters

An examination of the optical properties of smectic-A liquid crystals shows that they have uniaxial symmetry. Just as in the nematic phase, the smectic-A phase has a unique axis (again called the director and denoted by $\hat{n}$) along which the elongated rod-like molecules tend to align. In addition, x-ray diffraction from smectic-A liquid crystals displays one sharp ring demonstrating that this phase possesses one-dimensional translational periodicity. The structure is depicted in Fig. 1. The centers of mass of the molecules tend to lie on planes perpendicular to the director. The spacing between planes, d, is approximately a molecular length. There is no ordering of the centers of mass of the molecules within the planes.

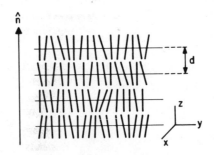

Fig. 1—Schematic representation of the structure of the smectic-A phase of liquid crystals.

As in the nematics, the orientational order of the molecules is described by the order parameter $\langle P_2(\cos\theta)\rangle$, where P_2 is the second-order Legendre polynomial, θ is the angle between the long axis of the molecule and the director, and $\langle\ \rangle$ denotes the average value. A simple phenomenological deduction of this orientational order parameter was presented in a previous chapter.[7] The identification of the order parameter required to describe the periodic layering of the molecules is not as straightforward, however. For the smectic-A structure we must examine the problem more formally.

In the case of nematics, Priestley[9] has described how the orientational distribution function could be expanded in a series of even-order Legendre polynomials:

$$f(\cos\theta) = \sum_{L(\text{even})} \frac{2L+1}{2} \langle P_L(\cos\theta)\rangle P_L(\cos\theta). \qquad [1]$$

The traditional order parameter, $\langle P_2(\cos\theta)\rangle$ appears in the first nontrivial term in the series. Succeeding terms contain the average values of higher-order Legendre polynomials, which can be thought of as order parameters of higher degree. The $\langle P_L \rangle$ thus describe features of increasing subtlety in the orientational ordering, and many are clearly required to give a good account of the true orientational distribution function.

How can this formal treatment of the distribution function (and resulting order parameters) be generalized to include the smectic-A structure? We find the clue in Kirkwood's treatment[10,11] of the melting of crystalline solids. In a crystal the density distribution function (the translational molecular distribution function) is periodic in three dimensions and can be expanded in a three-dimensional Fourier series. Kirkwood does this and then identifies the order parameters of the crystalline phase as the coefficients in the Fourier series. For simplicity let us consider a one-dimensionally periodic structure (such as the smectic-A but with the orientational order suppressed for the moment). The distribution function, which describes the tendency of the centers of mass of molecules to lie in layers perpendicular to the z-direction, can be expanded in a Fourier series:

$$f(z) = \sum_{n=0} \alpha_n \cos\left(\frac{2\pi n z}{d}\right), \qquad [2]$$

$$\int_0^d f(z)dz = 1, \qquad [3]$$

where d is the layer spacing and Eq. [3] expresses the normalization condition. Since the distribution is periodic we need only integrate over a single period. We now find the coefficients, α_n by multiplying both sides of Eq. [2] by $\cos(2\pi mz/d)$ and integrating:

$$\alpha_m = \frac{2}{d}\int_0^d \cos\left(\frac{2\pi m z}{d}\right) f(z) dz. \qquad [4]$$

The integral on the right hand side is immediately recognized as the definition of the average value, so that

$$\alpha_m = \frac{2}{d}\left\langle \cos\left(\frac{2\pi m z}{d}\right)\right\rangle. \qquad [5]$$

For the special case of $m = 0$, $\alpha_0 = 1/d$. Combining these results we obtain

$$f(z) = \frac{1}{d} + \frac{2}{d}\sum_{n=1}\left\langle \cos\left(\frac{2\pi nz}{d}\right)\right\rangle \cos\left(\frac{2\pi nz}{d}\right). \quad [6]$$

The coefficients in the series and hence the order parameters turn out to be the average values of the cosine functions of the series (in complete analogy to the situation in Eq. [1]). When the structure has perfect periodic order, all the $\langle\cos(2\pi nz/d)\rangle$ have the value unity; for the completely disordered system with all molecules randomly distributed in z, all $\langle\cos(2\pi nz/d)\rangle$ vanish. Again, many order parameters are required to make a good approximation to the distribution function, Eq. [6].

The smectic-A liquid crystals possess both orientational and translational order. The molecular distribution function must therefore describe both the tendency of the molecules to orient along $\hat{n}$ and to form layers perpendicular to $\hat{n}$. The distribution function is thus a function of both $\cos\theta$ and z, and can be expanded in a double series:

$$f(\cos\theta, z) = \sum_{L=0}\sum_{\substack{n=0 \\ \text{(even)}}} A_{Ln} P_L(\cos\theta)\cos\left(\frac{2\pi nz}{d}\right), \quad [7]$$

$$\int_{-1}^{1}\int_{0}^{d} f(\cos\theta, z)dz d(\cos\theta) = 1. \quad [8]$$

The coefficients A_{Ln} are found by multiplying both sides of Eq. [7] by $P_K(\cos\theta)\cos(2\pi mz/d)$, integrating and recognizing the definition of averages:

$$\langle X\rangle \equiv \int_{-1}^{1}\int_{0}^{d} X f(\cos\theta, z)dz d(\cos\theta). \quad [9]$$

The results are

$$A_{oo} = 1/2d,$$
$$A_{on} = \frac{1}{d}\left\langle\cos\left(\frac{2\pi nz}{d}\right)\right\rangle, (n \neq 0),$$
$$A_{Lo} = \frac{2L+1}{2d}\langle P_L(\cos\theta)\rangle, (L \neq 0), \quad [10]$$
$$A_{Ln} = \frac{2L+1}{2d}\left\langle P_L(\cos\theta)\cos\left(\frac{2\pi nz}{d}\right)\right\rangle, (L, n \neq 0).$$

In addition to the purely orientational and translational order parameters, the $\langle P_L(\cos\theta)\rangle$ and $\langle\cos(2\pi nz/d)\rangle$, we find the set of mixed-order parameters, $\langle P_L(\cos\theta)\cos(2\pi nz/d)\rangle$. These describe the correlation or coupling between the degrees of orientational and translational order. The three order parameters of lowest degree in Eq. [10] appear in all the published theories[1-5] of the smectic-A phase and have been given special symbols:

$$\eta \equiv \langle P_2(\cos\theta) \rangle,$$
$$\tau \equiv \langle \cos(2\pi z/d) \rangle, \qquad [11]$$
$$\sigma \equiv \langle P_2(\cos\theta) \cos(2\pi z/d) \rangle.$$

In the isotropic phase, $\eta = \tau = \sigma = 0$; in the nematic phase, $\eta \neq 0$, $\tau = \sigma = 0$; in the smectic-A phase $\eta \neq 0$, $\tau \neq 0$, $\sigma \neq 0$. For perfect order all three tend to unity. Part of the task of molecular theory is, of course, to calculate the temperature dependence of these order parameters. Again we point out that although the three quantities of Eq. [11] are sufficient to parametrize simple mean field models, a good approximation to the true distribution function, $f(\cos\theta,z)$ requires many terms in Eq. [7].

3. Phase Diagrams

Of special interest to the molecular theory are the transition temperatures at which the various liquid-crystal phases transform into each other and into the isotropic fluid. The collection of such temperatures in homologous series can be conveniently summarized in phase diagrams such as those schematically depicted in Fig. 2. The phase diagram for the homologous series of 4-ethoxybenzal-4-amino-n-alkyl-α-methyl cinnamates is displayed in Fig. 2a. The regions of stability of the smectic-A, nematic and isotropic phases are shown. Fig. 2b depicts the phase diagram for the homologous series of the cholesteryl esters of saturated aliphatic acids. Here the regions of stability of the smectic-A, cholesteric, and isotropic phases are shown. For the present purpose we can treat the cholesteric phase as thermodynamically similar to the nematic; the terms in the free energy, which differentiate between cholesteric and nematic structures, are very small and may be neglected in this context. Three major features of these diagrams should be noted: (a) the nematic (cholesteric) to isotropic transition temperature $T_{NI}(T_{CI})$ decreases strongly with increasing chain length, (b) the smectic-A to nematic (cholesteric) transition temperature $T_{AN}(T_{AC})$ first increases with chain length, then stays constant or gently decreases, and (c) $T_{NI}(T_{CI})$ and $T_{AN}(T_{AC})$ are converging with increasing chain length, so that for sufficiently long chains the smectic-A phase transforms directly into the isotropic fluid without passing through the nematic (cholesteric) phase.

Another experimental quantity of interest is the entropy of transition from smectic-A to nematic (cholesteric) structure. Just as in the

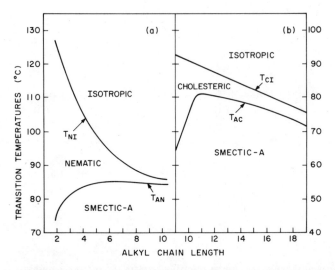

Fig. 2—(a) Schematic representation of the phase diagram of the homologous series of 4-ethoxybenzol-4-amino-n-alkyl-α-methyl cinnamates (after Ref. [3], data of Ref. [12]). (b) Schematic representation of the phase diagram of the homologous series of cholesteryl esters of saturated aliphatic acids (after Ref. [3], data of Ref. [13]).

case of the nematic to isotropic transition, the entropy changes are very small, in keeping with the fact that only one degree of freedom of the molecules (translational motion in the z-direction) is being influenced at $T_{AN}(T_{AC})$. For both of the examples shown in Fig. 2 the transition entropies range from about 0.2 to 1.2 cal/° mole, being low for short chains and increasing with increasing chain length.[12,13]

4. The Molecular Potential

The stability of the smectic-A structure is a direct consequence of the interactions between the constituent molecules. Even though we have virtually no detailed knowledge of their precise nature, we do know that there must be both orientation and distance dependence in the intermolecular pair potentials. That is, there must exist forces that cause the molecules to align parallel to each other and to form layers perpendicular to the director. Kobayashi[1,2] has suggested a simple form of pair interaction potential that contains the minimum necessary features:

$$V_{12} = U(r) + W(r)P_2(\cos\theta_{12}), \qquad [12]$$

where r is the separation between the centers of mass of the molecules and θ_{12} is the angle between their long axes. The functional dependence of U and W on r is not specified; $U(r)$ represents the short range central forces while $W(r)$ describes the orientational forces due to the anisotropic dispersion forces, quadrupole–quadrupole forces, etc.

An exact statistical theory of smectics based on the pair potential, Eq. [12], is extremely difficult to accomplish. Therefore we derive a mean-field approximation to the theory. For this purpose we require the mean-field version of the single molecule potential function. In a previous chapter[8] this problem was examined for the case of the nematic phase. A perfectly general form of V_{12} was assumed and expanded in a series of spherical harmonics. A new coordinate system was then chosen such that the polar axes coincided with the director. The single molecule potential was then obtained by averaging V_{12} over all possible positions and orientations of molecule 2 consistent with the structure of the nematic phase. The resulting single molecule potential had the form

$$V_1(\cos\theta) = \sum_{L}^{(\text{even})} \left\langle \sum_m U_{LLm}(r) \right\rangle \langle P_L \rangle P_L(\cos\theta), \qquad [13]$$

where the constants $\langle \sum_m U_{LLm}(r) \rangle$ are the averages of the distance dependent parts of the potential (averaged over all intermolecular separations), and where the $\langle P_L \rangle$ are the averages of the Legendre polynomials averaged with respect to the nematic orientational distribution function. The term in $L = 0$ is a constant and was discarded; the term in $L = 2$ led to the Maier–Saupe[6] version of the theory of nematics.

The analogous process applied to the Kobayashi potential, Eq. [12], in the case of the smectic-A structure is at the same time simpler and more complex. It is simpler because Eq. [12] exhibits a simple angular dependence and higher-order P_L do not enter the calculation. It is more complex, however, because we are now required to average over the positons and orientations of the second molecule in a way consistent with the smectic-A structure; that is, with a distribution function that depends on both angular and spatial coordinates, the $f(\cos\theta, z)$ discussed in Section 2 above. Applying these averaging procedures to Eq. [12] we obtain the single molecule potential as

$$V_1(\cos\theta, z) = \langle U(r) \rangle + \langle W(r)P_2 \rangle P_2(\cos\theta), \quad [14]$$

where the averages $\langle U(r) \rangle$ and $\langle W(r)P_2 \rangle$ are functions of z, the position of the centers of mass of the molecule of interest with respect to the layers, and where θ is the angle between the axis of this molecule and $\hat{n}$. In obtaining Eq. [14] we have used the relation, $P_2(\cos\theta_{12}) = P_2(\cos\theta_1)P_2(\cos\theta_2)$ + terms in $\varphi_2 - \varphi_1$. The terms in the azimuthal angle φ vanish in the averaging process since the smectic-A phase has cylindrical symmetry. It has been customary in the treatment[1-4] of the smectic-A phase to further simplify the potential by expanding the position dependent terms in a Fourier series. Taking $U(r)$ as an example:

$$U(r_{12}) = \frac{2}{\pi} \int_0^\infty \tilde{U}(x_{12}, y_{12}; s) \cos s z_{12} \, ds,$$

where $\tilde{U}$ is the Fourier transform of U,

$$\tilde{U}(s) = \tilde{U}(x_{12}, y_{12}; s) = \frac{1}{2\pi} \int_0^\infty U(r_{12}) \cos s z_{12} \, dz_{12}.$$

Taking the average of U over the smectic-A distribution of molecule 1,

$$\langle U(r_{12}) \rangle = \sum_n \langle U_n \rangle \left\langle \cos\left(\frac{2\pi n z_1}{d}\right) \right\rangle \cos\left(\frac{2\pi n z_2}{d}\right), \quad [15]$$

where the coefficient $\langle U_n \rangle = \langle \tilde{U}(2\pi n/d) \rangle (2/\pi)$. The several steps required to obtain this form of Eq. [15] are presented in the Appendix. Since there is no ordering of the molecules in the layers, the $\langle U_n \rangle$ are just constants. The same considerations as above apply to the $W(r)$ portion of the potential. Retaining only the first few terms gives the result

$$V_1(\cos\theta, z) = U_o + U_1 \tau \cos\left(\frac{2\pi z}{d}\right) + \ldots$$
$$+ \left[W_o \eta + W_1 \sigma \cos\left(\frac{2\pi z}{d}\right) + \ldots \right] P_2(\cos\theta), \quad [15']$$

where U_0, U_1, W_0, and W_1 are the Fourier coefficients of U and W, and where η, τ, and σ are the order parameters, the average values introduced earlier, Eq. [11]. U_0 is a constant and can be discarded. This version of the potential function is particularly instructive in demonstrating the cooperative nature of the formation of the smectic-A structure. The U_1 term shows the influence of the translational order τ in forcing the molecules into layers, the W_0 term shows the influence of the orientational order parameter η in forcing the molecules to align with $\hat{n}$, while the W_1 term shows how the degree of translational order can influence the orientational order (and vice versa) through the action of the mixed parameter σ.

Specific forms of the functions $U(r)$ and $W(r)$ were chosen by McMillan:[3,4]

$$W(r) = -\frac{v}{r_o^3 \pi^{3/2}} \exp\{-(r/r_o)^2\},$$
$$U(r) = \delta W(r),$$
[16]

where v and δ are constants characterizing the strengths of the two parts of the interaction, and where r_0 specifies the range of interaction; r_0 is of the order of the length of the molecules. The Fourier coefficients of Eq. [16] are

$$W_o = -v, \; U_1 = \delta W_1,$$
$$W_1 = -v\alpha = -2v \; e^{-(\pi r_0/d)^2}.$$
[17]

Substitution of Eq. [17] into Eq. [15'] gives the McMillan model of the potential in the form

$$V_M(\cos\theta, z) = -v\left\{\delta\alpha\tau\cos\left(\frac{2\pi z}{d}\right) + \left[\eta + \alpha\sigma\cos\left(\frac{2\pi z}{d}\right)\right]P_2(\cos\theta)\right\}.$$
[18]

The functional dependence chosen in Eq. [16] makes it convenient to discuss the variation of liquid-crystal behavior in homologous series. Lengthening the alkyl chains in a series increases the spacing d. This decreases the ratio r_0/d and hence increases the parameter α introduced in Eq. [17]. The properties of the short end of the homologous series can thus be computed using small values of α while those at the long end of the series will require larger values of α.

5. Statistical Thermodynamics

Having derived a particular form of the single molecule potential function in the mean-field approximation, we are now in a position to calculate the thermodynamic properties of the model. According to the rules of classical statistical mechanics the single molecule distribution function corresponding to the potential function, Eq. [18], is

$$f_M(\cos\theta, z) = Z^{-1}\exp[-\beta V_M(\cos\theta, z)],$$

$$Z = \int_0^1 \int_0^d \exp[-\beta V_M(\cos\theta, z)]dz d(\cos\theta),$$

[19]

where Z is the single molecule partition function and $\beta = 1/kT$ (k is Boltzmann's constant and T the temperature). The integrations can be restricted to $0 \leq \cos\theta \leq 1$ since V_M is even in $\cos\theta$, and $0 \leq z \leq d$ since V_M is periodic. The distribution function as it stands in Eq. [19] is not as yet useful for computing average values or thermodynamic quantities. The potential, V_M, Eq. [18], contains the as yet undetermined order parameters η, τ, and σ. The self-consistent determination of the order parameters and their temperature dependence can be realized by combining their definition, Eq. [11] with Eq. [19]:

$$\eta = \int_0^1 \int_0^d P_2(\cos\theta) f_M(\cos\theta, z) dz d(\cos\theta),$$

$$\tau = \int_0^1 \int_0^d \cos\left(\frac{2\pi z}{d}\right) f_M(\cos\theta, z) dz d(\cos\theta),$$

[20]

$$\sigma = \int_0^1 \int_0^d P_2(\cos\theta)\cos\left(\frac{2\pi z}{d}\right) f_M(\cos\theta, z) dz d(\cos\theta).$$

Each of the above equations contains one of the order parameters on the left and all three order parameters in the integrals on the right. The simultaneous solution of these three equations yields the temperature dependence of the order parameters.

The set of self-consistent equations above admits a number of simultaneous solutions. In addition to the smectic-A, nematic, and isotropic solutions, Eq. [20] also yields various less physical solutions. To find which of the possible solutions represents the physically observed states we must calculate the free energy and determine which solution gives the minimum in this quantity at different temperatures.

The energy, entropy, and free energy are obtained in exactly the same way as described in earlier chapters[7,8] on the simpler theory of the nematics:

$$E = \frac{1}{2}N\langle V_M \rangle = -\frac{1}{2}Nv(\eta^2 + \alpha\delta\tau^2 + \alpha\sigma^2),$$

[21]

$$S = -Nk\langle \ln f_M \rangle$$
$$= -\frac{Nv}{T}(\eta^2 + \alpha\delta\tau^2 + \alpha\sigma^2) + Nk\ln Z, \quad [22]$$

$$F = -NkT\ln Z + \frac{1}{2}Nv(\eta^2 + \alpha\delta\tau^2 + \alpha\sigma^2), \quad [23]$$

where N is the number of molecules present. Just as in the case of the mean-field theories of the nematics,[7,8] the free energy is found to contain an additional term beyond the usually expected $\ln Z$ term. Again, these terms arise because we have approximated a pair potential V_{12} by a temperature-dependent single molecule potential V_M. The form of Eq. [23] can be verified to be correct, however. Setting the partial derivatives of F with respect to η, τ, and σ to zero regains the self-consistency conditions, Eq. [20]. Further testing Eqs. [21] and [23] we see that they do satisfy the thermodynamic identity, $E = (\partial\beta F/\partial\beta)$.

6. Numerical Results

McMillan has obtained numerical solutions to Eqs. [20] through [23] for several sets of the potential parameters δ and α. In his first paper[3] he did not yet realize the need for the $U(r)$ term in the potential, Eq. [12], so that he was treating the case of $\delta = 0$. In his subsequent paper[4] he included the missing term. These two sets of results are now examined.

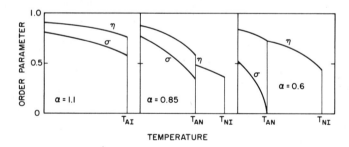

Fig. 3—Temperature dependence of the order parameters for $\delta = 0$ and several values of α (after Ref. [3]).

When $\delta = 0$, the term in τ drops out of the potential, Eq. [18], and the theory now involves only the order parameters η and σ. The translational order of the smectic-A phase is then described only by the mixed-order parameter σ. Fig. 3 shows the temperature dependence of the order parameters for three representative values of α. For $\alpha = 1.1$ (long chain length) η and σ change discontinuously at the same temperature; both the translational and orientational order vanish simultaneously. There is thus a first-order phase change from the smectic-A structure directly into the isotropic fluid; no nematic phase exists. For $\alpha = 0.85$, η and σ display discontinuities at T_{AN} but only σ vanishes. η then vanishes discontinuously at a higher temperature T_{NI}. The system thus displays a first-order transition from smectic-A to nematic followed by another first-order transiton from nematic to isotropic. For $\alpha = 0.6$ (short chain length), σ is seen to vanish continuously at a temperature T_{AN}. η shows a discontinuity in slope at this temperature then vanishes discontinuously at a higher temperature T_{NI}. There is thus a second-order phase transition from smectic-A to nematic followed by the usual first-order transition from nematic to isotropic.

The collection of phase transition temperatures obtained from the model with $\delta = 0$ and for numerous values of α is summarized in the phase diagram shown in Fig. 4a. These results are to be compared with the schematic phase diagram shown in Fig. 4b; this diagram is representative of many real systems and displays the features discussed in Section 3. Note that while there is qualitative agreement between the model and experiment, there are numerous significant discrepancies in the overall behavior of the phase transition lines.

The entropy changes of the smectic-A to nematic phase transitions have also been computed with $\delta = 0$ and for numerous values α. The results are summarized in Fig. 5, where we have plotted ΔS as a function of the ratio T_{AN}/T_{NI}; small values of T_{AN}/T_{NI} correspond to short chain length, large values of T_{AN}/T_{NI} to long chain length (see Fig. 4a). The results of this model are shown as the solid line; the points are experimental, taken from the homologous series described in Fig. 2. Again, there is qualitative agreement in that the general trend and order of magnitude are correct, but there are serious discrepancies.

In his second paper[4] McMillan included all the terms in the potential, Eq. [18] and studied the model for several values of δ and α. He was particularly interested in two substances, cholesteryl nonanoate and cholesteryl myristate. The nonanoate calculations were made using $\alpha \approx 0.41$; $\delta = 0$ and 0.65; for the myristate, $\alpha \approx 0.45$ was chosen with $\delta = 0$ and 0.65. With $\delta = 0.65$ the temperature dependences of the order parameters for both values of α were similar in appearance to those shown in Fig. 3b. In both examples η looked like the η of Fig. 3b while τ and σ resembled the σ of Fig. 3b. The model thus displayed successive first-order phase changes—smectic-A to cholesteric fol-

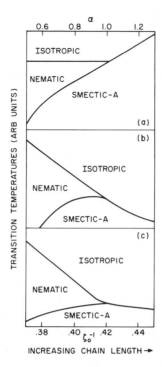

Fig. 4—Schematic phase diagrams for liquid crystals: (a) theory of McMillan,[3] (b) schematic representation of typical experimental diagrams and (c) theory of Lee et al (after Ref. [5]).

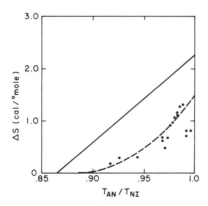

Fig. 5—Entropy of transition plotted as a function of the ratio T_{AN}/T_{NI}. Solid curve is theory of McMillan,[3] dashed curve is theory of Lee et al[5] and the points are experimental[12,13] for compounds of Fig. 2 (after Ref. [5]).

lowed by cholesteric to isotropic in agreement with experimental observations. A quantitative test of the theory was attempted by comparing the temperature dependence of τ^2 with the measured x-ray Bragg scattering intensity (to which it is proportional). The results are shown in Fig. 6. In the case of the myristate, excellent agreement

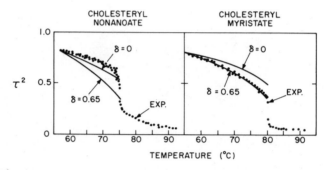

Fig. 6—Comparison of calculated and observed Bragg scattering intensity (after Ref. [4]).

with experiment is obtained using the value $\delta = 0.65$. For the nonanoate, however, the agreement between theory and experiment is poor, although the calculation with $\delta = 0$ is somewhat better than with $\delta = 0.65$. Phase diagrams such as those shown in Fig. 4 were not computed for finite values of δ. This is unfortunate, since it is important to know whether finite values of δ in the potential, Eq. [18], would have improved the appearance of the theoretical phase diagram, Fig. 4a.

7. Improved Theory

Motivated primarily by the disagreement between the theoretical and experimental phase diagrams, Fig. 4 (a and b), Lee et al[5] have derived a modified version of the mean-field theory of smectic-A liquid crystals. The authors begin by adopting the McMillan form of the Kobayashi pair potential, Eqs. [12] and [16]. The pair potential is then treated intact without going into the process of deriving the equivalent mean-field single molecule potential. Further, the pair potential is *not* expanded in a truncated Fourier series as in the manner leading to Eq. [15′].

The statistical thermodynamics is based on the variational principle. Guided by Eqs. [18] and [19], the following variational form was chosen for the single molecule distribution function (of the i-th molecule):

$$f(\cos\theta_i, z_i) = Z^{-1}\exp\{-\beta V(\cos\theta_i, z_i)\}$$

$$Z = \int_0^1 \int_0^d \exp\{-\beta V(\cos\theta_i, z_i)\} dz_i d(\cos\theta_i), \qquad [24]$$

$$V(\cos\theta_i, z_i) = -v\left[aP_2(\cos\theta_i) + bP_2(\cos\theta_i)\cos\left(\frac{2\pi z_i}{d}\right)\right.$$
$$\left. + c\delta\cos\left(\frac{2\pi z_i}{d}\right)\right], \qquad [25]$$

where a, b, and c are variational parameters. The values of these parameters as a function of the temperature are determined by minimizing the free energy. The free energy was taken to have the form

$$F = NkT \int_0^1 \int_0^d f(\cos\theta_1, z_1) \ln f(\cos\theta_1, z_1) dz_1 d(\cos\theta_1)$$
$$+ \frac{1}{2}N^2 \int_0^1 \int_0^d \int_0^1 \int_0^d f(\cos\theta_1, z_1) f(\cos\theta_2, z_2) V_{12}(\theta_{12}, z_{12})$$
$$\times dz_1 d(\cos\theta_1) dz_2 d(\cos\theta_2), \qquad [26]$$

where the first term comes from the entropy contribution (similar to Eq. [22]), and where the second term is the energy E written in terms of the average value of the pair potential. Although the energy is computed from a pair potential (in contrast to the usual mean-field definition, Eq. [21]) the free energy as written in Eq. [26] is still a mean-field approximation. The reason for this is the use of two single-molecule distribution functions in place of the proper (but unknown) pair distribution function in calculating E from V_{12}. The form of F is self-consistent, however, and, along with the variational conditions (Eq. [27] below), does satisfy the required thermodynamic identity, $E = (\partial\beta F/\partial\beta)$.

The temperature dependence of the variational parameters a, b, and c are obtained from minimization of the free energy, Eq. [26]:

$$\left(\frac{\partial F}{\partial a}\right) = \left(\frac{\partial F}{\partial b}\right) = \left(\frac{\partial F}{\partial c}\right) = 0. \qquad [27]$$

Of the many possible solutions to these equations we take only those giving the absolute minimum in F. The temperature dependence of the order parameters η, τ, and σ are then calculated by combining the equilibrium solutions of Eq. [27] with Eqs. [11], [24] and [25]. We note in passing that if the pair potential $V_{12}(\theta_{12}, z_{12})$ could be written in (or approximated by) a separable form $V_1(\theta_1, z_1)\ V_1(\theta_2, z_2)$, then $\eta = a$, $\sigma = b$, and $\tau = c$.

Numerical calculations were performed[5] using $\delta = 0.65$ and numerous values of the range parameter $\zeta_0 = 2\pi r_0/d$, where r_0 is the range of the interaction and d is the layer spacing (ζ_0 plays the same role in parameterizing chain length as McMillan's α). The collection of phase-transition temperatures obtained is displayed in Fig. 4c. The agreement between the theoretical diagram and the representative experimental phase diagram, Fig. 4b, is nothing less than remarkable. All the features described in Section 3 and shown in Fig. 4b are indeed present in the theoretical results. Numerical calculations of the entropy changes of the smectic-A to nematic phase transition were also made. Fig. 5 shows the results, ΔS plotted vs. the ratio T_{AN}/T_{NI}; the result of this calculation is depicted as the dashed line. The agreement with the experimental points is again remarkable. Note further that this theory predicts that the entropy change, ΔS (Fig. 5) will go to zero at $T_{AN}/T_{NI} = 0.88$; the first-order smectic-A to nematic phase change becomes second order. This feature is examined in more detail in the next section.

The comparison of the results of the two models in Figs. 4 and 5 shows that Lee et al[5] have indeed made a very significant improvement over the original considerations of McMillan.[3] Three important changes were made in the theory: (a) the theory was based on a variational principle with E being computed from a pair potential V_{12}; (b) the complete Kobayashi form of potential was used (finite δ); and (c) the pair potential was not approximated by a truncated Fourier series. With all three improvements made simultaneously one can only speculate as to which is the most important. (a) seems not to be a major factor; the resulting theory is still a mean-field approach, and using V_{12} in E should only influence small details. (c) is certainly an important factor. The influence of this improvement cannot be too great, however, since the variational form for the potential of mean force, Eq. [25], is in effect a truncated Fourier series of the true potential. (b) seems to be the most important. The influence of the purely translational term U in the Kobayashi potential, Eq. [12], must be the over-riding factor that produced the spectacular improvements depicted in Figs. 4 and 5.

8. The Possibility of Second-Order Transitions

The question of whether the normally first-order smectic-A to nematic phase transition can be second order in some materials is somewhat controversial. All of the published mean-field theories[1-5] of the smectic-A phase do exhibit second-order phase changes for certain values of the potential parameters. In both McMillan's theory[3] and that of Lee et al,[5] the second-order transition is predicted to occur at that end of homologous series having short chain lengths. More specifically, these models predict the second-order changes to occur when the ratio of transition temperatures T_{AN}/T_{NI} (or T_{AC}/T_{CI}) is at or below about 0.88 (see Fig. 5).

The experimental situation is contradictory. For the homologous series of 4-n-alkoxybenzylidene-4'-phenylazoanilines, Doane et al[14] have observed a possible second-order smectic-A to nematic transition at the extreme short end of the series. In the case of COC (cholesteryl olyel carbonate), Keyes et al[15] found the first-order phase transition to change to second order at a pressure of 2.66 kbar where $T_{AC}/T_{CI} = 0.88$. The phase transitions of CBAOB (p-cyanobenzylidene-amino-p-n-octyloxybenzene)[16,17] and CBOOA (p-cyanobenzylidene-p'-octyloxyaniline)[18,19] have also been believed to be of second order. Torza and Cladis,[20] on the other hand, have concluded from volumetric studies that the smectic-A to nematic phase transition is unambiguously first order in CBOOA. Lin, Keyes, and Daniels[21] concur in this conclusion based on high pressure studies of CBOOA. In any case, we must be mindful of the fact that the experimental verification of the order of the phase transition in these examples is most difficult. The entropy and volume discontinuities are very small and near critical fluctuations are present to complicate the interpretation. (The nature of fluctuations in the nematic to isotropic transition has been discussed in a previous chapter.[7])

The situation with respect to Landau-type phenomenological theories is also contradictory. Drawing an analogy between the smectic-A phase of liquid crystals and the superconducting phase of metals, de Gennes[22,23] has constructed a phenomenological theory from which he concludes that the smectic-A to nematic phase transition can be second order. Halperin and Lubensky,[24] on the other hand, have improved the analogy with superconductors and conclude that the transition will always be at least weakly first order.

My own conjecture in this matter is that the smectic-A to nematic phase transition is always first order. The basis of my argument is the analogy between this transition and the melting of a crystalline solid. When a crystal melts, three dimensional long-range translational order disappears; when a smectic-A goes nematic, one-dimensional long-range translational order disappears.

Kirkwood[10,11] has pointed out that the density distribution function of a crystalline solid (the translational molecular distribution function) can be expanded in a three-dimensional Fourier series. The coefficients in this series are then identified as the order parameters of the crystalline phase. All these order parameters vanish discontinuously at the first-order melting point. Empirically, there are no second-order melting transitions, nor do there seem to be any solid–liquid critical points. Though not a proven fact (as far as I am aware), it seems reasonable that crystal melting is always first order because all of the order parameters cannot vanish simultaneously and continuously before the free energy of the solid phase exceeds that of the liquid phase.

In the smectic-A phase, the single-molecule distribution function, Eqs. [7] and [10] can likewise be represented as a (one-dimensional) Fourier series in which all the coefficients may be considered order parameters. The disappearance of smectic-A order requires the simultaneous vanishing of all the order parameters. That they all can vanish simultaneously and continuously before the free energy of the smectic-A phase exceeds that of the nematic phase seems just as unlikely here as in the (empirically verified) case of the crystalline solids. It seems clear to me that the reason the various theoretical treatments mentioned above can exhibit second-order phase changes is that an insufficient number of order parameters is included. In all the treatments, either the potential, the potential of mean force, or the distribution function are expressed in terms of highly truncated Fourier series. Such truncation automatically limits the number of order parameters. Small numbers of order parameters can then vanish simultaneously and continuously under certain conditions providing the spurious second-order phase transitions.

Appendix

The average value of $U(r_{12})$, averaged over the distribution of molecule 1 is defined by

$$\langle U(r_{12}) \rangle = \int_0^\infty f(z_1) \, U(r_{12}) dz_1; \quad \int_0^\infty f(z_1) dz_1 = 1, \qquad [28]$$

where we have temporarily suppressed the angular dependence of f for clarity. Substituting the Fourier integral representation of U,

$$\langle U(r_{12})\rangle = \frac{2}{\pi}\int_0^\infty \int_0^\infty f(z_1)\langle \tilde{U}(s)\rangle \cos sz_{12}\, ds\, dz_1,$$
$$= \frac{2}{\pi}\int_0^\infty \langle \tilde{U}(s)\rangle \langle \cos sz_1\rangle \cos sz_2\, ds, \qquad [29]$$

where $\langle \tilde{U}(s)\rangle$ is the Fourer transform $\tilde{U}$ averaged over all the (random) positions of the molecules in the layers, and where the sine terms which come from $\cos sz_{12} = \cos s(z_2 - z_1)$ are omitted since they will vanish in the averaging. The next step is to evaluate $\langle \cos sz_1 \rangle$ using the expansion, Eq. [6]:

$$\langle \cos sz_1\rangle = \int_0^\infty f_1(z_1)\cos sz_1\, dz_1,$$
$$= \frac{2}{d}\sum_n \left\langle \cos\left(\frac{2\pi nz}{d}\right)\right\rangle \int_0^\infty \cos\left(\frac{2\pi nz_1}{d}\right)\cos sz_1\, dz_1,$$
$$= \sum_n \left\langle \cos\left(\frac{2\pi nz}{d}\right)\right\rangle \delta\left(s - \frac{2\pi n}{d}\right), \qquad [30]$$

where δ is the usual delta function. Substitution of the result of Eq. [30] into Eq. [29] gives

$$\langle U(r_{12})\rangle = \frac{2}{\pi}\int_0^\infty \langle \tilde{U}(s)\rangle \sum_n \left\langle \frac{\cos 2\pi nz}{d}\right\rangle \delta\left(s - \frac{2\pi n}{d}\right)\cos sz_2\, ds,$$
$$= \frac{2}{\pi}\sum_n \left\langle \tilde{U}\left(\frac{2\pi n}{d}\right)\right\rangle \left\langle \cos\frac{2\pi nz}{d}\right\rangle \cos\frac{2\pi nz_2}{d}, \qquad [31]$$

which is the desired result, Eq. [15].

References

[1] K. K. Kobayashi, "Theory of Translational and Orientational Melting with Application to Liquid Crystals," *J. Phys. Soc.* (Japan), **29**, p. 101 (1970).

[2] K. K. Kobayashi, "Theory of Translational and Orientational Melting with Application to Liquid Crystals," *Mol. Cryst. Liq. Cryst.*, **13**, p. 137 (1971).

[3] W. L. McMillan, "Simple Molecular Model for the Smectic-A Phase of Liquid Crystals," *Phys. Rev.*, **A4**, p. 1238 (1971).

[4] W. L. McMillan, "X-ray Scattering from Liquid Crystals. I. Cholesteryl Nonanoate and Myristate," *Phys. Rev.*, **A6**, p. 936 (1972).

[5] F. T. Lee, H. T. Tan, Y. M. Shih, and C. W. Woo, "Phase Diagram for Liquid Crystals," *Phys. Rev. Lett.*, **31**, p. 1117 (1973).

[6] W. Maier and A. Saupe, *Z. Naturforschg.*, **14a**, p. 882 (1959) and **15a**, p. 287 (1960).

[7] P. J. Wojtowicz, "Introduction to the Molecular Theory of Nematic Liquid Crystals," Chapter 3.

[8] P. J. Wojtowicz, "Generalized Mean Field Theory of Nematic Liquid Crystals," Chapter 4.

[9] E. B. Priestley, "Nematic Order: The Long Range Orientational Distribution Function," Chapter 6.

[10] J. G. Kirkwood and E. Monroe, "Statistical Mechanics of Fusion," *J. Chem. Phys.*, **9**, p. 514 (1941).

[11] J. G. Kirkwood, "Crystallization as a Cooperative Phenomenon," p. 67 in *Phase Transformations in Solids*, ed. by R. Smoluchowski, J. Wiley and Sons, Inc., New York, (1951).

[12] H. Arnold, *Z. Physik Chem.* (Leipzig) **239**, p. 283 (1968); **240**, p. 185 (1969).

[13] G. J. Davis and R. S. Porter, "Evaluation of Thermal Transitions in Some Cholesteryl Esters of Saturated Aliphatic Acids," *Mol. Cryst. Liq. Cryst.*, **10**, p. 1 (1970).

[14] J. W. Doane, R. S. Parker, B. Cvilk, D. L. Johnson, and D. L. Fishel, "Possible Second-Order Nematic-Smectic-A Phase Transition," *Phys. Rev. Lett.*, **28**, p. 1694 (1972).

[15] P. H. Keyes, H. T. Weston, and W. B. Daniels, "Tricritical Behavior in a Liquid-Crystal System," *Phys. Rev. Lett.*, **31**, p. 628 (1973).

[16] W. L. McMillan, "Measurement of Smectic-A Phase Order-Parameter Fluctuations near a Second Order Smectic-A Nematic Phase Transition," *Phys. Rev.*, **7A**, p. 1419 (1973).

[17] B. Cabane and W. G. Clark, "Orientational Order in the Vicinity of a Second Order Smectic-A to Nematic Phase Transition," *Solid State Comm.*, **13**, p. 129 (1973).

[18] L. Chueng, R. B. Meyer, and H. Gruler, "Measurement of Nematic Elastic Constants near a Second Order Nematic-Smectic-A Phase Change," *Phys. Rev. Lett.*, **31**, p. 349 (1973).

[19] M. Delaye, R. Ribotta, and G. Durand, "Rayleigh Scattering at a Second Order Nematic to Smectic-A Phase Transition," *Phys. Rev. Lett.*, **31**, p. 443 (1973).

[20] S. Torza and P. E. Cladis, "Volumetric Study of the Nematic-Smectic-A Transition of N-p-cyanobenzylidene-p-octyloxyaniline," *Phys. Rev. Lett.*, **32**, p. 1406 (1974).

[21] W. J. Lin, P. H. Keyes, and W. B. Daniels, "High Pressure Studies of Liquid Crystal Phase Transitions in CBOOA," *Phys. Lett.*, **A49**, p. 453 (1974).

[22] P. G. de Gennes, "An Analogy Between Superconductors and Smectics-A," *Solid State Comm.* **10**, p. 753 (1972).

[23] P. G. de Gennes, "Some Remarks on the Polymorphism of Smectics," *Mol. Cryst. Liq. Cryst.*, **21**, p. 49 (1973).

[24] B. I. Halperin and T. C. Lubensky, "On the Analogy Between Smectic-A Liquid Crystals and Superconductors," *Solid State Comm.*, **14**, p. 997 (1974).

Introduction to the Elastic Continuum Theory of Liquid Crystals

Ping Sheng

RCA Laboratories, Princeton, N. J. 08540

1. Introduction

In the preceding chapters of this book we have seen that in terms of molecular theories[1,2] one can calculate and successfully explain various properties of meosphase transitions. However, there exists a class of liquid-crystal phenomena involving the response of bulk liquid-crystal samples to external disturbances, with respect to which the usefulness of a molecular theory is not immediately obvious. These phenomena are usually distinguished by two characteristics: (1) the energy involved, per molecule, in producing these effects is small compared to the strength of intermolecular interaction; and (2) the characteristic distances involved in these phenomena are large compared to molecular dimensions. In describing these large-scale phenomena, it is more convenient to regard the liquid crystal as a continuous medium with a set of elastic constants than to treat it on a molecular basis. Based on this viewpoint, Zocher,[3] Oseen,[4] and Frank[5] developed a phenomenological continuum theory of liquid crystals that is very successful in explaining various magnetic (electric) field-induced effects. It is the purpose of the present chapter to develop this elastic continuum theory for nematic and cholesteric liquid crystals and to discuss and illustrate its use. In this paper, the derivation of the fundamental equation of the elastic continuum theory is followed by the application of the theory to four effects: (1) the twisted nematic cell, (2) the magnetic (electric) coherence length, (3) the Fréedericksz transition, and (4) the magnetic (electric) field-induced cholesteric–nematic transition.

2. The Fundamental Equation of the Continuum Theory of Liquid Crystals

In earlier chapters we have seen that liquid crystals are characterized by an orientational order of their constituent rod-like molecules.[1,2,6,7] In nematic liquid crystals this orientational order has uniaxial (cylindrical) symmetry, the axis of uniaxial symmetry being parallel to a unit vector $\hat{n}$, called the director. Let us now consider a very small spatial region inside a macroscopic sample of nematic liquid crystal that contains a sufficiently large number of molecules so that the long-range orientational order is well defined within that region. Such a spatial region can be characterized by a director pointing along the of local axis of uniaxial orientational symmetry. Let us imagine the division of the macroscopic sample into such small spatial regions. In each of the regions, we define an orientational director. In this manner the macroscopic sample of nematic liquid crystal can be characterized by a local director at every spatial "point," where we use the term point loosely to mean a small region of space as defined above. Obviously, this characterization of orientational order by a director field, $\hat{n}(\vec{r})$, is not limited to nematic liquid crystals, which we have used as an example in the above discussion. In fact, the director-field characterization can be applied equally well to cholesteric liquid crystals, since, locally, cholesterics also possess uniaxial symmetry in their orientational order. However, for convenience, we will continue to use nematic liquid crystals as our reference in the following discussion. The generalization of the theory will be made at appropriate places to permit application to cholesteric liquid crystals. In what follows, we first consider the free energy associated with a distortion in the director field. Next, we examine the free energy associated with the interaction of liquid crystals with external fields. The combination of all the free-energy contributions then yields the fundamental equation, which is an expression for the total free energy of a sample of nematic or cholesteric liquid crystal in an external field.

The starting point for the development of continuum theory is the consideration of the equilibrium state. In nematic liquid crystals, parallel alignment of all the local directors represents the equilibrium state, or the state of minimum free energy. However, when we perturb the system by pinning the surface directors to the walls of the container, applying an external field, or introducing thermal fluctuations, the local directors will no longer be spatially invariant. A quantitative formulation of the above statements is that the quantities dn_α/dx_β, (where x is the spatial variable and the subscripts $\alpha,\beta = 1,2,3$ denote the components along the three orthogonal axes of the Cartesian coordinate system) are zero for the equilibrium state but are nonzero for some, or all, values of α and β when the system is distorted. In other words, we can think of dn_α/dx_β as the distortion parameters, and the equilibrium state is given by the uniformly aligned state for which $dn_\alpha/dx_\beta = 0$ everywhere. Since the distorted state represents a state with higher free energy than the equilibrium state, we can write the free-energy density of the distorted state as

$$f(\text{distorted}) = f_0(\text{equilibrium}) + \Delta f, \qquad [1]$$

where f and f_o are the free-energy densities of the distorted and equilibrium states, respectively; Δf (>0) is a function of the n_α and the dn_α/dx_β that vanishes when all $dn_\alpha/dx_\beta = 0$. Since, in general, $dn_\alpha/dx_\beta \ll$ (molecular dimension)$^{-1}$ for the phenomena of interest, we can expand Δf as a power series in the n_α and the dn_α/dx_β and retain only the lowest-order nonvanishing terms of the series. Let $\mathcal{F}_D$ (subscript D stands for distortion) denote such an approximation to Δf. $\mathcal{F}_D$ must satisfy several requirements. First, since we are expanding in powers of the dn_α/dx_β around $dn_\alpha/dx_\beta = 0$, which is a free-energy minimum, the lowest-order nonvanishing terms must be quadratic in the dn_α/dx_β (i.e. proportional to terms of the form $(dn_\alpha/dx_\beta) \cdot (dn_\gamma/dx_\delta)$). Second, since the "head" and the "tail" of a nematic director represent the same physical state, $\mathcal{F}_D$ must be even in the n_α. Third, $\mathcal{F}_D$ must be a scalar quantity. In addition, we will discard terms of the form $\nabla \cdot \vec{u}(\vec{r})$ where $\vec{u}(\vec{r})$ is any arbitrary vector field, since they represent surface contributions to the distortion free-energy density and are assumed to be small (by Gauss's Theorem, $\int \nabla \cdot \vec{u}(\vec{r}) dV = \int d\hat{\sigma} \cdot \vec{u}(\vec{r})$, where $d\hat{\sigma}$ is a surface element with unit vector perpendicular to the surface). With the above constraints, it can be shown[5] that $\mathcal{F}_D$ contains only three linearly independent terms. They are (1)

$[\nabla \cdot \hat{n}(\vec{r})]^2$, (2) $[\hat{n}(\vec{r}) \cdot \nabla \times \hat{n}(\vec{r})]^2$, and (3) $[\hat{n}(\vec{r}) \times \nabla \times \hat{n}(\vec{r})]^2$. It is obvious that all three terms satisfy the requirements stated above. Interested readers are referred to Refs. [5] and [8] for proofs that these three terms are indeed unique. In Fig. 1 we show the physical distor-

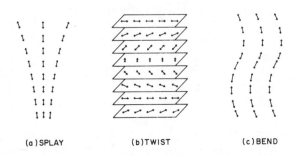

Fig. 1—Three types of distortion in a director field: (a) gives $\nabla \cdot \hat{n}(\vec{r}) \neq 0$, (b) gives $\hat{n}(\vec{r}) \cdot \nabla \times \hat{n}(\vec{r}) \neq 0$, and (c) gives $\hat{n}(\vec{r}) \times \nabla \times \hat{n}(\vec{r}) \neq 0$. Each director is shown with double arrows in order to indicate that the "head" and the "tail" directions represent exactly the same physical state in a nematic sample.

tions of the director field associated with the three terms. The first term is called "splay," the second term "twist," and the third term "bend." $\mathcal{F}_D$ can now be written as

$$\mathcal{F}_D = \frac{1}{2}\{K_{11}[\nabla \cdot \hat{n}(\vec{r})]^2 + K_{22}[\hat{n}(\vec{r}) \cdot \nabla \times \hat{n}(\vec{r})]^2 + K_{33}[\hat{n}(\vec{r}) \times \nabla \times \hat{n}(\vec{r})]^2\}, \qquad [2]$$

where the constants K_{11}, K_{22}, K_{33} are, respectively, the splay, twist, and bend elastic constants and are named collectively as the Frank elastic constants. The factor ½ is included so that the K's may agree with their historical definitions. Since $\mathcal{F}_D$ must be positive in order to give stability for the uniformly aligned state, all the K's must be positive. As for their values, we note that the theoretical determination of the K's from molecular parameters represents a task of linking the continuum theory to the microscopic theories of liquid crystals and is beyond the scope of the present paper. However, from dimensional analysis we can get an order-of-magnitude estimate of what the values of the K's should be. Since the K's are in units of energy/length, we must look for the characteristic energy and length in the problem. The only energy in the problem is the intermolecular interaction energy, which is estimated* to be ~0.01 eV, and the only suitable length is the separation between two molecules, which is ~10 Å. Therefore, $K \simeq 10^{-7}$ dyne, in order-of-magnitude agreement with measured values[8] of 10^{-7}–10^{-6} dyne. The K's are also temperature dependent. In fact, it can be shown[9] that the temperature dependence is of the form $K \sim <P_2(\cos \Theta)>^2$, where $<>$ denotes averaging over that small volume of the sample that is characterized by a local director $\hat{n}(\vec{r})$, P_2 is the Legendre polynomial of second order, and Θ is the angle between any molecule (inside the volume where the average is taken) and the local director $\hat{n}(\vec{r})$. $<P_2(\cos \Theta)>$ is just the local order parameter measured with $\hat{n}(\vec{r})$ as the axis of symmetry. The dependence of the K's on the square of the local order parameter is plausible if one thinks of the K's as the macroscopic analog of the anisotropic intermolecular interaction constants, the difference being that, in place of molecules, we have small volumes of the sample with well-defined long-range orientational order. In such an analogy $<P_2(\cos \Theta)>$ plays the role of a (temperature-dependent) dipole strength of the molecules.

Suppose now the sample of nematic liquid crystal is placed under the influence of a magnetic or an electric field. Because the liquid crystal molecules are generally diamagnetic, electrically polarizable, and anisotropic in their magnetic and electric properties, the application of a field usually contributes an amount of free-energy density which is opposite in sign to that of the distortion free-energy density (because fields help align molecules). Let us first discuss the magnetic field contribution. Consider again a small region of the sample characterized by a local director $\hat{n}(\vec{r})$. The diamagnetic susceptibility per unit volume in such a small volume is usually anisotropic. Let $\chi_\parallel$ denote the susceptibility per unit volume parallel to $\hat{n}(\vec{r})$ and $\chi_\perp$ denote the susceptibility per unit volume perpendicular to $\hat{n}(\vec{r})$. The difference, $\Delta\chi \equiv \chi_\parallel - \chi_\perp$, is a measure of the local anisotropy. As shown in Ref. [6], $\Delta\chi$ is equal to $N < P_2(\cos \Theta) > (\zeta_\parallel - \zeta_\perp)$, where N is the number of molecules per unit volume, $< P_2(\cos \Theta) >$ is the orientational order within the small region under consideration, and $\zeta_\parallel(\zeta_\perp)$ is the diamagnetic susceptibility of a single rod-like molecule

---

* The intermolecular interaction energy responsible for the nematic ordering can be estimated from the latent heat of isotropic–nematic phase transition $\Delta E \simeq 300$ cal/mol $\simeq 0.01$ eV/molecule (see, for example, G. H. Brown, J. W. Doane, and D. D. Neff, *A Review of the Structure and Physical Properties of Liquid Crystals*, p. 43, CRC Press, Cleveland, Ohio, 1971).

parallel (perpendicular) to its long axis. In the following we will assume that $\chi_\parallel > \chi_\perp$ (i.e., positive anisotropy) and that the value of $<P_2(\cos \Theta)>$ is uniform throughout the volume of the sample, implying that $\Delta\chi$, K_{11}, K_{22}, and K_{33} have no spatial dependence. In

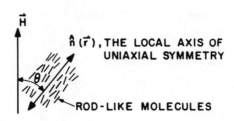

Fig. 2—Orientation of the local axis of uniaxial symmetry with respect to the external magnetic field direction.

Fig. 2 we show the relative directions of magnetic field $\vec{H}$ and local director $\hat{n}(\vec{r})$. The induced diamagnetic moments per unit volume parallel and perpendicular to $\hat{n}(\vec{r})$ are, respectively,

$$M_\parallel = H\chi_\parallel \cos\theta,$$
$$M_\perp = H\chi_\perp \sin\theta.$$

The work done by the field per unit volume is

$$W_{\text{magnetic}} = \int_0^H (-M_\perp' \sin\theta - M_\parallel' \cos\theta) dH'$$
$$= -\frac{H^2}{2}(\chi_\perp + \Delta\chi \cos^2\theta).$$

Discarding the spatially invariant term, $-H^2\chi_\perp/2$, we obtain the magnetic field contribution to the free-energy density

$$\mathcal{F}_M = -\frac{\Delta\chi}{2}[\vec{H}\cdot\hat{n}(\vec{r})]^2. \qquad [3]$$

Using similar arguments as above, we obtain the electric-field contribution to the free-energy density

$$\mathcal{F}_E = -\frac{\Delta\epsilon}{8\pi}[\vec{E}\cdot\hat{n}(\vec{r})]^2, \qquad [4]$$

where $\Delta\epsilon \equiv \epsilon_\parallel - \epsilon_\perp$ is the difference between the local dielectric constants in directions parallel and perpendicular to the local director. Values of $\Delta\chi$ and $\Delta\epsilon$ typically range from 10^{-7}–10^{-6} cgs units for $\Delta\chi$ and 0.1–1 for $\Delta\epsilon$.

At this point we make a slight generalization so that the theory can be applied to cholesteric liquid crystals as well. The basic difference between cholesteric and nematic liquid crystals lies in the fact that the equilibrium state of cholesterics is characterized by a nonvanishing twist in the director field. If we denote the cholesteric helical axis as the x-axis, the equilibrium state is characterized by

$$n_x = 0,$$
$$n_y = \cos(\pi x/\lambda_0),$$
$$n_z = \sin(\pi x/\lambda_0),$$

where λ_0 is the pitch of the helix. Using this representation of the director field, the twist term, $[\hat{n}(\vec{r}) \cdot \nabla \times \hat{n}(\vec{r})]^2$, is calculated to be $(\pi/\lambda_0)^2$. This suggests that the twist part of free-energy density for the cholesteric liquid crystals should be expanded around $|\hat{n}(\vec{r}) \cdot \nabla \times \hat{n}(\vec{r})| = \pi/\lambda_0$, where $||$ is the absolute value sign. In fact, it can be shown[5] that the appropriate form is $[|\hat{n}(\vec{r}) \cdot \nabla \times \hat{n}(\vec{r})| - \pi/\lambda_0]^2$.

We are now in a position to combine all the free-energy density terms to give a total free-energy density $\mathcal{F}$ of the system under external fields:

$$\mathcal{F} = \mathcal{F}_D + \mathcal{F}_M + \mathcal{F}_E$$
$$= \frac{1}{2}\left\{K_{11}[\nabla \cdot \hat{n}(\vec{r})]^2 + K_{22}\left[\left|\hat{n}(\vec{r}) \cdot \nabla \times \hat{n}(\vec{r})\right| - \frac{\pi}{\lambda_0}\right]^2\right.$$
$$\left. + K_{33}[\hat{n}(\vec{r}) \times \nabla \times \hat{n}(\vec{r})]^2 - \Delta\chi[\vec{H}\cdot\hat{n}(\vec{r})]^2 - \frac{1}{4\pi}\Delta\epsilon[\vec{E}\cdot\hat{n}(\vec{r})]^2\right\}. \quad [5]$$

The total free energy of the sample is given by

$$F = \int_{\substack{\text{volume} \\ \text{of the sample}}} \mathcal{F}\, d^3r. \quad [6]$$

Eqs. [5] and [6] are the fundamental equations of the elastic continuum theory of nematic and cholesteric liquid crystals (for nematics λ_0 in Eq. [5] is set equal to ∞). In the following section we use the fundamental equations to solve four examples as illustrations for their applications.

3. Applications of the Elastic Continuum Theory

The basic principle involved in the application of the fundamental equations to the solution of actual problems is that the equilibrium state of the director field is always given by that director configuration that minimizes the free energy of the system with specified boundary conditions.

Before getting into actual calculations we first simplify Eq. [5] by setting $K_{11} = K_{22} = K_{33} = K$. This greatly facilitates the mathematics but does not affect the qualitative behavior of the results. Neglecting the term π/λ_0 for the moment, we have

$$\begin{aligned}\mathcal{F} &= \frac{K}{2}\Big\{[\nabla \cdot \hat{n}(\vec{r})]^2 + [\hat{n}(\vec{r}) \cdot \nabla \times \hat{n}(\vec{r})]^2 + [\hat{n}(\vec{r}) \times \nabla \times \hat{n}(\vec{r})]^2 \\ &\quad - \frac{\Delta \chi}{K}[\vec{H} \cdot \hat{n}(\vec{r})]^2 - \frac{1}{4\pi}\frac{\Delta \epsilon}{K}[\vec{E} \cdot \hat{n}(\vec{r})]^2\Big\} \\ &= \frac{K}{2}\Big\{[\nabla \cdot \hat{n}(\vec{r})]^2 + [\nabla \times \hat{n}(\vec{r})]^2 - \frac{\Delta \chi}{K}[\vec{H} \cdot \hat{n}(\vec{r})]^2 \\ &\quad - \frac{1}{4\pi}\frac{\Delta \epsilon}{K}[\vec{E} \cdot \hat{n}(\vec{r})]^2\Big\}.\end{aligned}\qquad [7]$$

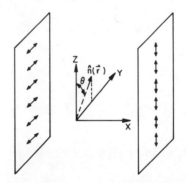

Fig. 3—The geometry of a 90°-twisted nematic cell.

3.1 Twisted Nematic Cell

In Fig. 3 we show a planar cell containing nematic liquid crystal, two of whose bounding walls are rubbed or otherwise treated so that the directors near the walls are pinned in the directions shown. Since the

ELASTIC CONTINUUM THEORY

directors at the two walls are perpendicular to each other, the local nematic directors must undergo a 90° twist in passing from one wall to the other. The question is how this twist is distributed across the cell, i.e., should the distribution be uniform or nonuniform? To answer this question, we must calculate the free energy of an arbitrary twist pattern with the specified boundary conditions. The correct twist pattern is then given by that director configuration that minimizes the free energy of the system. Assuming all the directors lie in the y-z plane, we write

$$n_x = 0,$$
$$n_y = \sin\theta(x),$$
$$n_z = \cos\theta(x).$$

From this representation of the director field, it is easily calculated that $\nabla \cdot \hat{n}(\vec{r}) = 0$ and $\nabla \times \hat{n}(\vec{r}) = (d\theta(x)/dx)[\sin\theta \hat{j} + \cos\theta \hat{k}]$, where we use $\hat{i}, \hat{j}, \hat{k}$ to denote the unit vectors in the x, y, z directions, respectively. Using Eqs. [6] and [7], we obtain

$$F = \int d^3r \frac{K}{2}[\nabla \times \hat{n}(\vec{r})]^2$$

$$= \frac{K}{2}\int d^3r \left[\frac{d\theta(x)}{dx}\right]^2$$

$$= \frac{KA}{2} \int_{\text{thickness of the cell}} dx \left[\frac{d\theta(x)}{dx}\right]^2, \qquad [8]$$

where A is the area of the cell in the y-z plane. To minimize F, we recall that if one wants to minimize the value of an integral

$$I = \int_a^b G\left(y(x), \frac{dy(x)}{dx}, x\right) dx$$

by varying the functional form of $y(x)$, the optimal function $y(x)$ must satisfy the equation

$$\frac{\partial G}{\partial y} - \frac{d}{dx}\frac{\partial G}{\partial\left(\frac{dy}{dx}\right)} = 0, \qquad [9]$$

which is called the Euler–Lagrange equation. In our present case θ corresponds to y, and $G = [d\theta(x)/dx]^2$; the application of Eq. [9]

yields $d^2\theta(x)/dx^2 = 0$, or $d\theta(x)/dx = C$, where C is an integration constant that can be determined by the boundary condition $C\times$ (cell thickness) $= \pm\pi/2$ ($+\pi/2$ is indistinguishable from $-\pi/2$ for nematics). This is the result we are looking for. It tells us that the twist will be uniformly distributed across the cell. However, because $d\theta/dx$ can be either + or −, the twist can be either left handed or right handed. In an actual 90°-twisted nematic cell, both senses of the twist are usually present, a fact that is indicated by the existence of visible disclination lines separating regions of opposite senses of the twist.

3.2 Magnetic Coherence Length

The geometry of this problem is shown in Fig. 4. A semi-infinite sample of nematic liquid crystal with positive anisotropy is bound on one

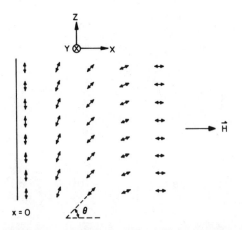

Fig. 4—Distortion of the director field when the molecules are pinned to the wall perpendicular to the external magnetic field direction. The nematic molecules are assumed to be diamagnetic with positive anisotropy.

side by a wall that is treated so that the directors near the wall are pinned along the z-direction. A magnetic field is applied along the x-axis, so that far away from the wall the directors would lie along the field direction. There is a transition region near the wall where the directors gradually change from one direction to the other. The problem is to find the characteristic length of that transition region. From

Fig. 4 we have

$$n_x = \cos\theta(x),$$
$$n_y = 0,$$
$$n_z = \sin\theta(x).$$

Straightforward calculation gives

$$[\nabla \cdot \hat{n}(\vec{r})]^2 = \sin^2\theta \left[\frac{d\theta(x)}{dx}\right]^2,$$

$$[\nabla \times \hat{n}(\vec{r})]^2 = \cos^2\theta \left[\frac{d\theta(x)}{dx}\right]^2,$$

$$-\frac{\Delta\chi}{K}[\vec{H}\cdot\hat{n}(\vec{r})]^2 = -\frac{\Delta\chi}{K}[H\cos\theta(x)]^2.$$

Substitution into Eq. [7] yields

$$\frac{F}{A} = \frac{K}{2}\int_0^\infty dx \left(\left[\frac{d\theta(x)}{dx}\right]^2 - \frac{\Delta\chi}{K}[H\cos\theta(x)]^2\right), \qquad [10]$$

where A is the area of the bounding wall. Application of the Euler–Lagrange equation to Eq. [10] gives an equation for the $\theta(x)$ that minimizes the free energy of the system:

$$\frac{K}{H^2\Delta\chi}\frac{d^2\theta(x)}{dx^2} - \sin\theta\cos\theta = 0. \qquad [11]$$

The combination $K/(H^2\Delta\chi)$ has the dimension of (length)^2. We define

$$\xi_M \equiv \frac{1}{H}\sqrt{\frac{K}{\Delta\chi}} \qquad [12]$$

as the characteristic distance of the problem. Eq. [11] can now be rewritten as

$$\frac{\sin\theta\cos\theta}{\xi_M^2}\frac{d\theta}{dx} = \frac{d^2\theta}{dx^2}\frac{d\theta}{dx} = \frac{1}{2}\frac{d}{dx}\left(\frac{d\theta}{dx}\right)^2$$

$$\frac{\sin\theta\, d\sin\theta}{\xi_M^2\, dx} = \frac{1}{2}\frac{d}{dx}\left(\frac{d\theta}{dx}\right)^2$$

or

$$\frac{1}{2}\frac{1}{\xi_M^2}\frac{d\sin^2\theta}{dx} = \frac{1}{2}\frac{d}{dx}\left(\frac{d\theta}{dx}\right)^2 \qquad [13]$$

Integration of Eq. [13] yields

$$\left(\frac{d\theta}{dx}\right)^2 = \left(\frac{\sin\theta}{\xi_M}\right)^2 + C.$$

The constant of integration is fixed by the condition that as $x \to \infty$, $\theta \to 0$ and $d\theta/dx \to 0$. Therefore $C = 0$, and

$$\frac{d\theta}{dx} = \pm \frac{\sin\theta}{\xi_M}.$$

Choosing the $-$ sign for $x > 0$ and integrating once more, we have

$$\ln\left(\tan\frac{\theta}{2}\right) = -\frac{x}{\xi_M},$$

or

$$\theta(x) = 2\arctan[\exp\{-x/\xi_M\}], \text{ where } x \geq 0. \tag{14}$$

Eq. [14] is plotted in Fig. 5. As we can see, ξ_M is the characteristic length scale below which the magnetic field does not have much in-

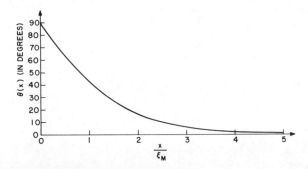

Fig. 5—Tilt angle (deviation from the field direction) of the director field plotted as a function of distance from the wall. The distance is measured in units of magnetic coherence length, which is the characteristic length of magnetic phenomena in nematic liquid crystals.

fluence on the relative orientations of the directors. Another way of saying the same thing is that ξ_M defines the scale of magnetic phenomena. ξ_M is usually called the "magnetic coherence length." For $K \sim 10^{-6}$ dyne, $\Delta\chi \sim 10^{-7}$ cgs units, and $H \sim 10\text{kOe}$, ξ_M is about 3 μm.

Suppose now that in place of the magnetic field an electric field is applied. If impurity conduction and other dynamical effects are neglected, the problem is qualitatively the same, and a quantity ξ_E can be obtained which is the exact analog of ξ_M. Substitution of $\Delta\epsilon/4\pi$ for $\Delta\chi$ and E for H in Eq. [12] gives

$$\xi_E \equiv \frac{1}{E}\sqrt{\frac{4\pi K}{\Delta\epsilon}}. \qquad [15]$$

Setting $\xi_M = \xi_E$, we can compare the relative effectiveness of magnetic and electric fields in orienting the nematic directors. The relation is

$$E = \sqrt{\frac{4\pi\Delta\chi}{\Delta\epsilon}} H. \qquad [16]$$

Taking $H = 1$ Oe, $\Delta\chi \sim 10^{-7}$ cgs unit, and $\Delta\epsilon \sim 0.1$, we have

$$E = \sqrt{\frac{4\pi\Delta\chi}{\Delta\epsilon}} H \simeq \sqrt{10^{-5}} H \simeq \frac{1}{300}\frac{\text{statvolt}}{\text{cm}} = 1 \text{ V/cm}.$$

Therefore, one oersted of magnetic field is equivalent to the order of one volt/cm of electric field in terms of effectiveness in orienting the nematic liquid crystals.

3.3 Fréedericksz Transition

Consider a planar cell of nematic liquid crystal with directors on both surfaces anchored perpendicular to the walls as shown in Fig. 6. It was first observed by Fréedericksz[10] in 1927 that such a cell would undergo an abrupt change in its optical properties when the strength of an external magnetic field, applied normal to the director (z-direction in Fig. 6), exceeded a well-defined threshold. (In the original experiment one wall of the cell was concave in shape so as to give some variation in the cell thickness.) Fréedericksz further noted that the strength of the magnetic field at threshold was inversely proportional to the cell thickness d.[11] This Fréedericksz transition is now a well-studied phenomenon, and has found applications in liquid-crystal display devices. The transition is essentially due to the magnetic alignment of the bulk sample directors at sufficiently high field strength. However, both the abruptness of its onset and the relationship between the threshold field strength and the cell thickness are of theoretical and practical interest. Here we will use the continuum theory to calculate the various properties of the Fréedericksz transition. From Fig. 6 we get

$$n_x = \cos\theta(x),$$
$$n_y = 0,$$
$$n_z = \sin\theta(x).$$

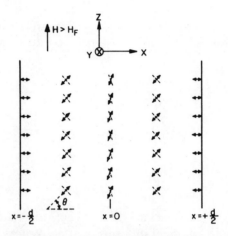

Fig. 6—Local nematic directors in a Fréedericksz cell when $H > H_F$. Dotted lines indicate the equivalent tilt configuration of the directors. Tilt angle θ is defined as shown.

From these expressions one obtains

$$[\nabla \cdot \hat{n}(\vec{r})]^2 + [\nabla \times \hat{n}(\vec{r})]^2 = \left[\frac{d\theta(x)}{dx}\right]^2,$$

and

$$-\frac{\Delta\chi}{K}[H \cdot \hat{n}(\vec{r})]^2 = -\frac{\Delta\chi}{K}H^2\sin^2\theta.$$

Substitution into Eq. [7] and application of the Euler–Lagrange equation results in

$$\xi_M^2 \frac{d^2\theta}{dx^2} + \sin\theta\cos\theta = 0. \qquad [17]$$

Using the same manipulations as those for Eq. [11], we get

ELASTIC CONTINUUM THEORY

$$\left(\frac{d\theta}{dx}\right)^2 = C - \frac{\sin^2\theta}{\xi_M^2}. \quad [18]$$

The constant of integration C is obtained by noting that, from the symmetry of the problem, $d\theta/dx = 0$ at $x = 0$. Defining $\theta(x = 0)$ as θ_M, we get $C = \sin^2\theta_M/\xi_M^2$, and

$$\frac{d\theta}{dx} = \pm \frac{1}{\xi_M}\sqrt{\sin^2\theta_M - \sin^2\theta}, \quad [19]$$

where the $+$ sign corresponds to the solution in region $x < 0$ and the $-$ sign corresponds to the solution in region $x > 0$. Since the solution is symmetric about $x = 0$, we will choose the $+$ sign in the following calculations. Integration of Eq. [19] yields

$$\int_{-\frac{d}{2}}^{x} \frac{dx}{\xi_M} = \int_0^{\theta(x)} \frac{d\theta'}{\sqrt{\sin^2\theta_M - \sin^2\theta'}},$$

or

$$\frac{1}{\xi_M}\left(\frac{d}{2} + x\right)\sin\theta_M = \int_0^{\theta(x)} \frac{d\theta'}{\sqrt{1 - \left[\frac{\sin\theta'}{\sin\theta_M}\right]^2}}, \quad [20]$$

where $-d/2 \leq x \leq 0$. The solution of Eq. [20] will proceed in two steps. First θ_M will be determined as a function of ξ_M (or of H, since $\xi_M \equiv \sqrt{K/\Delta\chi}/H$). Then this $\theta_M(H)$ can be substituted back into Eq. [20] for the solution of $\theta(x)$ as a function of H.

From the definition of θ_M we get

$$\frac{d}{2\xi_M}\sin\theta_M = \int_0^{\theta_M} \frac{d\theta'}{\sqrt{1 - \left(\frac{\sin\theta'}{\sin\theta_M}\right)^2}}. \quad [21]$$

This equation can be solved graphically as in Fig. 7 by plotting, as a function of $\sin\theta_M$, the left- and right-hand sides on the same graph and locating the points of intersection. By expanding the integral on the right-hand side of Eq. [21], denoted here as $L(\sin\theta_M)$, for small values of θ_M, we get the slope

$$\left.\frac{dL(\sin\theta_M)}{d\sin\theta_M}\right|_{\theta_M = 0} = \frac{\pi}{2}.$$

Therefore, for $d/(2\xi_M) < \pi/2$ the only solution of Eq. [21] is $\theta_M = 0$.

However, when $d/(2\xi_M) > \pi/2$, a second solution with $\theta_M \neq 0$ is obtained that gives lower free energy than the $\theta_M = 0$ solution. The critical magnetic field H_F for the transition is found by equating $d/(2\xi_M)$

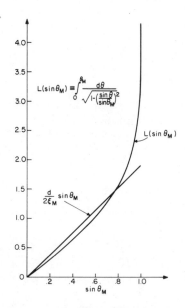

Fig. 7—Graphical solution of Eq. [21].

and $\pi/2$. Substitution of $\sqrt{K/\Delta\chi}/H$ for ξ_M gives

$$H_F = \sqrt{\frac{K}{\Delta\chi}}\frac{\pi}{d}, \qquad [22]$$

which agrees with Fréedericksz's observation that the threshold field strength varies inversely with the thickness of the sample. The form of Eq. [22] can be understood by a simple plausibility argument. In section 3.2 we have seen that the magnetic coherence length ξ_M can be thought of as that length below which the magnetic field does not have much influence on the relative orientations of the directors. By applying this interpretation of ξ_M to our present example it is clear that only when $\xi_M < d/2$ would it be possible for the magnetic field to have significant influence on the orientations of the directors. Therefore, we would estimate

$$H_F \simeq \sqrt{\frac{K}{\Delta\chi}}\frac{2}{d},$$

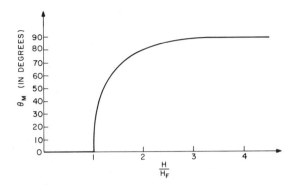

Fig. 8—Tilt angle of the directors at the center of Fréedericksz cell, θ_M, plotted as a function of reduced field H/H_F. For H/H_F slightly greater than 1, θ_M behaves as $\sim(H/H_F - 1)^{1/2}$.

which differs with the exact result only by a factor of $\pi/2$. In Fig. 8, θ_M is plotted as a function of H/H_F. For $H \sim H_F$, Eq. [21] can be expanded around $\theta_M = 0$ to give $\theta_M \propto (H - H_F)^{1/2}$.

Having obtained $\theta_M(H)$, we can now determine $\theta(x)$ as a function of H. By writing $d/2\xi_M = \pi H/2H_F$, Eq. [20] is put in the form

$$\frac{\pi}{2} \frac{H}{H_F}\left(1 + \frac{2x}{d}\right) \sin\theta_M = \int_0^{\theta(x)} \frac{d\theta'}{\sqrt{1 - \left(\frac{\sin\theta'}{\sin\theta_M}\right)^2}}. \quad [23]$$

The right-hand side can be numerically integrated on computer, and the results $\theta(x)$ are plotted in Fig. 9 for three different values of H.

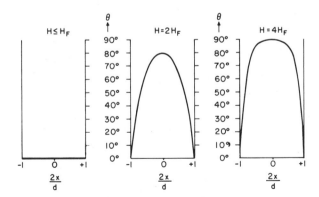

Fig. 9—Tilt angle of the directors plotted as a function of position in a Fréedericksz cell for three different magnetic field strengths.

The Fréedericksz transition can be induced by an electric field as well as by a magnetic field. The threshold electric field in that case is given by

$$E_F = \frac{\pi}{d}\sqrt{\frac{4\pi K}{\Delta \epsilon}}, \quad [24]$$

or

$$V_F = E_F d = \pi\sqrt{\frac{4\pi K}{\Delta \epsilon}}. \quad [25]$$

For $K \sim 10^{-6}$ dyne, $\Delta\chi \sim 10^{-7}$ cgs unit, $\Delta\epsilon \sim 0.1$, the critical magnetic field H_F is ~ 10 Oe for $d \sim 1$ cm, and the critical voltage is ~ 10 V, independent of cell thickness.

To conclude the discussion of the Fréedericksz transition, we note that for $H > H_F$ ($V > V_F$) there are two equivalent tilt configurations of the directors, denoted by the solid and the dotted lines in Fig. 6. In practice, for $H > H_F$ (or $V > V_F$) both tilt configurations are usually present, and regions of different tilt patterns are separated by visible disinclination lines.

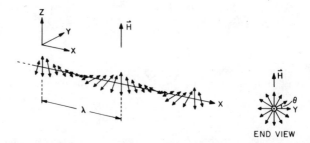

Fig. 10—Two views of the local directors in a cholesteric liquid crystal. In order to induce the cholesteric–nematic transition, a magnetic field H is applied perpendicular to the cholesteric helical axis. The angle θ used in the calculation is defined as shown.

3.4 Field-Induced Cholesteric–Nematic Transition

Consider a sample of cholesteric liquid crystal placed in a magnetic field H, with the field direction perpendicular to the cholesteric helical axis as shown in Fig. 10. From Eqs. [5], and [6], the free energy of the system over one period (or pitch) of the helix, λ, can be written as

ELASTIC CONTINUUM THEORY

$$\frac{F(\lambda)}{A} = \frac{K_{22}}{2} \int_0^\lambda dx \left\{ \left[\left| \hat{n}(r) \cdot \nabla \times \hat{n}(r) \right| - \frac{\pi}{\lambda_0} \right]^2 - \frac{\Delta \chi}{K_{22}} H^2 \sin^2\theta \right\}, \quad [26]$$

where A is the area of the sample in the y-z plane, assumed to be a constant, λ_0 is the pitch of the cholesteric helix at $H = 0$, θ is the angle between a local director and the y-axis as defined in Fig. 9, and $x = 0$ is defined by any point at which $\theta = 0$ (or π). With

$$n_x = 0,$$
$$n_y = \cos\theta(x),$$
$$n_z = \sin\theta(x),$$

we have

$$\hat{n}(r) \cdot \nabla \times \hat{n}(r) = \frac{d\theta(x)}{dx}$$

and

$$\frac{F(\lambda)}{A} = \frac{K_{22}}{2} \int_0^\lambda dx \left\{ \left[\frac{d\theta}{dx} - \frac{\pi}{\lambda_0} \right]^2 - \frac{\Delta \chi H^2 \sin^2\theta}{K_{22}} \right\}. \quad [27]$$

Here, we have to remember to take the absolute value of $d\theta/dx$. Application of the Euler–Lagrange equation yields

$$\xi_M^2 \frac{d^2\theta}{dx^2} + \sin\theta\cos\theta = 0,$$

which, as seen previously, can be put in the form

$$\left(\frac{d\theta}{dx} \right)^2 = \frac{1}{\xi_M^2} \left(\frac{1}{k^2} - \sin^2\theta \right), \quad [28]$$

where k^2 is an integration constant. At $H = 0$, it follows from Eq. [27] that $(d\theta/dx)$ equals a constant, (π/λ_0). Therefore, k must behave as $\sim H$ for $H \to 0$ in order to cancel the H^2 from $1/\xi_M^2$ in Eq. [28]. Writing Eq. [28] in the form

$$\frac{d\theta}{dx} = \pm \frac{1}{k\xi_M} \sqrt{1 - k^2\sin^2\theta}, \quad [28a]$$

we note that for finite values of H, $d\theta/dx$ is no longer a constant.

Plotting the z-component of the local director, $n_z = \sin\theta(x)$, as a function of position along the helical axis (x-axis) reveals that the sinusoidal pattern for $n_z(x)$ at $H = 0$ becomes distorted at finite values of H as shown in Fig. 11. The distortion makes $n_z(x)$ more square-wave-like and lengthens the pitch of the helix. Both of these effects can be understood on the basis that alignment along the magnetic

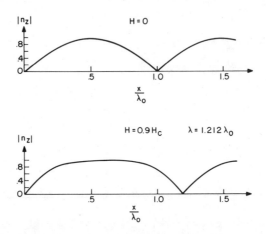

Fig. 11—Component of the cholesteric local director along the external field direction (z) for two different magnetic field strengths. Note that at finite field strength the helical pitch is lengthened and the sinusoidal shape of the curve at $H = 0$ is distorted, becoming more square-wave-like.

field direction lowers the energy of the system. The $\pm$ signs for $d\theta/dx$ indicate the two possible senses of the helical twist. Since they are equivalent, we choose the $+$ sign in the following calculation. From the expression for $d\theta/dx$ we can get an expression for the pitch λ:

$$\lambda = \int_0^\lambda dx = \int_0^\pi d\theta \frac{dx}{d\theta} = \int_0^\pi \frac{\xi_M k}{\sqrt{1 - k^2 \sin^2\theta}} d\theta$$

$$= 2\xi_M k \int_0^{\frac{\pi}{2}} \frac{d\theta}{\sqrt{1 - k^2 \sin^2\theta}}. \qquad [29]$$

At this point, it becomes necessary to know k^2 as a function of H. To do that, we must substitute Eq. [28a] back into Eq. [27] and minimize the average free-energy density by varying k^2. Let us rewrite Eq. [27] as

ELASTIC CONTINUUM THEORY

$$g = \frac{2F(\lambda)}{q_0^2 K_{22} A \lambda} = \frac{1}{\lambda}\int_0^\pi d\theta \frac{dx}{d\theta}\left(\left[\frac{1}{q_0\xi_M k}\sqrt{1-k^2\sin^2\theta} - 1\right]^2 - \frac{\sin^2\theta}{q_0^2 \xi_M^2}\right), \quad [30]$$

where g is a dimensionless average free-energy density and $q_0 \equiv \pi/\lambda_0$. Substitution of Eq. [28a] for $dx/d\theta$ and expansion of the terms in the integrand gives

$$g = 1 - \frac{2\pi}{\lambda q_0} + \frac{2}{k\lambda \xi_M q_0^2}\int_0^\pi d\theta\sqrt{1-k^2\sin^2\theta} - \frac{1}{k^2 q^2 \xi_M^2}. \quad [31]$$

Detailed steps leading from Eq. [30] to Eq. [31] are given in the Appendix. Differentiation of g with respect to k^2 yields

$$\frac{dg}{dk^2} = \frac{2\pi}{\lambda^2 q_0}\frac{d\lambda}{dk^2} - \frac{2}{k\lambda^2 q_0}\frac{1}{\xi_M q_0}\frac{d\lambda}{dk^2}\int_0^\pi d\theta$$

$$\times \sqrt{1-k^2\sin^2\theta} + \frac{1}{k^4 q_0^2 \xi_M^2} - \frac{1}{k^4 q_0^2 \xi_M^2}$$

$$= \frac{d\lambda}{dk^2}\frac{2}{\lambda^2 q_0}\left(\pi - \frac{1}{k\xi_M q_0}\int_0^\pi d\theta\sqrt{1-k^2\sin^2\theta}\right) \quad [32]$$

The desired equation for determining k as a function of H is obtained by setting $dg/dk^2 = 0$:

$$\xi_M q_0 = \sqrt{\frac{K_{22}}{\Delta\chi}}\frac{\pi}{H\lambda_0} = \frac{2}{\pi}\frac{1}{k}\int_0^{\frac{\pi}{2}} d\theta\sqrt{1-k^2\sin^2\theta}. \quad [33]$$

Define

$$E_1(k) \equiv \int_0^{\frac{\pi}{2}} \frac{d\theta}{\sqrt{1-k^2\sin^2\theta}},$$

$$E_2(k) \equiv \int_0^{\frac{\pi}{2}} d\theta\sqrt{1-k^2\sin^2\theta},$$

where E_1 and E_2 are the complete elliptic integrals of the first kind and the second kind, respectively. Eqs. [29] and [33] can be put in the form

$$\lambda = 2\xi_M k E_1(k),$$

and

$$\frac{\pi \xi_M}{\lambda_0} = \frac{2}{k\pi} E_2(k),$$

Combining the two equations yields

$$\frac{\lambda}{\lambda_0} = \frac{4}{\pi^2} E_1(k) E_2(k). \qquad [34]$$

$E_1(k)$ diverges at $k = 1$. Therefore, λ/λ_0 diverges at a field given by Eq. [33]:

$$\frac{\xi_M \pi}{\lambda_0} = \frac{2}{\pi} E_2(1) = \frac{2}{\pi},$$

which defines a critical magnetic field

$$H_c = \sqrt{\frac{K_{22}}{\Delta \chi} \frac{\pi^2}{2\lambda_0}}. \qquad [35]$$

If one takes $K_{22} \sim 10^{-6}$ dyne, $\Delta \chi \sim 10^{-6}$ cgs units, $\lambda_0 \sim 10^{-4}$ cm, H_c is $\sim$50 kOe. A similar threshold can be obtained if the magnetic field is replaced by an electric field:

$$E_c = \sqrt{\frac{4\pi K_{22}}{\Delta \epsilon} \frac{\pi^2}{2\lambda_0}}. \qquad [36]$$

Eqs. [29], [33], [34], and [35] were first obtained by de Gennes.[12] In terms of H_c, Eq. [33] can be put in the form

$$\frac{H_c}{H} k = E_2(k). \qquad [33a]$$

For $H > H_c$ this equation has no solution. When $H < H_c$, the values of k ranges from 0 to 1 as plotted in Fig. 12. In Fig. 13 we show a plot of λ/λ_0 vs. H/H_c. At $H = H_c$ the pitch diverges and the cholesteric phase transforms into the nematic phase. Experimentally this curve is well verified.[13,14]

Finally, it should be noted that if the field is initially applied parallel to the helical axis, the cholesteric helix would usually rotate at $H < H_c$ so as to make the field perpendicular to the helical axis. Therefore, the geometry shown in Fig. 10 is always the situation seen experimentally just before the field strength reaches the cholesteric–nematic transition threshold.[14]

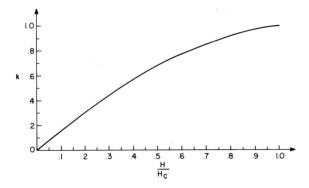

Fig. 12—Solution of Eq. [33a].

4. Concluding Remarks

The above discussion of the continuum theory of liquid crystals is by no means complete. There exist many more effects that can be described by the continuum theory, either in its present or modified form. In view of the diverse applications of the theory, the selection

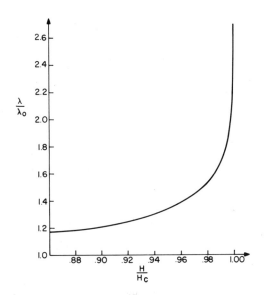

Fig. 13—Ratio of the helical pitch in finite field to the pitch in zero field, λ/λ_0, plotted as a function of reduced field H/H_c. The divergence at $H/H_c = 1$ is logarithmic in nature.

of the four examples discussed in this chapter is based on the consideration that they all have practical relevance to liquid-crystal display devices. It is hoped that their description by the continuum theory can, on the one hand, demonstrate the power and the flavor of the theory and, on the other hand, complement the discussion of the device physics aspects of these effects in other papers of this series.

Appendix

In this appendix we show the steps leading from Eq. [30] to Eq. [31]. From Eqs. [28a] and [30] we have

$$g = \frac{1}{\lambda} \int_0^\pi d\theta \frac{k\xi_M}{\sqrt{1 - k^2\sin^2\theta}} \left\{ 1 - \frac{2\sqrt{1 - k^2\sin^2\theta}}{kq_0\xi_M} \right.$$
$$\left. + \frac{1 - k^2\sin^2\theta}{k^2 q_0^2 \xi_M^2} - \frac{\sin^2\theta}{q_0^2 \xi_M^2} \right\}$$
$$= \frac{k\xi_M}{\lambda} \int_0^\pi d\theta \frac{d\theta}{\sqrt{1 - k^2\sin^2\theta}} - \frac{2\pi}{\lambda q_0} + \frac{1}{k\lambda\xi_M q_0^2}$$
$$\times \int_0^\pi d\theta \sqrt{1 - k^2\sin^2\theta} - \frac{1}{q_0^2 \lambda \xi_M^2} \int_0^\lambda dx \; \sin^2\theta. \quad [37]$$

By writing $\sin^2\theta = (1/k^2) - \xi_M^2(d\theta/dx)^2$, the fourth term in Eq. [37] can be simplified as

$$-\frac{1}{q_0^2 \lambda \xi_M^2} \int_0^\lambda dx \; \sin^2\theta = -\frac{1}{k^2 q_0^2 \xi_M^2} + \frac{1}{q_0^2 \lambda} \int_0^\pi \left(\frac{d\theta}{dx}\right) d\theta$$
$$= -\frac{1}{k^2 q_0^2 \xi_M^2} + \frac{1}{k\lambda\xi_M q_0^2} \int_0^\pi d\theta \sqrt{1 - k^2\sin^2\theta} \quad [38]$$

Therefore,

$$g = 1 - \frac{2\pi}{\lambda q_0} + \frac{2}{k\lambda\xi_M q_0^2} \int_0^\pi d\theta \sqrt{1 - k^2\sin^2\theta} - \frac{1}{k^2 q_0^2 \xi_M^2}. \quad [39]$$

References

[1] P. J. Wojtowicz, "Introduction to the Molecular Theory of Nematic Liquid Crystals," Chapter 3.
[2] P. J. Wojtowicz, "Generalized Mean Field Theory of Nematic Liquid Crystals," Chapter 4.
[3] H. Zocher, "The Effect of a Magnetic Field on the Nematic State," *Trans. Faraday Soc.*, **29**, p. 945 (1933).

[4] C. W. Oseen, "The Theory of Liquid Crystals," *Trans. Faraday Soc.*, **29**, p. 883 (1933).
[5] F. C. Frank, "On the Theory of Liquid Crystals," *Faraday Soc. Disc.*, **25**, p. 19 (1958).
[6] E. G. Priestley, "Nematic Order: The Long Range Orientational Distribution Function," Chapter 6.
[7] P. Sheng, "Hard Rod Model of the Nematic-Isotropic Phase Transition, Chapter 5.
[8] P. G. de Gennes, *Lecture Notes on Liquid Crystal Physics*, Part I (1970).
[9] A. Saupe, "Temperaturabhängigkeit und Grösse der Deformationskonstanten nematischer Flüssigkeiten," *Z. Naturforsch.*, **15a**, 810 (1960).
[10] V. Frëedericksz and A. Repiewa, "Theoretisches und Experimentelles zur Frage nach der Natur der Anisotropen Flussigkeiten," *Z. Physik*, **42**, p. 532 (1927).
[11] V. Frëedericksz and V. Zolina, "Forces Causing the Orientation of an Anisotropic Liquid," *Trans. Faraday Soc.*, **29**, p. 919 (1933).
[12] P. G. de Gennes, "Calcul de la Distorsion d'une Structure Chlosteric par un Champ Magnetique," *Solid State Comm.*, **6**, p. 163 (1968).
[13] G. Durand, L. Leger, F. Rondelez, and M. Veyssie, "Magnetically Induced Cholesteric-to-Nematic Phase Transition in Liquid Crystals," *Phys. Rev. Lett.*, **22**, p. 227 (1969).
[14] R. B. Meyer, "Distortion of a Cholesteric Structure by a Magnetic Field," *Appl. Phys. Lett.*, **14**, p. 208 (1969).

9

Electrohydrodynamic Instabilities in Nematic Liquid Crystals

Dietrich Meyerhofer

RCA Laboratories, Princeton, N. J. 08540

1. Introduction

The best-known electrohydrodynamic instabilities in liquid crystals are the Williams domains.[1] They are observed when an electric field is applied to a thin layer of a nematic liquid crystal having negative dielectric anisotropy and sufficient electrical conductivity. They manifest themselves as a set of parallel straight lines separated by a constant distance that is approximately equal to the cell thickness. They appear above a well-defined threshold voltage and exist in their original form over only a small voltage range. At higher voltages, the pattern becomes more complicated and leads to heavily scattering turbulence (dynamic scattering).[2]

The structure of the domain instability is well-understood qualitatively.[3] The liquid flows in cylindrical motion at right angles to the domain walls, which represent the vortices of the motion. The hydrodynamic motion is not visible directly, but becomes manifest because of the anisotropy in index of refraction. The pattern is only visible for light polarized perpendicular to the domain walls, which means that the director (parallel to the optic axis) is located in the plane perpendicular to the walls. The flow and alignment pattern in this plane are sketched in Fig. 1, following Penz.[3]

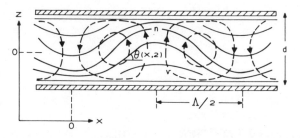

Fig. 1—Cross section through a liquid crystal cell at right angles to the domain lines. The solid lines indicate the director orientation n, the dashed lines represent the flow v.

It should be noted here that hydrodynamic instabilities are also possible in isotropic liquids. Examples are the interaction of thermal gradients and gravitational forces (Bénard's problem[4]) and electric-field-driven flow under space charge limited current conditions.[5] The latter is the likely cause of flow that has been observed in liquid crystals above the nematic–isotropic transition[6] (N-L point). In contrast to this, the instabilities discussed here depend on the anisotropic nature of the material parameters.

The nature of the Williams domain instability was explained by Helfrich who calculated the threshold voltage of domain formation for the case of dc voltage and one-dimensional geometry.[7] The calculations were extended to ac fields by Dubois-Violette[8] and to two-dimensional geometry by Penz and Ford.[9] They show good qualitative agreement with measured thresholds.

In this chapter we will extend the threshold calculations to show how the instability arises and to combine both frequency dependence and two-dimensional features. We will compare the results with measured values of threshold voltage and domain spacing as function of ac frequency.[10] Good quantitative agreement will be demonstrated.

2. Nature of the Instability and the Balance of Forces

We now calculate the threshold of domain formation following the concepts of Helfrich.[7] We consider a thin planar cell with electrodes on the two surfaces. We assume that the dielectric anisotropy is negative ($\Delta\epsilon = \epsilon_\parallel - \epsilon_\perp < 0$) and that the liquid crystal is aligned uniformly parallel to the surface by suitable surface treatment. The geometry is that of Fig. 1; the director lies in the x-z plane and is parallel to the x-axis at the surfaces. There is no variation in the y-direction and all variations in the x-direction are periodic with period Λ. In this geometry, the electric field alone does not distort the liquid and any instability must be of hydrodynamic nature.

To calculate the threshold, we assume a small fluctuation of the director $\theta(x,z)$ and calculate whether the fluctuation grows or decays in time. In the absence of applied fields, there is only an elastic torque (Γ_elast), which tends to restore the uniform alignment. An applied electric field in the z-direction produces two kinds of forces. First, there is a purely dielectric restoring torque (Γ_diel) due to $\Delta\epsilon < 0$. Second, the anisotropy in the conductivity produces a charge separation (similar to the Hall effect) through which the applied field exerts a force on the mass of the liquid causing it to flow. The nonuniformity of the flow (shear) produces a torque on the director (Γ_visc) that is in such a direction as to increase the fluctuation. The magnitude of Γ_diel and Γ_visc both increase as the square of the applied field so that, if $|\Gamma_\text{visc}|$ is larger than $|\Gamma_\text{diel}|$, there will exist a field at which $|(\Gamma_\text{visc} + \Gamma_\text{diel})|$ is larger than $|\Gamma_\text{elast}|$ and the alignment will become unstable.

In calculating the torque, we make the assumption that $\theta \ll \pi/2$. This will lead to linearized equations and all torques will be proportional to θ. This permits an accurate calculation of the threshold, but does not allow the determination of the distortion above threshold for which higher order terms in θ are required.

3. Dielectric Response

Apply an external field $E^\circ = V/d$. Thus, the field in the sample is given by $\mathbf{E} = (E_x, 0, E^\circ + E_z')$ where $E_x, E_z' \ll E^\circ$. Because $\nabla \times \mathbf{E} = -\partial B/\partial t = 0$,

$$\frac{\partial E_x}{\partial z} = \frac{\partial E_z}{\partial x} = \frac{\partial E_z'}{\partial x}. \qquad [1]$$

The director $\mathbf{n}$ lies in the (x, z) plane at angle θ to the x axis, so $\mathbf{n} =$

$(\cos\theta, 0, \sin\theta) \approx (1, 0, \theta)$. The dielectric and conductivity tensors (ϵ, σ) are both uniaxial and parallel to **n**, and are given by

$$\epsilon_{ij} = \epsilon_\perp \delta_{ij} + \Delta\epsilon n_i n_j \quad [2]$$
$$\Delta\epsilon = \epsilon_\parallel - \epsilon_\perp.$$

with an identical relationship for σ. If we keep only first order terms in θ, E_x, E_z', we obtain

$$D_x = \epsilon_\parallel E_x + \Delta\epsilon\theta E_z = \epsilon_\parallel E_x + \Delta\epsilon\theta E° \quad [3]$$
$$D_z = \Delta\epsilon\theta E_x + \epsilon_\perp E_z = \epsilon_\perp E° + \epsilon_\perp E_z',$$
$$j_x = \sigma_\parallel E_x + \Delta\sigma\theta E° \quad [4]$$
$$j_z = \sigma_\perp E° + \sigma_\perp E_z'.$$

The free charge is given by

$$\rho = \nabla \cdot \mathbf{D} = \epsilon_\parallel \frac{\partial E_x}{\partial x} + \Delta\epsilon E° \frac{\partial \theta}{\partial x} + \epsilon_\perp \frac{\partial E_z'}{\partial z}, \quad [5]$$

and from the charge continuity equation

$$\frac{-\partial \rho}{\partial t} = \nabla \cdot \mathbf{j} = \sigma_\parallel \frac{\partial E_x}{\partial x} + \Delta\sigma E° \frac{\partial \theta}{\partial x} + \sigma_\perp \frac{\partial E_z'}{\partial z}. \quad [6]$$

Eqs. [1], [5], and [6] define E_x, E_z', and ρ in terms of E_o and $d\theta/dx$.

4. Hydrodynamic Effects

The hydrodynamic equations describe the relationships between the applied forces and the fluid velocities. As given by Leslie,[11] in linear approximation, they are

$$F_i + \sum_j \frac{\partial \sigma_{ij}'}{\partial x_j} = 0 \quad [7]$$

for the isothermal, steady-state case. Here, **F** is the applied body force

$$\mathbf{F} = \rho \mathbf{E} = \rho(0,0,E°) \quad [8]$$

and σ' is the viscous stress tensor (σ'_{ij} is as defined by Penz[9]). For an incompressible fluid, there is the additional requirement

$$\nabla \cdot \mathbf{v} = 0, \qquad [9]$$

where $\mathbf{v}$ is the fluid velocity.

The viscous tensor as obtained from thermodynamic considerations and symmetry properties is defined in terms of velocity gradients and director orientations as

$$\sigma_{ij}' = -p\delta_{ij} + \alpha_1 \sum_{kl} n_k n_l A_{kl} n_i n_j + \alpha_2 n_j N_i + \alpha_3 n_i N_j$$
$$+ \alpha_4 A_{ij} + \alpha_5 \sum_k n_j n_k A_{ki} + \alpha_6 \sum_k n_i n_k A_{kj}, \qquad [10]$$

where

$$A_{ij} = \frac{1}{2}\left(\frac{\partial v_i}{\partial x_j} + \frac{\partial v_j}{\partial x_i}\right),$$
$$N = -\frac{1}{2}[\nabla \times \mathbf{v}] \times \mathbf{n}, \qquad [11]$$

p is the hydrostatic pressure, and the α_i's are the viscosity coefficients. The velocity gradients are assumed small, so that the only non-zero terms are

$$A_{xx} = \frac{\partial v_x}{\partial x}, \; A_{xz} = A_{zx} = \frac{1}{2}\left(\frac{\partial v_x}{\partial z} + \frac{\partial v_z}{\partial x}\right).$$
$$A_{zz} = \frac{\partial v_z}{\partial z}, \; N_z = \frac{1}{2}\left(\frac{\partial v_x}{\partial z} - \frac{\partial v_z}{\partial x}\right)$$

and

$$\sigma_{xx}' = (\alpha_1 + \alpha_4 + \alpha_5 + \alpha_6)\frac{\partial v_x}{\partial x} - p$$
$$\sigma_{xz}' = \frac{1}{2}(-\alpha_3 + \alpha_4 + \alpha_6)\frac{\partial v_z}{\partial x} + \frac{1}{2}(\alpha_3 + \alpha_4 + \alpha_6)\frac{\partial v_x}{\partial z} \qquad [12]$$
$$\sigma_{zx}' = \frac{1}{2}(-\alpha_2 + \alpha_4 + \alpha_5)\frac{\partial v_z}{\partial x} + \frac{1}{2}(\alpha_2 + \alpha_4 + \alpha_5)\frac{\partial v_x}{\partial z}$$
$$\sigma_{zz}' = \alpha_4 \frac{\partial v_x}{\partial z} - p.$$

Inserting Eqs. [8] and [12] into Eq. [7], we obtain

$$-F_x = 0 = \frac{\partial \sigma_{xx}'}{\partial x} + \frac{\partial \sigma_{xz}'}{\partial z} = -\frac{\partial p}{\partial x} + (\alpha_1 + \alpha_4 + \alpha_5 + \alpha_6)\frac{\partial^2 v_x}{\partial x^2}$$
$$+ \frac{1}{2}(-\alpha_3 + \alpha_4 + \alpha_6)\frac{\partial^2 v_z}{\partial x \partial z} + \frac{1}{2}(\alpha_3 + \alpha_4 + \alpha_6)\frac{\partial^2 v_x}{\partial z^2} \quad [13]$$

$$-F_z = -\rho E_0 = \frac{\partial \sigma_{zx}'}{\partial x} + \frac{\partial \sigma_{zz}'}{\partial z} = -\frac{\partial p}{\partial z} + \alpha_4 \frac{\partial^2 v_z}{\partial z^2} + \frac{1}{2}(-\alpha_2 + \alpha_4$$
$$+ \alpha_5)\frac{\partial^2 v_z}{\partial x^2} + \frac{1}{2}(\alpha_2 + \alpha_4 + \alpha_5)\frac{\partial^2 v_x}{\partial x \partial z}. \quad [14]$$

Eqs. [9], [13], and [14] can be solved for v_x, v_z, and p in terms of ρ which was previously defined in terms of $d\theta/dx$.

5. The Boundary Value Problem in the Conduction Regime

The boundary conditions of the two-dimensional problem are that θ, E_x, v_x, and v_z all must be zero at the electrodes ($z = \pm d/2$). Penz and Ford[9] have solved the various equations for these boundary conditions in the dc case. The solution had to be obtained numerically, which obscures the physical processes.

We can obtain a much simpler solution by relaxing one boundary condition and letting v_x be finite at the electrode. While this can not be the case in practice, due to viscous forces, we will show below that the approximation is a good one. Penz[3] demonstrated experimentally that the distortion of the director above the Williams domain threshold may be described as

$$\theta = \theta^\circ \sin q_x x \cos q_z z, \quad [15]$$

where $q_z = \pi/d$, $q_x = 2\pi/\Lambda$. This is the pattern plotted in Fig. 1. By inspection, and by making use of Eqs. [1] and [9], we can write the simplest trial functions

$$\begin{aligned} E_x &= E_x' \sin q_x x \cos q_z z \\ E_z' &= SE_x' \cos q_x x \sin q_z z \\ v_x &= Sv_z' \sin q_x x \sin q_z z \\ v_z &= v_z' \cos q_x x \cos q_z z \end{aligned} \quad [16]$$

where $S = q_z/q_x = \Lambda/2d$. The corresponding flow pattern is indicated in Fig. 1 (the figure anticipates v_z' being negative).

Inserting the values of E_x and E_z' in Eqs. [5] and [6] and solving for ρ, one obtains

$$\rho + \frac{\epsilon_\parallel + S^2\epsilon_\perp}{\sigma_\parallel + S^2\sigma_\perp} \frac{\partial \rho}{\partial t} = \left[\Delta\epsilon - \frac{\epsilon_\parallel + S^2\epsilon_\perp}{\sigma_\parallel + S^2\sigma_\perp} \Delta\sigma \right] E^\circ \theta^\circ q_x \cos q_x x \cos q_z z. \quad [17]$$

Let the applied field be sinusoidal,

$$E^\circ = \sqrt{2} E_0 \cos \omega t, \quad [18]$$

and assume that θ and $\mathbf{v}$ are independent of time. This defines the *conduction regime* of Dubois-Violette et al[8] in which Williams domains can exist. Thus Eq. [17] is solved to obtain

$$\rho = -\tau \sigma_H \sqrt{2} E_0 \frac{\cos \omega t + \omega \tau \sin \omega t}{1 + \omega^2 \tau^2} q_x \theta^\circ \cos q_x x \cos q_z z, \quad [19]$$

where

$$\tau = \frac{\epsilon_\parallel + S^2\epsilon_\perp}{\sigma_\parallel + S^2\sigma_\perp}; \quad \sigma_H = \Delta\sigma - \frac{\Delta\epsilon}{\tau}$$

in the notation of Dubois-Violette. τ is the equivalent dielectric relaxation time.

Inserting Eqs. [18] and [19] into Eq. [5], one obtains

$$E_x' = -\frac{\tau \sigma_H \dfrac{\cos \omega t + \omega \tau \sin \omega t}{1 + \omega^2 \tau^2} + \Delta\epsilon \cos \omega t}{\epsilon_\parallel + S^2 \epsilon_\perp} \sqrt{2} E_0 \theta^\circ, \quad [20]$$

the desired relationship between E_x and θ.

Next, we insert Eq. [16] into Eqs. [13] and [14] and eliminate p to obtain

$$q_x^3 \{ \eta_1 + (\alpha_1 + \eta_1 + \eta_2)S^2 + \eta_2 S^4 \} v_z' \sin q_x x \cos q_z z = -E_0 \frac{\partial \rho}{\partial x}, \quad [21]$$

where $\eta_1 = \frac{1}{2}(-\alpha_2 + \alpha_4 + \alpha_5)$, $\eta_2 = \frac{1}{2}(\alpha_3 + \alpha_4 + \alpha_6)$ are the Helfrich viscosity parameters (Penz and Ford[9] use the same notation, but interchange η_1 and η_2). The left-hand side of Eq. [21] is time independent, so the right-hand side can be averaged over time (after inserting Eq. [19]) to obtain

$$-\tau\sigma_H E_0^2(q_x^2\theta° \sin q_x x \cos q_z z)\frac{1}{1+\omega^2\tau^2}.$$

Therefore,

$$v_z' = \frac{-\theta° E_0^2}{q_x[(1+S^2)(\eta_1 + S^2\eta_2) + S^2\alpha_1]} \frac{\tau\sigma_H}{1+\omega^2\tau^2}. \qquad [22]$$

6. The Torque Balance Equation

We can now calculate the various torques acting on the director. Because of the two-dimensional geometry, only the y-component of the torque is non-zero. The dielectric and elastic torques are obtained most easily by taking the functional derivative of the free energy with respect to θ.[8]

$$\Gamma_y = \frac{-\partial F}{\partial\theta} \qquad [23]$$

The elastic free energy is given by Frank[12] as

$$F_{elast} = \frac{1}{2}[k_{11}(\nabla\cdot\mathbf{n})^2 + k_{22}(\mathbf{n}\cdot\nabla\times\mathbf{n})^2 + k_{33}(\mathbf{n}\times\nabla\times\mathbf{n})^2]$$
$$= \frac{k_{11}}{2}\left(\frac{\partial n_x}{\partial x} + \frac{\partial n_z}{\partial z}\right)^2 + \frac{k_{33}}{2}\left(\frac{\partial n_x}{\partial z} - \frac{\partial n_z}{\partial x}\right)^2. \qquad [24]$$

Inserting the values for $\mathbf{n}$, and keeping only linear terms,

$$\Gamma_{elast,y} = -\left(k_{11}\frac{\partial^2\theta}{\partial z^2} + k_{33}\frac{\partial^2\theta}{\partial x^2}\right)$$
$$= q_x^2(k_{33} + S^2 k_{11})\theta_0 \sin q_x x \cos q_z z. \qquad [25]$$

This is a restoring torque ($\Gamma/\theta > 0$), i.e., it tends to decrease θ_0.

The dielectric free energy is

$$F_{diel} = \frac{1}{2}\mathbf{E}\cdot\mathbf{D}. \qquad [26]$$

Making use of Eq. [2] and the exact value of $\mathbf{n} = (\cos\theta, 0, \sin\theta)$, one obtains

$$\Gamma_{diel,y} = -\Delta\epsilon E_z(E_x + E_z\theta). \qquad [27]$$

ELECTROHYDRODYNAMIC INSTABILITIES

Inserting Eqs. [15] and [20] into Eq. [27] and averaging over time, one obtains

$$\Gamma_{\text{diel},y} = -\Delta\epsilon E_0^2 \left\{ \frac{-\tau\sigma_H(1+\omega^2\tau^2)^{-1}+\Delta\epsilon}{\epsilon_{\parallel}+S^2\epsilon_{\perp}} + 1 \right\} \theta_0 \sin q_x x \cos q_z z$$

$$= -\Delta\epsilon E_0^2 \left\{ \frac{\tau\sigma_H}{\epsilon_{\parallel}+S^2\epsilon_{\perp}} \frac{\omega^2\tau^2}{1+\omega^2\tau^2} \right.$$

$$\left. + (1+S^2)\frac{\sigma_{\perp}}{\sigma_{\parallel}+S^2\sigma_{\perp}} \right\} \theta_0 \sin q_x x \cos q_z z \quad [28]$$

for $\Delta\epsilon < 0$, $\Gamma/\theta > 0$, and this is also a restoring torque as predicted.

The viscous torque that the flow exerts on the director is given by[9,13]

$$\Gamma_{\text{visc},y} = \sigma_{xz}' - \sigma_{zx}'$$

$$= \frac{1}{2}(\alpha_2 - \alpha_3 + \alpha_6 - \alpha_5)\frac{\partial v_z}{\partial x} + \frac{1}{2}(\alpha_3 - \alpha_2 + \alpha_6 - \alpha_5)\frac{\partial v_x}{\partial z}$$

$$= -K_1\frac{\partial v_z}{\partial x} + K_2\frac{\partial v_x}{\partial z} \quad [29]$$

$$= q_x v_z'(K_1 + S^2 K_2) \sin q_x x \cos q_z z,$$

$$= -\frac{E_0^2(K_1 + S^2 K_2)}{\eta_1 + (\alpha_1+\eta_1+\eta_2)S^2 + \eta_2 S^4} \frac{\tau\sigma_H}{1+\omega^2\tau^2} \theta_0 \sin q_x x \cos q_z z,$$

where Eq. [22] has been used. K_1 and K_2 are the combinations of viscosity parameters Helfrich[7] calls the shear-torque coefficients. If the Parodi relationship[13] ($\alpha_6 - \alpha_5 = \alpha_2 + \alpha_3$) is used, they become

$$K_1 = -\alpha_2, \quad K_2 = \alpha_3. \quad [30]$$

Since K_1 is larger than K_2 and positive, $\Gamma_{\text{visc}}/\theta < 0$ and this can be seen to be the driving torque.

All three torques have the same spatial dependence and are proportional to θ. This confirms that our set of trial functions, Eqs. [16], are consistent. The threshold for domain formation may now be calculated by setting the sum of the torques equal to zero. After some rearrangement,

$$\frac{E_0^2}{q_x^2} = [(k_{33}+S^2 k_{11})(1+\omega^2\tau^2)] \left[\frac{(K_1+S^2 K_2)(1+S^2)}{(1+S^2)(\eta_1+S^2\eta_2)+S^2\alpha_1} \right.$$

$$\left. \cdot \frac{\sigma_{\parallel}\epsilon_{\perp}-\sigma_{\perp}\epsilon_{\parallel}}{\sigma_{\parallel}+S^2\sigma_{\perp}} + \Delta\epsilon\frac{(1+S^2)\sigma_{\perp}}{\sigma_{\parallel}+S^2\sigma_{\perp}} + \omega^2\tau^2 \Delta\epsilon\frac{(1+S^2)\epsilon_{\perp}}{\epsilon_{\parallel}+S^2\epsilon_{\perp}} \right]^{-1}, \quad [31]$$

where

$$\tau = \frac{\epsilon_\parallel + S^2 \epsilon_\perp}{\sigma_\parallel + S^2 \sigma_\perp}.$$

Eq. [31] has been cast in this form to allow comparison with previous work. If $\omega = 0$, Eq. [31] of Penz and Ford[9] is obtained (α_1 is very small or zero); if $\omega = 0$ and $S = 0$, Eq. [5.2] of Helfrich;[7] and if $S = 0$, Eq.

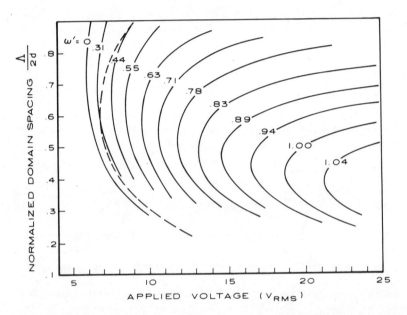

Fig. 2—Plot of Eq. (32) for MBBA (Table 1): $S = \Lambda/2d$ versus V. The parameter is the reduced frequency ω'. The dashed curve is the domain spacing given by Penz and Ford.[9] Instability occurs at the point on the curve where the voltage is a minimum.

[III,9] of Dubois-Violette[8] (except for a different definition of the viscosity factor). We note, however, that Penz and Ford's Eq. [31] has a different meaning, because it simply relates E_0 and the various values of S that form the more complicated solutions to the exact boundary value problem (q_z is not defined *a priori*). In the present case, the form of the solution has been given in Eqs. [15] and [16] and q_z has been defined as π/d. Therefore, we can rewrite Eq. [31]

$$V_{th}^2 = \frac{\pi^2}{S^2}(k_{33} + S^2 k_{11})(1 + \omega^2 \tau^2)\left[\frac{K_1 + S^2 K_2}{\eta_1 + S^2 \eta_2} \cdot \frac{\sigma_\| \epsilon_\perp - \sigma_\perp \epsilon_\|}{\sigma_\| + S^2 \sigma_\perp}\right.$$
$$\left. + \Delta\epsilon \frac{(1 + S^2)\sigma_\perp}{\sigma_\| + S^2 \sigma_\perp} + \omega^2 \tau^2 \Delta\epsilon \frac{(1 + S^2)\epsilon_\perp}{\epsilon_\| + S^2 \epsilon_\perp}\right]^{-1}, \qquad [32]$$

which relates the applied voltage V to the domain spacing Λ. The threshold voltage is determined by minimizing V with respect to S. Eq. [32] demonstrates that the relationship is independent of cell thickness, which also appears to be the case for the exact result.[9]

Table 1—Material Parameters of MBBA[9]

k_{11}	$= 6.10 \times 10^{-12}$ newton
k_{33}	$= 7.25 \times 10^{-12}$ newton
$\epsilon_\|$	$= 4.72$
$\epsilon_\perp$	$= 5.25$
η_1	$= 103.5 \times 10^{-3}$ kg/m/sec
η_2	$= 23.8 \times 10^{-3}$ kg/m/sec
α_2	$= -77.5 \times 10^{-3}$ kg/m/sec
α_3	$= -1.2 \times 10^{-3}$ kg/m/sec
$\sigma_\|/\sigma_\perp$	$= 1.5$

7. Numerical Results and Comparison with Experiment

Eq. [32] is plotted in Fig. 2 for various values of the normalized frequency $\omega' = \omega/\omega_c = \omega\tau_c$, $\tau_c = \epsilon_\|/\sigma_\|$. The numerical values used are the same as those used by Penz and Ford[9] for MBBA (Table 1). For comparison, Fig. 2 also shows the complete solution of these authors ($\omega = 0$). As can be seen, our approximation leads to a somewhat lower threshold, because we have neglected the viscosity force of the walls on the liquid. Note that the threshold for the one-dimensional case ($S = 0$, $q_x = \pi/d$ in Eq. [31]), 2.57 V, is considerably in error. From the curves of Fig. 2, the values of threshold voltage and the corresponding domain spacing are obtained as a function of voltage. They are plotted in Fig. 3 as a function of frequency. Also shown are experimental results of MBBA taken from the paper of Meyerhofer and Sussman[10] for a 10-μm-thick cell of MBBA with the alignment parallel to the surface. Since the conductivity had not been measured independently, the experimental data were normalized to agree with the calculated threshold at the highest frequency. The agreement is very satisfactory and shows that the inconsistency discussed in Ref. [10], that resulted when the actual values of Λ are inserted in the Orsay equation,

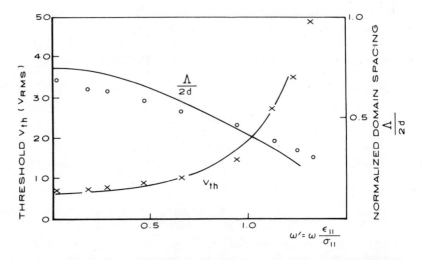

Fig. 3—Threshold and domain spacing as function of ω'. The lines are calculated from the curves of Fig. 2, and the points are the experimental results.[10] The only adjustment of the data was to determine $\tau_c = \epsilon_{\parallel}/\sigma_{\parallel}$ by fitting the measured threshold to the calculated one at one frequency, because the value of $\sigma_{\parallel}$ was not available.

is due to the one-dimensional nature of that theory, and it disappears when the two-dimensional calculation is performed. The good agreement is further demonstrated in Fig. 4, where the threshold voltage is plotted versus domain size (Fig. 2 of Ref. [10]) and the calculated curve has no adjustable parameters.

8. Range of Applicability

The calculations we have performed apply to the "conduction regime" because of the assumption that θ is constant in time. The range of this regime has been discussed in detail by Dubois-Violette et al[8] and shown experimentally by Meyerhofer and Sussman.[10] The low-frequency limit varies as d^{-2}; below this frequency both θ and ρ are time dependent. The upper frequency limit, near ω_c, is independent of thickness; above this frequency the dielectric regime applies, where ρ is constant and θ varies.

The conduction regime as described by Figs. 2 to 4 occurs in suitably doped nematic liquid crystals of negative dielectric anisotropy. This can be seen qualitatively by studying the denominator of Eq. [32]. The first (viscous) term must be larger than the second (dielectric) term for the denominator to be positive and instability to occur. This restricts negative values of $\Delta\epsilon$ to the range -2 to 0 for materials

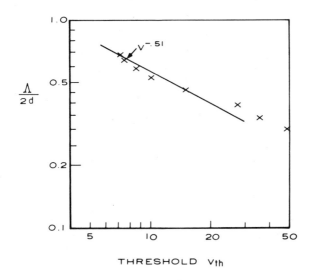

Fig. 4—The data of Fig. 3 plotted as domain spacing versus threshold with frequency as parameter. The crosses are the experimental results.[10] There are no adjustable parameters.

similar to MBBA and PAA ($K_1/\eta_1 = 0.75$, $\sigma_\parallel/\sigma_\perp = 1.5 > \epsilon_\parallel/\epsilon_\perp$). For such materials, the threshold increases strongly with frequency, leading to a cutoff of the conduction regime near $\omega\tau_c = 1$.

Eq. [32] shows that instability can also take place for $\Delta\epsilon > 0$, as discussed by Dubois-Violette[8] and Penz.[14] In that case, the threshold voltage becomes independent of frequency at high frequency, but this is not a conduction regime because V_{th} is then only dependent on k and ϵ. Furthermore, this threshold is higher than that of the dielectric Freedericksz transition[15] for the case of the entire sample deforming uniformly ($S = 0$, $V_{th} = \pi\sqrt{k_{11}/\Delta\epsilon}$), and so will not occur. In fact,

Williams domains will occur only over the range of $\Delta\epsilon$ from 0 to approximately 0.5. These considerations are in agreement with experimental observations.[16]

References

[1] R. Williams, "Domains in Liquid Crystals," *J. Chem. Phys.*, **39**, p. 384 (1963).

[2] G. H. Heilmeier, L. A. Zanoni, and L. A. Barton, "Dynamic Scattering: A New Electrooptic Effect in Certain Classes of Nematic Liquid Crystals," *Proc. IEEE*, **56**, p. 1162 (1968).

[3] P. A. Penz, "Voltage-Induced Velocity and Optical Focusing in Liquid Crystals," *Phys. Rev. Lett.*, **24**, p. 1405 (1970); "Order Parameter Distribution for the Electrohydrodynamic Mode of a Nematic Liquid Crystal," *Mol. Cryst. Liq. Cryst.*, **15**, p. 141 (1971).

[4] S. Chandrasekhar, *Hydrodynamic and Hydromagnetic Stability*, Clarendon Press, Oxford, England (1961).

[5] N. Filici, "Phénomènes hydro et Aerodynamiques dans la Conduction des diélectriques Fluids," *Rev. Gen. Elect.*, **78**, p. 717 (1969).

[6] H. Gruler and G. Meier, "Correlation between Electrical Properties and Optical Behaviour of Nematic Liquid Crystals," *Mol. Cryst. Liq. Cryst.*, **12**, p. 289 (1971).

[7] W. Helfrich, "Conduction-Induced Alignment of Nematic Liquid Crystals: Basic Model and Stability Considerations," *J. Chem. Phys.*, **51**, p. 4092 (1969).

[8] E. Dubois-Violette, P. G. deGennes, and O. Parodi, "Hydrodynamic Instabilities of Nematic Liquid Crystals under AC Electric Fields," *J. Phys. (Paris)*, **32**, p. 305 (1971).

[9] P. A. Penz and G. W. Ford, "Electromagnetic Hydrodynamics of Liquid Crystals," *Phys. Rev.*, **A6**, p. 414 (1972).

[10] D. Meyerhofer and A. Sussman, "Electrohydrodynamic Instabilities in Nematic Liquid Crystals in Low-Frequency Fields," *Appl. Phys. Lett.*, **20**, p. 337 (1972).

[11] F. M. Leslie, "Some Constitutive Equations for Anisotropic Fluids," *Quart. J. Mech. Appl. Math.*, **19**, p. 357 (1966); and "Some Constitutive Equations for Liquid Crystals," *Arch. Ration. Mech. Analysis*, **28**, p. 265 (1968).

[12] F. C. Frank, "On the Theory of Liquid Crystals," *Discuss. Faraday Soc.*, **25**, p. 19 (1958).

[13] O. Parodi, "Stress Tensor for a Nematic Liquid Crystal," *J. Phys. (Paris)*, **31**, p. 581 (1970).

[14] P. A. Penz, "Electrohydrodynamic Solutions for Nematic Liquid Crystals with Postivie Dielectric Anisotrophy," *Mol. Cryst. Liq. Cryst.*, **23**, p. 1 (1973).

[15] V. Freedericksz and W. Zwetkoff, "Uber die Einwirkung des Elektrischen Feldes auf Anisotrope Flussigkeiten. II. Orientierung der Flussigkeit im Elektrischen Felde," *Acta Physiocochim. USSR*, **3**, p. 895 (1935).

[16] H. Gruler and G. Meier, "Electric Field-Induced Deformations in Oriented Liquid Crystals of the Nematic Type," *Mol. Cryst. Liq. Cryst.*, **16**, p. 299 (1972); W. H. deJeu, C. J. Gerritsma, and T. W. Lathouwers, "Instabilities in Electric Fields of Nematic Liquid Crystals with Positive Dielectric Anisotropy: Domains, Loop Domains, and Reorientation," *Chem. Phys. Lett.*, **14**, p. 503 (1972).

10

The Landau-de Gennes Theory of Liquid Crystal Phase Transitions

Ping Sheng and E. B. Priestley

RCA Laboratories, Princeton, N. J. 08540

1. Introduction

A physical system in which phase transition(s) can occur is usually characterized by one or more long range order parameters (order parameter for short). For example, in nematic liquid crystals the order parameter is the quantity $S \equiv \langle P_2(\cos \theta) \rangle$ as defined in previous chapters;[1-3] in ferromagnets the order parameter is the magnetization in a single domain; and in liquid-gas systems the order parameter is the density difference between the liquid and gas phases. In each of the above cases the state of the system, at any fixed temperature, can be described by an equilibrium value of the order parameter and fluctuations about that value. A phase transition can be accompanied by either a continuous or a discontinuous change in the equilibrium value of the order parameter when the system transforms from one phase to the other. (For simplicity we will consider temperature as the only thermodynamic variable in this paper; the pressure depedence of the various phenomena will be neglected).

From the above discussion it is apparent that an essential element in the theory of phase transitions is the determination, at every temperature, of the equilibrium value and fluctuation amplitude of the order parameter. In principle this can be accomplished if one can calculate the free energy $F[\varrho,T]$ of the system when it is in a state characterized by a value ϱ of the order parameter and a temperature T. $F[\varrho,T]$ is sometimes called the Landau free energy function to distinguish it from the total free energy of the system. The equilibrium value of the order parameter is that value $\varrho_{eq}(T)$ which minimizes $F[\varrho,T]$. The relative probability for the system to fluctuate to a state characterized by ϱ is proportional to $\exp\{-\beta(F[\varrho,T] - F[\varrho_{eq}(T),T])\}$, where $\beta \equiv 1/k_BT$, k_B being the Boltzmann constant. So far we have neglected spatially nonuniform fluctuations of the system, i.e., fluctuations that produce spatial variation of the order parameter. To include such fluctuations in the theory, one usually first calculates a free energy density at each temperature as a function of the order parameter and its spatial derivatives. The free energy for an arbitrary spatial variation of the order parameter, $\varrho(\vec{r})$, where $\vec{r}$ denotes spatial position, is then obtained from the free energy density by a volume integration. Thus, in the general case, the theoretical description of a phase transition is equivalent to the determination of the free energy density as a function of the order parameter, its spatial derivatives, and the temperature. However, a rigorous determination of the free energy density function is extremely difficult since it is almost equivalent to the complete solution of the phase transition problem.

In 1937 Landau[4] made an elegant and far-reaching speculation about the functional dependence of the free-energy density on the order parameter and its spatial derivatives near a second-order phase transition point. Briefly, Landau speculated that near a second-order transition the free-energy density function can be expanded as a power series in the order parameter and its spatial derivatives, with temperature dependent coefficients. Landau further argued that, sufficiently close to the transition, only the leading terms of the series are important, so that the expansion of the free-energy density function becomes a simple low-order polynomial. Despite certain shortcomings,[5] Landau's theory of phase transitions has proven to be as good as mean field theory in providing a semi-quantitative description of the specific heat, the order parameter, and the entropy in the vicinity of a second-order phase transition. Moreover, the Landau theory is mathematically simpler than mean field theory, and the inclusion of spatial variations of the order parameter gives it a new dimension not found in mean field theory. Of course, the Landau theory is useful only in a limited temperature range close to the transition point, and it contains more phenomenological parameters than does mean field theory. In these respects it is somewhat less satisfying than mean field theory.

Although originally intended as a theory of second-order phase transitions, the Landau theory can easily be generalized to include first-order phase transitions.[6] de Gennes[7] was the first to successfully apply Landau's theory to the first-order liquid-crystal phase transitions. It is the purpose of the present chapter to develop this Landau–de Gennes theory of liquid-crystal phase transitions and to discuss and illustrate its use. In the following sections, the derivation and discussion of the basic equations will be followed by application of the theory to the calculation of thermodynamic properties and fluctuation phenomena of liquid-crystal phase transitions, and by a description of some of the theory's more novel predictions and their experimental verifications.

2. Derivation of the Fundamental Equations of the Landau–de Gennes Theory

2.1 The Partition Function

We begin by considering a macroscopic system whose equilibrium state is characterized by a spatially invariant, dimensionless, scalar order parameter σ. Any disturbance in the system, such as thermal fluctuations, produces spatial variations of the order parameter. However, for low-energy (long wavelength) fluctuations, the spatial variations occur on a scale much larger than the molecular dimension. Therefore, if we limit our consideration only to the equilibrium state and low-energy fluctuations about the equilibrium state, we can imagine dividing the system into, say, M small, cubic, spatial regions each of volume ΔV. Each of these regions contains a sufficient number of molecules so that long-range order is well defined inside the region, yet is small enough compared to the wavelength of low-energy fluctuations that spatial variation of the order parameter within the region is negligible. The partition function for one such region, say region α, is

$$\mathfrak{z}(\alpha, T) = \sum_i \exp[-\beta E_i(\alpha)], \qquad [1]$$

where $E_i(\alpha)$ is the energy of region α (excluding its interaction with the rest of the system) when it is in state i (a state is defined in quantum systems by the eigenstate of the system and in classical systems by a set of numbers giving the spatial positions and momenta of all the particles). In general, the summation extends over all possible states of region α; however, we can make a simplifying approximation since we are only interested in low-energy fluctuations, which produce negligible spatial variation of the order parameter in region α. Thus out of all the possible states i of region α, we only sum over those states p which can be characterized by a single value of the order parameter. The resulting approximation to $\mathfrak{z}(\alpha,T)$ will be denoted by $Z(\alpha,T)$. We now divide the possible values of the order parameter into small intervals of width $\Delta\sigma(\alpha)$ and group together all those states having values of the order parameter in the same interval. By denoting the number of states in the interval centered at $\sigma(\alpha)$ as $N[\sigma(\alpha)]$, it follows that

$$\mathfrak{z}(\alpha, T) \simeq Z(\alpha, T) = \sum_p \exp[-\beta E_p(\alpha)]$$

$$= \sum_{\sigma(\alpha)} \left\{ \sum_{j=1}^{N[\sigma(\alpha)]} \exp(-\beta E_j[\sigma(\alpha)]) \right\}, \quad [2]$$

where $E_j[\sigma(\alpha)]$ is the energy of one of the states having an order parameter value within the interval centered around $\sigma(\alpha)$. Note that $N[\sigma(\alpha)]$ depends on the width of the interval $\Delta\sigma(\alpha)$ and, though region α is small, it still contains a sufficiently large number of molecules that $N[\sigma(\alpha)]$ must be very large in any finite interval $\Delta\sigma(\alpha)$. It is therefore meaningful to define a density of states $\rho[\sigma(\alpha)]$ such that

$$\rho[\sigma(\alpha)] = \lim_{\Delta\sigma(\alpha)\to 0} \frac{N[\sigma(\alpha)]}{\Delta\sigma(\alpha)}. \quad [3]$$

By reducing $\Delta\sigma(\alpha)$ to a differential, $d\sigma(\alpha)$, and by assuming that the energy of any state p in region α can be expressed as a continuous function, $E[\sigma(\alpha)]$, of the order parameter, Eq. [2] can be accurately replaced by

$$Z(\alpha, T) = \int d\sigma(\alpha) \rho[\sigma(\alpha)] \exp\{-\beta E[\sigma(\alpha)]\}, \quad [4]$$

where the integral extends over all possible values of the order parameter.

We now define a Landau free energy function $\hat{f}[\sigma(\alpha),T]$ of region α, when it is characterized by the value $\sigma(\alpha)$, as

$$\hat{f}[\sigma(\alpha), T] \equiv E[\sigma(\alpha)] - k_B T \ln \rho[\sigma(\alpha)]. \quad [5]$$

In terms of $\hat{f}[\sigma(\alpha),T]$, Eq. [4] can be expressed as

$$Z(\alpha, T) = \int d\sigma(\alpha) \exp\{-\beta\hat{f}[\sigma(\alpha), T]\}, \qquad [6]$$

and the total free energy of region α, $F(\alpha,T)$, is given by the usual relation

$$F(\alpha, T) = -k_B T \ln Z(\alpha, T). \qquad [7]$$

Let us now consider the interaction energy between two neighboring regions, say regions α and $\alpha + 1$. When the order parameter characterizing a physical system deviates from spatial uniformity, there is always a restoring force tending to bring the system back into spatial uniformity. It is therefore plausible to treat the interaction between two neighboring regions as being elastic. That is, the interaction energy I between regions α and $\alpha + 1$, when they are characterized by the order parameter values $\sigma(\alpha)$ and $\sigma(\alpha + 1)$, is, to a first approximation, a function of the *difference* between $\sigma(\alpha)$ and $\sigma(\alpha + 1)$. I has the property

$$I[\sigma(\alpha) - \sigma(\alpha + 1), T] = \begin{cases} 0 & \text{for } \sigma(\alpha) = \sigma(\alpha + 1) \\ \text{positive and increases with } |\sigma(\alpha) - \sigma(\alpha + 1)| & \text{for } \sigma(\alpha) \neq \sigma(\alpha + 1), \end{cases} \qquad [8]$$

where $|\ |$ denotes the absolute value. The combined Landau free energy of regions α and $\alpha + 1$, when they are characterized by $\sigma(\alpha)$ and $\sigma(\alpha + 1)$, is then given by $\hat{f}[\sigma(\alpha),T] + \hat{f}[\sigma(\alpha + 1),T] + I[\sigma(\alpha) - \sigma(\alpha + 1),T]$. The complete partition function for the two regions, including the interaction, can be written as

$$Z(\alpha, \alpha + 1, T) = \int d\sigma(\alpha) \int d\sigma(\alpha + 1) \exp\{-\beta(\hat{f}[\sigma(\alpha), T] + \hat{f}[\sigma(\alpha + 1), T] + I[\sigma(\alpha) - \sigma(\alpha + 1), T])\}. \qquad [9]$$

It should be noted that if $I = 0$, Eq. [9] gives the expected result $F(\alpha,\alpha + 1,T) = -k_B T \ln Z(\alpha,\alpha + 1,T) = F(\alpha,T) + F(\alpha + 1,T)$ for two noninteracting regions. For the purpose of simplifying the counting, we will define the quantities $\Gamma(\alpha,T)$, $\Gamma(\alpha + 1,T)$ such that

$$\Gamma(\alpha + 1, T) = \Gamma(\alpha, T) = \frac{1}{2} I[\sigma(\alpha) - \sigma(\alpha + 1), T]. \quad [10]$$

In other words, the interaction energy between two neighboring regions is split equally and counted twice: once as belonging to region α and once as belonging to region $\alpha + 1$. Eq. [9] can now be put in the form

$$Z(\alpha, \alpha + 1, T) = \int d\sigma(\alpha) \int d\sigma(\alpha + 1) \exp\{-\beta(\hat{f}[\sigma(\alpha), T] +$$
$$\hat{f}[\sigma(\alpha + 1), T] + \Gamma(\alpha, T) + \Gamma(\alpha + 1, T))\}. \quad [9a]$$

We can easily generalize Eq. [9a] to obtain the partition function $Z(T)$ of the whole system, where each region interacts elastically with its neighbors:

$$Z(T) = \int d\sigma(1) \int d\sigma(2) \ldots \int d\sigma(M) \exp\left[-\beta \sum_{\alpha=1}^{M} (\hat{f}[\sigma(\alpha), T] + \Gamma(\alpha, T))\right]$$
$$= \int D\{\sigma(1), \sigma(2), \ldots \sigma(M)\} \exp\left[-\beta \sum_{\alpha=1}^{M} (\hat{f}[\sigma(\alpha), T] + \Gamma(\alpha, T))\right], \quad [11]$$

where $\Gamma(\alpha, T)$ is equal to half the interaction energy of region α with its neighbors and is understood to depend on the difference in order parameter values between region α and its neighboring regions. $\{\sigma(1), \sigma(2), \ldots, \sigma(M)\}$ denotes a set of M numbers the value of each of which is bounded by the maximum and minimum values of the order parameter, and the integral over $D\{\sigma(1), \sigma(2), \ldots \sigma(M)\}$ means integration over all possible sets of M numbers with the above constraint. In Fig. 1 we illustrate schematically the equivalence of integration over each $\sigma(\alpha)$ individually and integration over all possible sets $\{\sigma(1), \sigma(2), \ldots, \sigma(M)\}$.

It is convenient at this point to change notation somewhat. We will henceforth label each region by the spatial coordinate $\vec{r}_\alpha$ of its center. Since the volume of each region is ΔV, we can define a free energy density,

$$f[\sigma(\vec{r}_\alpha), T] \equiv \frac{\hat{f}[\sigma(\alpha), T]}{\Delta V}, \quad [12]$$

and an interaction energy density,

$$\gamma(\vec{r}_\alpha, T) \equiv \frac{\Gamma(\alpha, T)}{\Delta V}. \quad [13]$$

Since individual regions are small compared to the entire system, it is the usual practice to regard each region as a spatial "point" in the system. Thus $\sigma(\vec{r}_\alpha) \to \sigma(\vec{r})$ and $f[\sigma(\vec{r}_\alpha),T] \to f[\sigma(\vec{r}),T]$, where $\sigma(\vec{r})$

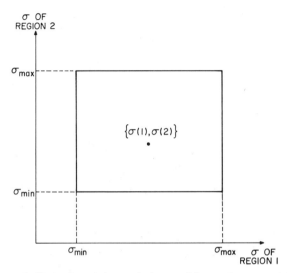

Fig. 1—Schematic illustration of the equivalence of integrating over $d\sigma(1)\,d\sigma(2)$ and summing over all possible sets of numbers $\{\sigma(1), \sigma(2)\}$. A set of numbers $\{\sigma(1), \sigma(2)\}$ corresponds to a point inside the square as shown; summing over all possible sets assures that the area of the entire square is included. Generalization to any arbitrary number of variables is straightforward.

and $f[\sigma(\vec{r}),T]$ are defined at any arbitrary spatial "point" $\vec{r}$. Recall that, to a first approximation, $\Gamma(\alpha,T)$ depends only on differences in the order parameter values between region α and its neighboring regions, so that $\gamma(\vec{r}_\alpha,T)$ can be replaced* by $\gamma[\vec{\nabla}\sigma(\vec{r}),T]$. It therefore follows that

$$\sum_{\alpha=1}^{M} \{\hat{f}[\sigma(\alpha),T] + \Gamma(\alpha,T)\} = \sum_{\alpha=1}^{M} \{f[\sigma(\vec{r}_\alpha),T] + \gamma(\vec{r}_\alpha,T)\}\Delta V$$

$$\longrightarrow \int_{\substack{\text{volume} \\ \text{of the} \\ \text{sample}}} d^3\vec{r}\,\{f[\sigma(\vec{r}),T] + \gamma[\vec{\nabla}\sigma(\vec{r}),T]\}, \qquad [14]$$

and each set of numbers $\{\sigma(1), \sigma(2) \ldots \sigma(M)\}$ is uniquely replaced by

---

* Theoretically the interaction term should depend on all orders of spatial derivatives of the order parameter. However, later development will show that one needs only retain the first order spatial derivatives of $\sigma(\vec{r})$ to obtain the Landau expression. Therefore, for simplicity, the function γ will be assumed to depend only on $\vec{\nabla}\sigma(\vec{r})$.

a spatial function $\sigma(\vec{r})$. Eq. [11] can now be rewritten as

$$Z(T) = \int \boldsymbol{D}\sigma(\vec{r}) \exp\{-\beta \int d^3\vec{r} (f[\sigma(\vec{r}), T] + \gamma[\vec{\nabla}\sigma(\vec{r}), T])\} \quad [15]$$

where we have used Feynman's path integral notation[8] $\boldsymbol{D}\sigma(\vec{r})$ to denote integration over all possible functions $\sigma(\vec{r})$. Eq. [15] is the central result of this subsection. Some of the implications of Eq. [15] are discussed in Appendix A.

2.2 The Landau Expansion

Calculation of the physical properties of a spatially uniform, macroscopic system requires information about $f[\sigma,T]$ only in the immediate vicinity of its minimum at temperature T, due to the sharpness of the peak in the function $\exp\{-\beta V f[\sigma,T]\}$ (See Appendix A). This implies that for a physical system having a second-order phase transiton at $T = T_c$, with $\sigma = 0$ for $T > T_c$ and $\sigma \neq 0$ for $T < T_c$ (such as the ferromagnetic phase transition), many features of the transition can be deduced if $f[\sigma,T]$ is known only in the neighborhood of $\sigma = 0$, $T = T_c$. Landau speculated that the first few derivatives of the function $f[\sigma,T]$ with respect to σ exist, and that they have finite values when evaluated at $\sigma = 0$, $T = T_c$. For a second-order phase transition the value of σ varies continuously and can be arbitrarily small near $T = T_c$; therefore, the Landau assumption enables one to express $f[\sigma,T]$ near $T = T_c$ as[4]

$$f[\sigma, T] = f_0[T] + \lambda(T)\sigma + \frac{1}{2}A(T)\sigma^2 +$$
$$\frac{1}{3}B(T)\sigma^3 + \frac{1}{4}C(T)\sigma^4 + \ldots . \quad [16]$$

Various properties of the expansion coefficients λ, A, B and C can be obtained from quite general considerations.

For definiteness we will examine the expansion in relation to two physical systems: (1) the CuZn (β-brass) binary alloy and (2) a ferromagnet such as iron. It is well known[9] that CuZn has a second-order, order–disorder transition at $T_c = 742°K$. The crystal structure at $0°K$ can be described by two interpenetrating simple cubic lattices each with N_0 sites. Let us suppose that lattice 1 is occupied by Cu atoms and lattice 2 by Zn atoms. As the temperature is raised above

0°K, some Zn atoms will be found on lattice 1 and some Cu atoms on lattice 2 but, so long as the temperature is less than T_c, $N_{Cu}(1)/N_0 > \frac{1}{2}$, where we have denoted the number of copper atoms on lattice 1 by $N_{Cu}(1)$. For temperatures in excess of T_c, complete randomization occurs and $N_{Cu}(1)/N_0 = N_{Zn}(1)/N_0 = \frac{1}{2}$. A suitable order parameter for the system can be defined as

$$\sigma = \frac{N_{Cu}(1) - N_{Zn}(1)}{N_0}, \qquad [17]$$

which is zero for $T \geq T_c$ and in the range 0 to 1 for $T < T_c$. It is also obvious that the state characterized by $-\sigma = [N_{Cu}(2) - N_{Zn}(2)]/N_0$ is physically equivalent to the state characterized by $+\sigma$ because the difference between the two can be ascribed to the interchange of labelings for lattices 1 and 2. It follows therefore that the free-energy density for $-\sigma$ must be equal to that for $+\sigma$ and hence $\lambda(T) = B(T) = 0$. In fact, all the terms having odd powers of σ must necessarily have vanishing coefficients. This is a general result not limited to the CuZn system alone. Consider the ferromagnet, iron. In this case the order parameter is given by a vector $\vec{m}$ defined as

$$\vec{m} = \vec{M}/M_0, \qquad [18]$$

where $\vec{M}$ is the magnetization vector and M_0 is the magnitude of the saturation magnetization at 0°K. The equilibrium state is characterized by $\vec{m} = 0$ for temperatures above the Curie point and by $0 < |\vec{m}| \leq 1$ for temperatures below the Curie point. In exact analogy with the case of a scalar order parameter, the free energy density can be expanded in terms of $\vec{m}$. However, since the free energy density is a scalar quantity, the Landau expansion of $f[\vec{m},T]$ about $\vec{m} = 0$ can only contain scalar combinations of $\vec{m}$. It is therefore obvious that $\lambda(T) = 0$. $B(T)$ must also vanish since it is impossible to construct a scalar from three vectors. As before, we see that the coefficients of the odd order terms in the expansion vanish. The above arguments can be similarly applied to other examples of second-order transitions.[10,11] In fact, for a second-order phase transition, one can always define an order parameter such that its equilibrium value is zero for $T \geq T_c$ and nonzero for $T < T_c$ and such that $\lambda(T) = B(T) = 0$ in the Landau expansion of $f[\sigma,T]$.

For a second order phase transition, Eq. [16] thus takes the form

$$f[\sigma, T] \simeq f_0[T] + \frac{1}{2}A(T)\sigma^2 + \frac{1}{4}C(T)\sigma^4 + \ldots \qquad [19]$$

At this point Landau[4] assumed further that near T_c the coefficient $C(T)$ is a slowly varying function of T compared to $A(T)$ and therefore can be replaced by a constant $C > 0$ in the temperature range of interest. Since the order parameter is zero at equilibrium in the high temperature phase, it is necessary that $\sigma = 0$ be a minimum of $f[\sigma,T]$ for $T > T_c$ and thus $A(T)$ must be positive. However, in the low temperature phase nonzero values of σ must correspond to the minimum in $f[\sigma,T]$. This minimum occurs at $\sigma = \pm[-A(T)/C(T)]^{1/2}$ provided $A(T)$ is negative. Therefore $A(T)$ must be positive for $T > T_c$ and negative for $T < T_c$, which implies that $A(T = T_c)$ must be zero and, in the neighborhood of T_c, $A(T)$ can be approximated by

$$A(T) = a(T - T_c), \qquad [20]$$

where a is a positive constant. We can now rewrite Eq. [19] as

$$f[\sigma, T] \simeq f_0[T] + \frac{1}{2}a(T - T_c)\sigma^2 + \frac{1}{4}C\sigma^4; \quad a, C > 0. \qquad [21]$$

Eq. [21] is illustrated schematically in Fig. 2.

So far we have considered only spatially uniform systems. Generalization of Eq. [21] to include spatial variations of the order parameter involves: (1) replacing σ by $\sigma(\vec{r})$, and (2) including the contribution to the free-energy density due to the interaction term $\gamma[\vec{\nabla}\sigma(\vec{r}),T]$. Following Landau,[4] we expand $\gamma[\vec{\nabla}\sigma(\vec{r}),T]$ in a power series in $\vec{\nabla}\sigma(\vec{r})$ and retain only the leading terms. Again, the expansion can contain only scalar combinations of $\vec{\nabla}\sigma(\vec{r})$ and, since $\gamma[\vec{\nabla}\sigma(\vec{r}),T] = 0$ when $\vec{\nabla}\sigma(\vec{r}) = 0$, we obtain*

$$\gamma[\vec{\nabla}\sigma(\vec{r}), T] \simeq \frac{1}{2}D(T)[\vec{\nabla}\sigma(\vec{r})]^2. \qquad [22]$$

In order that the spatially uniform state be the state of lowest free energy, $D(T)$ must be positive. Furthermore, near the critical temperature, $D(T)$ can be approximated by a constant D. Combining the leading terms in the expansions of $f[\sigma(\vec{r}),T]$ and $\gamma[\vec{\nabla}\sigma(\vec{r}),T]$, we obtain

---

*Theoretically, expansion of the interaction term can contain second order scalar combinations such as $\sigma(\vec{r})\nabla^2\sigma(\vec{r})$ and $\nabla^2\sigma(\vec{r})$. However, $\nabla^2\sigma(\vec{r})$ integrated over volume can be converted into a surface integral and can be neglected since the surface contribution is assumed to be small. The volume integral of $\sigma(\vec{r})\nabla^2\sigma(\vec{r})$ is equivalent to the integral of $[\vec{\nabla}\sigma(\vec{r})]^2$. Therefore, even in the most general case, the leading term in the expansion is proportional to $[\vec{\nabla}\sigma(\vec{r})]^2$.

LANDAU–deGENNES THEORY

$$f[\sigma(\vec{r}), T] + \gamma[\vec{\nabla}\sigma(\vec{r}), T] \simeq \mathcal{F}_L$$
$$\equiv f_0[T] + \frac{1}{2}a(T - T_c)\sigma^2(\vec{r}) + \frac{1}{4}C\sigma^4(\vec{r}) + \frac{1}{2}D[\vec{\nabla}\sigma(\vec{r})]^2, \quad [23]$$

where $\mathcal{F}_L$ is usually referred to as the Landau free energy density and a, C, and D are all positive constants.

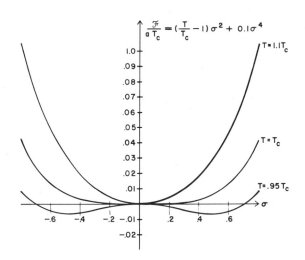

Fig. 2—Illustration of the behavior of Eq. [21]. The Landau free energy density is symmetric about $\sigma = 0$; for $T \geq T_c$ the disordered state ($\sigma = 0$) is the stable state while for $T < T_c$ the ordered state ($\sigma \neq 0$) is the state of lowest free energy.

2.3 Generalization of the Landau Expansion to Liquid Crystals

We must first identify a suitable order parameter for describing liquid crystal phase transitions. In previous chapters, we have seen that for nematic and cholesteric liquid crystals, the molecular ordering at every spatial "point" $\vec{r}$ (where the term "point" has the meaning defined in Section (2.1)) is characterized by a director $\hat{n}(\vec{r})$ pointing along the local axis of uniaxial symmetry, and by a quantity $S(\vec{r})$ giving the local orientational order of the rod-like molecules. $S(\vec{r})$ is defined by

$$S(\vec{r}) = \ll P_2(\cos\theta) \gg_{\vec{r}}, \quad [24]$$

where θ is the angle between the long axis of any molecule in the small region of space associated with "point" $\vec{r}$ and the local director $\hat{n}(\vec{r})$, P_2 is the second Legendre polynomial, and $\ll \gg_{\vec{r}}$ denotes spatial averaging over the configurations of the molecules at "point" $\vec{r}$ ($\ll \gg_{\vec{r}}$ should not be confused with thermal averaging $<>$, which gives the equilibrium value of a quantity. $S(\vec{r})$ should therefore be distinguished from its equilibrium value $<S(\vec{r})>$, which is spatially invariant and temperature-dependent). In Chapter 6 it was shown that in a coordinate system where $\hat{n}(\vec{r})$ coincides with the external z-axis (3-axis), the local order parameter of a nematic or cholesteric liquid crystal is given by*

$$\mathbf{Q}(\vec{r}) = S(\vec{r}) \begin{pmatrix} -\frac{1}{2} & 0 & 0 \\ 0 & -\frac{1}{2} & 0 \\ 0 & 0 & 1 \end{pmatrix}, \qquad [25]$$

where $\mathbf{Q}(\vec{r})$ denotes that the quantity $Q(\vec{r})$ is a tensor. For arbitrary orientation of $\hat{n}(\vec{r})$ relative to the external coordinate system, the components of $\mathbf{Q}(\vec{r})$ can be expressed as

$$Q_{ij}(\vec{r}) = \frac{1}{2} S(\vec{r})[3n_i(\vec{r})n_j(\vec{r}) - \delta_{ij}], \qquad [26]$$

where $i,j = 1, 2, 3$ denote the components along the three orthogonal axes of the Cartesian coordinate system and $\delta_{ij} = 1$ for $i = j$ and zero otherwise. Comparing the order parameters for a ferromagnet and a liquid crystal, we see that $S(\vec{r})$ corresponds to the magnitude of $\vec{m}$, and the matrix in Eq. [25] corresponds to the unit vector pointing along the direction of magnetization; this matrix can be thought of as a "unit tensor." To see that the order parameter for a liquid crystal cannot be a vector, we observe that $\hat{n}(\vec{r})$ and $-\hat{n}(\vec{r})$ correspond to physically equivalent states and $Q(\vec{r})$ must therefore be proportional to an even order combination of $\hat{n}(\vec{r})$. The two lowest even-order combinations of a unit vector are a scalar and a second rank tensor and, since the liquid-crystal order cannot be completely described by a scalar, we are left with the tensor as our only choice.

Let us now consider the Landau free-energy density expression for liquid crystals. All isotropic-nematic (cholesteric) phase transitions

---

* Our definition of $\mathbf{Q}(\vec{r})$ differs from that of de Gennes by a factor of 2.

are first order, and exhibit a discontinuous jump at $T = T_c$ in the equilibrium value of $\mathbf{Q}(\vec{r})$ from $\mathbf{Q}(\vec{r}) = 0$ in the isotropic phase to some finite value in the low temperature phase. Consequently, we expect the Landau expansion about $\mathbf{Q}(\vec{r}) = 0$ to provide a better description of phenomena such as fluctuations in the high-temperature isotropic phase than it does in the low-temperature phase, since the expansion would not be accurate for the values of the order parameter in the low-temperature phase. We will first develop the Landau–de Gennes expansion about the isotropic phase of a nematic liquid crystal and later generalize the expression to include the effects of external fields and the special symmetry of cholesteric liquid crystals. Since the expansion can contain only scalar combinations of $\mathbf{Q}(\vec{r})$ and its spatial derivatives, the term linear in $\mathbf{Q}(\vec{r})$ again vanishes because the scalar $\sum_{i=1}^{3} Q_{ii}(\vec{r})$ (the trace of $\mathbf{Q}(\vec{r})$) is identically zero. The coefficient of the term linear in the spatial derivative of $\mathbf{Q}(\vec{r})$ must also be zero because there is no way of forming a scalar quantity from the derivative. However, unlike the situation described for ferromagnetism and other second order phase transitions, the term cubic in $\mathbf{Q}(\vec{r})$ does not have to vanish because it is possible to construct a scalar from three tensors, and also because $\mathbf{Q}(\vec{r})$ and $-\mathbf{Q}(\vec{r})$ correspond to physically different states, as illustrated in Fig. 3. Thus the free

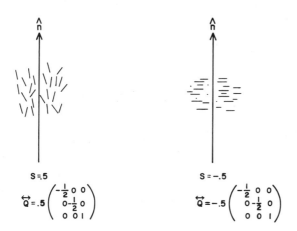

Fig. 3—Schematic illustration of the physical difference in molecular ordering for $S = 0.5$ and $S = -0.5$. The short lines represent projections of the molecular long axes on the plane of the paper. Thus a molecule whose long axis is normal to the paper is represented by a dot.

energy density is no longer required to be symmetric about $\mathbf{Q}(\vec{r}) = 0$. Recalling that the trace of a matrix or of a product of matrices is always a scalar, we may write

$$\mathcal{F}_L = f_0[T] + \frac{1}{2}a(T - T_c^*)Q_{ij}(\vec{r})Q_{ji}(\vec{r}) +$$

$$\frac{1}{3}BQ_{ij}(\vec{r})Q_{jk}(\vec{r})Q_{ki}(\vec{r}) + \frac{1}{4}C_1[Q_{ij}(\vec{r})Q_{ji}(\vec{r})]^2 +$$

$$\frac{1}{4}C_2 Q_{ij}(\vec{r})Q_{jk}(\vec{r})Q_{kl}(\vec{r})Q_{li}(\vec{r}) +$$

$$\frac{1}{2}L_1 \partial_i Q_{jk}(\vec{r}) \partial_i Q_{jk}(\vec{r}) + \frac{1}{2}L_2 \partial_i Q_{ij}(\vec{r}) \partial_k Q_{kj}(\vec{r}), \qquad [27]$$

where T_c^* is a temperature slightly below T_c, $a > 0$, B, C_1, C_2, L_1, L_2 are constants; $i, j, k, l = 1,2,3$ denote the components along the three orthogonal axes of the coordinate system; $\partial_i \equiv \partial/\partial x_i$ is the partial derivative with respect to spatial coordinate x_i; and summation over repeated indices is implied. Two points regarding Eq. [27] are worthy of special attention. First, because the order parameter is a tensor, $\mathcal{F}_L$ contains two fourth-order terms and two spatial derivative terms, in contrast with Eq. [23] where the order parameter is a scalar. Second, a new phenomenological parameter T_c^* has been defined as the temperature at which the curvature of $\mathcal{F}_L$ at $\mathbf{Q}(\vec{r}) = 0$ changes sign. Since $T_c^* < T_c$, we can find a temperature range, $T_c^* < T < T_c$, in which $\mathbf{Q}(\vec{r}) = 0$ is not the position of the absolute minimum of $\mathcal{F}_L$ (since otherwise $\mathbf{Q}(\vec{r}) = 0$ would be the equilibrium state for $T < T_c$, which is a contradiction) and yet the curvature of $\mathcal{F}_L$ at $\mathbf{Q}(\vec{r}) = 0$ is positive. In other words, $\mathbf{Q}(\vec{r}) = 0$ is a relative minimum of $\mathcal{F}_L$ for $T_c^* < T < T_c$. Since the existence of a relative minimum of $\mathcal{F}_L$ is a necessary condition for supercooling, it follows that T_c^* can be interpreted physically as that temperature below which supercooling becomes impossible.

Eq. [27] can be put into a more physically interpretable form by substituting Eq. [26] for $Q_{ij}(\vec{r})$ and noting that

$$n_i(\vec{r})\partial_j n_i(\vec{r}) = \frac{1}{2}\partial_j[n_1^2(\vec{r}) + n_2^2(\vec{r}) + n_3^2(\vec{r})] = \frac{1}{2}\partial_j[1] = 0$$

and

$$[\hat{n}(\vec{r}) \cdot \vec{\nabla}]\hat{n}(\vec{r}) = \frac{1}{2}\vec{\nabla}[\hat{n}(\vec{r}) \cdot \hat{n}(\vec{r})] - \hat{n}(\vec{r}) \times [\vec{\nabla} \times \hat{n}(\vec{r})]$$

$$= -\hat{n}(\vec{r}) \times [\vec{\nabla} \times \hat{n}(\vec{r})].$$

We obtain

$$\mathcal{F}_L = f_0[T] + \frac{3}{4}a(T - T_c^*)S^2(\vec{r}) + \frac{1}{4}BS^3(\vec{r}) +$$

$$\frac{9}{16}CS^4(\vec{r}) + \frac{3}{4}L_1[\vec{\nabla}S(\vec{r})]^2 + \frac{9}{4}L_1 S^2(\vec{r})(\partial_i n_j(\vec{r}))(\partial_i n_j(\vec{r})) +$$

$$\frac{1}{8}L_2[\vec{\nabla}S(\vec{r})]^2 + \frac{3}{8}L_2[\hat{n}(\vec{r}) \cdot \vec{\nabla}S(\vec{r})]^2 + \frac{9}{8}L_2 S^2(\vec{r})[\vec{\nabla} \cdot \hat{n}(\vec{r})]^2 +$$

$$\frac{3}{2}L_2 S(\vec{r})[\vec{\nabla} \cdot \hat{n}(\vec{r})][\hat{n}(\vec{r}) \cdot \vec{\nabla}S(\vec{r})] +$$

$$\frac{3}{4}L_2 S(\vec{r})[\hat{n}(\vec{r}) \times (\vec{\nabla} \times \hat{n}(\vec{r}))] \cdot \vec{\nabla}S(\vec{r}) +$$

$$\frac{9}{8}L_2 S^2(\vec{r})[\hat{n}(\vec{r}) \times (\vec{\nabla} \times \hat{n}(\vec{r}))]^2, \qquad [28]$$

where $C \equiv C_1 + (C_2/2)$. Eq. [28] can be further reduced by noting that

$$(\partial_i n_j(\vec{r}))(\partial_i n_j(\vec{r})) = [\vec{\nabla} \cdot \hat{n}(\vec{r})]^2 + [\hat{n}(\vec{r}) \cdot (\vec{\nabla} \times \hat{n}(\vec{r}))]^2 + [\hat{n}(\vec{r}) \times (\vec{\nabla} \times \hat{n}(\vec{r}))]^2 - \vec{\nabla} \cdot [\hat{n}(\vec{r})(\vec{\nabla} \cdot \hat{n}(\vec{r})) + \hat{n}(\vec{r}) \times \vec{\nabla} \times \hat{n}(\vec{r})],$$
$$[29]$$

where the last term represents the surface contribution to the free-energy density and therefore can be neglected (since the volume integral of $\vec{\nabla} \cdot \vec{V}(\vec{r})$, where $\vec{V}(\vec{r})$ is an arbitrary vector field, can be converted to a surface integral by Gauss's Theorem). Substitution of Eq. [29] into Eq. [28] yields

$$\mathcal{F}_L = f_0[T] + \frac{3}{4}a(T - T_c^*)S^2(\vec{r}) + \frac{1}{4}BS^3(\vec{r}) +$$

$$\frac{9}{16}CS^4(\vec{r}) + \frac{3}{4}\left(L_1 + \frac{1}{6}L_2\right)[\vec{\nabla}S(\vec{r})]^2 + \frac{3}{8}L_2[\hat{n}(\vec{r}) \cdot \vec{\nabla}S(\vec{r})]^2 +$$

$$\frac{9}{4}S^2(\vec{r})\left\{\left(L_1 + \frac{1}{2}L_2\right)[\vec{\nabla} \cdot \hat{n}(\vec{r})]^2 + L_1[\hat{n}(\vec{r}) \cdot \vec{\nabla} \times \hat{n}(\vec{r})]^2 +$$

$$\left(L_1 + \frac{1}{2}L_2\right)[\hat{n}(\vec{r}) \times \vec{\nabla} \times \hat{n}(\vec{r})]^2\right\} + \frac{3}{2}L_2 S(\vec{r})[\vec{\nabla} \cdot \hat{n}(\vec{r})] \times$$

$$[\hat{n}(\vec{r}) \cdot \vec{\nabla}S(\vec{r})] + \frac{3}{4}L_2 S(\vec{r})[\hat{n}(\vec{r}) \times \vec{\nabla} \times \hat{n}(\vec{r})] \cdot \vec{\nabla}S(\vec{r}). \qquad [30]$$

There are four types of terms in Eq. [30]. The first four terms concern only the value of the orientational order $S(\vec{r})$. The next two terms account for spatial variation of $S(\vec{r})$. Next there is a term concerned with the spatial variation of $\hat{n}(\vec{r})$; we have expressed this term in the familiar form of splay, twist, and bend distortions[12] of the director field $\hat{n}(\vec{r})$. It should be noted that to second order in the Landau expansion there are only two independent elastic constants, L_1 and L_2, whereas in the nematic phase there are known to be three independent elastic constants.[13] The last two terms in Eq. [30] represent the interaction between spatial variations of $S(\vec{r})$ and spatial variations of $\hat{n}(\vec{r})$. Clearly, the mathematics can be quite complicated if $S(\vec{r})$ and $\hat{n}(\vec{r})$ are allowed to vary simultaneously.

Let us now examine the expansion coefficients B, C, L_1, and L_2 in more detail. We will consider first the spatially uniform state which can be described by the first four terms of Eq. [30]. Following Landau, we choose $C > 0$, which requires that $B < 0$ if the equilibrium value of S is to be positive in the low-temperature phase. We illustrate the behavior of $\mathcal{F}_L$ as a function of S for a spatially uniform system in Fig. 4. Next we consider a state for which $S(\vec{r})$ is constant but $\hat{n}(\vec{r})$ is allowed to vary from point to point. Since the spatially uniform state must be stable against any distortion, it is clear from Eq. [30] that

$$L_1 > 0 \qquad [31]$$

and

$$L_1 + \frac{1}{2} L_2 > 0. \qquad [32]$$

If now we fix $\hat{n}(\vec{r})$ and let $S(\vec{r})$ vary, the same reasoning leads to the inequalities

$$L_1 + \frac{1}{6} L_2 > 0 \text{ for } \hat{n}(\vec{r}) \perp \vec{\nabla} S(\vec{r}) \qquad [33]$$

and

$$L_1 + \frac{2}{3} L_2 > 0 \text{ for } \hat{n}(\vec{r}) \parallel \vec{\nabla} S(\vec{r}). \qquad [34]$$

In order to clarify the physical meaning of L_1 and L_2, imagine a disturbance of the equilibrium state $\mathbf{Q}(\vec{r}) = 0$ of the high temperature phase such that $S(\vec{r} = 0) \neq 0$ along a certain director $\hat{n}$. If $L_1 = L_2 = 0$, we see from Eq. [30] that the disturbance would be a delta function at $\vec{r} = 0$ since the rest of the system, including the immediate region

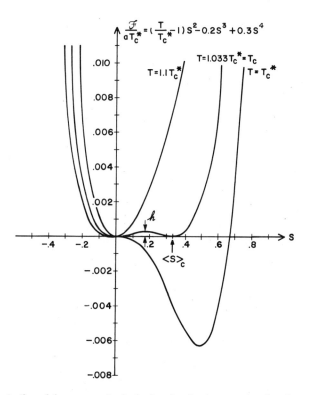

Fig. 4—Illustration of the asymmetry in the Landau free energy density when a non-zero third-order term is included. For $T > T_c$, $S = 0$ is the absolute minimum. As T approaches T_c from above a second minimum appears and for $T = T_c$ the physical states corresponding to the two minima have the same free energy density and are separated by a barrier of height h. For $T_c^* < T < T_c$ the $S \neq 0$ minimum represents the stable state and the $S = 0$ minimum represents the metastable (supercooled) state. For temperatures below T_c^* the metastable state becomes unstable.

surrounding $\vec{r} = 0$, would still prefer to be in its lowest free energy state, $S(\vec{r}) = 0$. However, as soon as $L_1, L_2 \neq 0$, the delta function becomes energetically unfavorable because of the divergence of the spatial derivatives, and any disturbance at $\vec{r} = 0$ must decay in a continuous manner to the equilibrium value $S(\vec{r}) = 0$ over some region of space surrounding $\vec{r} = 0$. The characteristic decay length is determined by the competition between the first four terms of Eq. [30], which tend to minimize the decay length so as to reduce their free-energy contribution, and the spatial derivative terms, which tend to maximize the decay length so as to reduce $|\vec{\nabla}S(\vec{r})|$. The decay of a disturbance of the sort described above ($\hat{n}(\vec{r})$ fixed, $S(\vec{r} = 0) \neq 0$ along $\hat{n}$) is illustrated schematically in Fig. 5. Note that for $L_2 \neq 0$, the

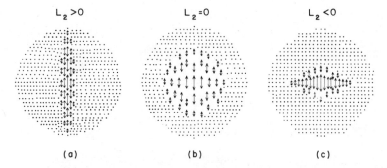

Fig. 5—Illustration of the spatial anistropy in the decay of a disturbance in the local order for different values of L_2. Regions of order are represented by double-headed arrows whose length is proportional to the magnitude of the local order and whose direction is parallel to the local director. The dots represent the surrounding isotropic field. In (a), $L_2 > 0$ and the disturbance (fluctuation) relaxes faster in the direction perpendicular to the director, whereas in (c), $L_2 < 0$ and the relaxation is faster in the direction parallel to the director. In (b), $L_2 = 0$ and the decay pattern is isotropic.

decay length of the disturbance parallel to $\hat{n}$ is different from that perpendicular to $\hat{n}$. Specifically, spatial variation of $S(\vec{r})$ in a direction parallel to $\hat{n}$ contributes a free energy density

$$\frac{3}{4}\left(L_1 + \frac{2}{3}L_2\right)[\vec{\nabla}S(\vec{r})]^2,$$

whereas spatial variation of $S(\vec{r})$ perpendicular to $\hat{n}$ contributes a free energy density

$$\frac{3}{4}\left(L_1 + \frac{1}{6}L_2\right)[\vec{\nabla}S(\vec{r})]^2.$$

Therefore, if $L_2 > 0$, the elastic constant governing the spatial variation of $S(\vec{r})$ in the direction parallel to $\hat{n}$ is larger than that in the direction perpendicular to $\hat{n}$. Indeed, the elastic constant along a direction which makes an arbitrary angle ψ with $\hat{n}$ is proportional to

$$1 + \frac{1}{6}\frac{L_1}{L_2}(1 + 3\cos^2\psi).$$

The decay pattern of a disturbance for $L_2 > 0$ is shown in Fig. 5a, while that for $L_2 < 0$ is shown in Fig. 5c.

Now, a word about the magnitude of the expansion coefficients a, B, C, L_1, and L_2. Taking the intermolecular interaction energy ($\simeq$ 0.01 eV) as the characteristic energy,[12] the intermolecular separation ($\simeq 10$ Å) as the characteristic length, and $T_c \simeq 300°$K as the characteristic temperature of nematic and cholesteric phases, we can estimate the magnitudes of these coefficients using dimensional analysis. The coefficient a has units of energy/(volume·°K). Combining the characteristic values in the appropriate way leads to $a \simeq 0.005$ J/(cm^3 °K), compared to the measured value[14] of 0.042 J/(cm^3 °K) for MBBA. B and C have units of energy/volume which gives $|B|, |C| \simeq 1.6$ J/cm^3, in order of magnitude agreement with measured values[14] $\simeq 0.5$ J/cm^3. Finally, we obtain $|L_1|, |L_2| \simeq 10^{-7}$ dyne which is in reasonable agreement with the experimental value[15] of 10^{-6} dyne.

In the remainder of this section we will generalize Eq. [30] to include the effects of external magnetic and electric fields and also to take account of the finite pitch in the equilibrium state of cholesteric liquid crystals.

Suppose that a static magnetic field $\vec{H}$ is applied to the system. Since nematic and cholesteric liquid crystals are normally diamagnetic, the energy density induced by the magnetic field is given by

$$W_M = -\frac{1}{2}H_i\chi_{ij}(\vec{r})H_j, \qquad [35]$$

where χ_{ij} are the components of the susceptibility tensor in units of susceptibility per unit volume. It was shown in Chapter 6 that

$$\chi(\vec{r}) = \frac{2}{3}(\Delta\chi)_{max}\mathbf{Q}(\vec{r}) + \bar{\chi}\mathbf{I}, \qquad [36]$$

where $\mathbf{I}$ is the unit matrix, $(\Delta\chi)_{max} \equiv N(\zeta_\| - \zeta_\perp)$ (N being the number of molecules per unit volume, and $\zeta_\|(\zeta_\perp)$ the susceptibility of a single rod-like molecule parallel (perpendicular) to its long axis), and $\bar{\chi}$ is the susceptibility per unit volume when $\mathbf{Q}(\vec{r}) = 0$. Using Eq. [13]

of Chapter 6, we see that $\bar{\chi} = N(2\zeta_\perp + \zeta_\parallel)/3$. Thus, Eq. [35] can be rewritten as

$$W_M = -\frac{1}{3}(\Delta\chi)_{\max} H_i Q_{ij}(\vec{r}) H_j - \frac{1}{6} NH^2(2\zeta_\perp + \zeta_\parallel). \quad [37]$$

Since the last term in Eq. [37] is independent of the order parameter and its gradients, it can be absorbed into the term $f_0[T]$ of $\mathcal{F}_L$. Therefore, the magnetic field contribution to the free energy density is

$$\mathcal{F}_M = -\frac{1}{3}(\Delta\chi)_{\max} H_i Q_{ij}(\vec{r}) H_j$$

$$= -\frac{1}{6}(\Delta\chi)_{\max} S(\vec{r})\{3[\vec{H}\cdot\hat{n}(\vec{r})]^2 - H^2\}. \quad [38]$$

Using similar arguments, we obtain the electric field contribution to the free-energy density

$$\mathcal{F}_E = -\frac{1}{24\pi}(\Delta\epsilon)_{\max} S(\vec{r})\{3[\vec{E}\cdot\hat{n}(\vec{r})]^2 - E^2\}, \quad [39]$$

where $(\Delta\epsilon)_{\max}$ is the maximum possible value of $(\epsilon_\parallel - \epsilon_\perp)$, $\epsilon_\parallel(\epsilon_\perp)$ being the dielectric constant parallel (perpendicular) to the local director.

It must be stressed that, unlike our expansion of $\mathcal{F}_L$, Eq. [30], which is valid only for vanishingly small values of the order parameter, Eqs. [38] and [39] are valid for arbitrary values of the order parameter. In fact, the magnetic and electric energy density expressions of the elastic continuum theory (Eqs. [3] and [4] of Chapter 8) are special cases of Eqs. [38] and [39]. For example, consider the magnetic energy density of the low temperature, anisotropic phase of a system in which only spatial variation of $\hat{n}(\vec{r})$ is important. $S(\vec{r})$ in Eq. [38] can then be replaced by its equilibrium value $<S>$, where $<\ >$ denotes thermal averaging. The term proportional to $<S>H^2$ can be neglected because of its spatial invariance, and one obtains Eq. [3] of Chapter 8 directly,

$$\mathcal{F}_M = -\frac{1}{2}\Delta\chi[\vec{H}\cdot\hat{n}(\vec{r})]^2,$$

where $\Delta\chi \equiv (\Delta\chi)_{\max} <S>$.

The expression for the free-energy density can also be generalized to include the special symmetry properties of cholesteric liquid crystals. Cholesteric order is indistinguishable from nematic order on a microscopic scale (see Chapters 1 and 11). However, when we examine the spatial variation of the local order, we find that cholesteric liquid crystals always exhibit helical ordering on a macroscopic scale (compared to molecular dimensions). Moreover, for any given cholesteric liquid crystal, the helical ordering has a definite handedness. In other words, cholesteric liquid crystals are not symmetric under the operation of spatial inversion, since a helix is always either left or right handed and the handedness changes upon inversion. Since the two cholesteric states related by spatial inversion are physically different,[16] the free-energy density is no longer required to be invariant under spatial inversion. This means that the free-energy density can contain pseudoscalar terms as well as the usual scalar terms. de Gennes found that the pseudoscalar term required by cholesteric order is given by[7]

$$\mathcal{F}_C = \pm 2 q_0 L_1 \epsilon_{ijk} Q_{il}(\vec{r}) \partial_k Q_{jl}(\vec{r}), \quad [40]$$

where the + and − signs refer to the two senses of helical pitch; $q_0 = \pi/\lambda_0$ (> 0), λ_0 being the pitch of the cholesteric phase as $T \to T_c$; and ϵ_{ijk} is the Levi–Cevita antisymmetric tensor of the third rank, which has the property that

$$\epsilon_{ijk} = \begin{cases} 1 & ijk = 123, 231, 312 \\ -1 & ijk = 213, 321, 132 \\ 0 & \text{otherwise} \end{cases}$$

Eq. [40] can be reduced to the form

$$\mathcal{F}_C = \mp \frac{9}{2} q_0 L_1 S^2(\vec{r}) \{\hat{n}(\vec{r}) \cdot [\vec{\nabla} \times \hat{n}(\vec{r})]\}, \quad [41]$$

where the − sign refers to a left-handed helix ($\hat{n}(\vec{r}) \cdot [\vec{\nabla} \times \hat{n}(\vec{r})] > 0$) and the + sign refers to a right-handed helix ($\hat{n}(\vec{r}) \cdot [\vec{\nabla} \times \hat{n}(\vec{r})] < 0$). Eq. [41] can be combined with the twist term of Eq. [30]

$$(9/4) L_1 S^2(\vec{r}) [\hat{n}(\vec{r}) \cdot \vec{\nabla} \times \hat{n}(\vec{r})]^2$$

to yield

$$(9/4) L_1 S^2(\vec{r}) [|\hat{n}(\vec{r}) \cdot \vec{\nabla} \times \hat{n}(\vec{r})| - q_0]^2 - (9/4) L_1 S^2(\vec{r}) q_0^2,$$

where | | is the absolute value sign.

The total Landau–de Gennes free-energy density for nematic or cholesteric liquid crystals in an external field is given by

$$\mathcal{F} = \mathcal{F}_L + \mathcal{F}_M + \mathcal{F}_E + \mathcal{F}_C = f_0[T] +$$

$$\frac{3}{4}a\left(T - T_c^* - \frac{3L_1 q_0^2}{a}\right)S^2(\vec{r}) + \frac{1}{4}BS^3(\vec{r}) + \frac{9}{16}CS^4(\vec{r}) +$$

$$\frac{3}{4}\left(L_1 + \frac{1}{6}L_2\right)[\vec{\nabla}S(\vec{r})]^2 + \frac{3}{8}L_2[\hat{n}(\vec{r})\cdot\vec{\nabla}S(\vec{r})]^2 +$$

$$\frac{9}{4}S^2(\vec{r})\left\{\left(L_1 + \frac{1}{2}L_2\right)[\vec{\nabla}\cdot\hat{n}(\vec{r})]^2 + L_1[|\hat{n}(\vec{r})\cdot\vec{\nabla}\times\hat{n}(\vec{r})| - q_0]^2 +\right.$$

$$\left.\left(L_1 + \frac{1}{2}L_2\right)[\hat{n}(\vec{r})\times\vec{\nabla}\times\hat{n}(\vec{r})]^2\right\} +$$

$$\frac{3}{2}L_2 S(\vec{r})\left\{[\vec{\nabla}\cdot\hat{n}(\vec{r})][\hat{n}(\vec{r})\cdot\vec{\nabla}S(\vec{r})] + \frac{1}{2}[\hat{n}(\vec{r})\times\vec{\nabla}\times\hat{n}(\vec{r})]\cdot\vec{\nabla}S(\vec{r})\right\} -$$

$$\frac{1}{6}(\Delta\chi)_{\max}S(\vec{r})\{3[\vec{H}\cdot\hat{n}(\vec{r})]^2 - H^2\} -$$

$$\frac{1}{24\pi}(\Delta\epsilon)_{\max}S(\vec{r})\{3[\vec{E}\cdot\hat{n}(\vec{r})]^2 - E^2\}, \qquad [42]$$

where $q_0 = 0$ for nematic liquid crystals. It should be noted that for cholesteric liquid crystals T_c^* is replaced by $T_c^{**} = T_c^* + (3L_1 q_0^2/a)$ in Eq. [42]; we can interpret T_c^{**} as the temperature below which supercooling of cholesterics is impossible. Eq. [42] can also be expressed in terms of $Q_{ij}(\vec{r})$,

$$\mathcal{F} = f_0[T] + \frac{1}{2}a(T - T_c^*)Q_{ij}(\vec{r})Q_{ji}(\vec{r}) +$$

$$\frac{1}{3}BQ_{ij}(\vec{r})Q_{jk}(\vec{r})Q_{ki}(\vec{r}) + \frac{1}{4}C[Q_{ij}(\vec{r})Q_{ji}(\vec{r})]^2 +$$

$$\frac{1}{2}L_1\partial_i Q_{jk}(\vec{r})\partial_i Q_{jk}(\vec{r}) + \frac{1}{2}L_2\partial_i Q_{ij}(\vec{r})\partial_k Q_{kj}(\vec{r}) -$$

$$\frac{1}{3}(\Delta\chi)_{\max}H_i Q_{ij}(\vec{r})H_j - \frac{1}{12\pi}(\Delta\epsilon)_{\max}E_i Q_{ij}(\vec{r})E_j$$

$$\pm 2q_0 L_1 \epsilon_{ijk} Q_{il}(\vec{r})\partial_k Q_{jl}(\vec{r}). \qquad [43]$$

In terms of $\mathcal{F}$ the partition function for the system is

$$Z(T) = \int D\mathbf{Q}(\vec{r}) \ \exp\{-\beta \int d^3\vec{r} \ \mathcal{F}[\mathbf{Q}(\vec{r}), \ \partial_i \mathbf{Q}(\vec{r}), \ T]\}, \quad [44]$$

where $D\mathbf{Q}(\vec{r})$ denotes integration over all possible tensor fields $\mathbf{Q}(\vec{r})$, and the dependence of $\mathcal{F}$ on $\mathbf{Q}(\vec{r})$, the spatial derivatives of $\mathbf{Q}(\vec{r})$, and T is explicitly displayed. Eqs. [42] through [44] are the basic equations of the Landau–de Gennes theory of liquid-crystal phase transitions.

3. Thermodynamic Properties of Liquid Crystal Phase Transitions

In this section we use Landau–de Gennes theory to calculate the thermodynamic properties of liquid-crystal phase transitions in terms of the phenomenological parameters a, B, C, and T_c^*. In particular, we will derive expressions for T_c, the transition temperature; $<S>_c$, the equilibrium order parameter value in the low temperature phase at the transition; and Δs, the transition entropy per molecule. We will also calculate the temperature dependence of $<S>$ (the equilibrium value of S) for T close to T_c; and h, the height of the free-energy barrier between $<S> = 0$ and $<S> = <S>_c$ at $T = T_c$.

To calculate the thermodynamic properties listed above, it is sufficient to consider a spatially uniform system in which the order parameter value is spatially invariant. This means that the spatial derivative terms in the Landau free-energy density, which are important for the calculation of fluctuation phenomena as shown in the next section, can be neglected for the present purpose. Therefore, from Eq. [42] we get

$$\mathcal{F}[S, T] = f_0[T] + \frac{3}{4} a(T - T_c^*) S^2 + \frac{1}{4} BS^3 + \frac{9}{16} CS^4, \quad [45]$$

where we have set $H = E = q_0 = 0$. The partition function is then given by

$$Z(T) = \int dS \ \exp\{-\beta V \mathcal{F}[S, T]\}. \quad [46]$$

From $Z(T)$ one can calculate the free energy density of the system $\mathcal{F}^0(T)$:

$$\mathcal{F}^0(T) = \lim_{V \to \infty} \frac{F(T)}{V} = \lim_{V \to \infty} \frac{-k_B T \ln Z(T)}{V}. \quad [47]$$

Thermodynamic properties of the system can be derived directly from $\mathfrak{F}^0(T)$.

In Appendix A it is shown that, if $Z(T)$ is in the form of Eq. [46], then $\mathfrak{F}^0(T)$ is equal to the value of the absolute minimum of the Landau free energy density $\mathfrak{F}$ at temperature T. Moreover, the (temperature-dependent) equilibrium value of the order parameter, $\langle S \rangle$, is equal to the value of S which minimizes $\mathfrak{F}$ at each T. Therefore, in order to obtain $\mathfrak{F}^0(T)$, we differentiate $\mathfrak{F}$ with respect to S and set the result equal to zero for $S = \langle S \rangle$:

$$\frac{3}{2} a(T - T_c^*)\langle S \rangle + \frac{3}{4} B \langle S \rangle^2 + \frac{9}{4} C \langle S \rangle^3 = 0. \qquad [48]$$

Eq. [48] has three solutions:

$$\langle S \rangle = 0,$$

$$\langle S \rangle = -\frac{1}{6}\frac{B}{C} \pm \sqrt{\frac{1}{324}\frac{B^2}{C^2} + \frac{2}{3}\frac{a}{C}\left(T_c^* + \frac{B^2}{27aC} - T\right)}. \qquad [49]$$

The above solutions for $\langle S \rangle$ must be substituted into Eq. [45] to determine which one gives the lowest value of $\mathfrak{F}$. Straightforward calculation yields

$$\mathfrak{F}^0(T) = f_0[T] + \frac{3}{4} a(T - T_c^*)\langle S \rangle^2 + \frac{1}{4} B \langle S \rangle^3 + \frac{9}{16} C \langle S \rangle^4, \qquad [50]$$

where

$$\langle S \rangle = \begin{cases} 0, \text{ for } T \geq T_c^* + \dfrac{B^2}{27aC} \\[2mm] -\dfrac{1}{6}\dfrac{B}{C} + \sqrt{\dfrac{1}{324}\dfrac{B^2}{C^2} + \dfrac{2}{3}\dfrac{a}{C}\left(T_c^* + \dfrac{B^2}{27aC} - T\right)}, \\[2mm] \qquad\qquad\qquad\qquad \text{ for } T \leq T_c^* + \dfrac{B^2}{27aC} \end{cases} \qquad [51]$$

Eq. [51] shows that the system transforms from the high-temperature phase, $\langle S \rangle = 0$, to the low-temperature phase, $\langle S \rangle \neq 0$, at a temperature

$$T_c = T_c^* + \frac{B^2}{27aC} .$$ [52]

The order parameter value of the low temperature phase at $T = T_c$, $\langle S \rangle_c$, can be obtained directly from Eq. [51]:

$$\langle S \rangle_c = -\frac{1}{6}\frac{B}{C} + \frac{1}{18}\frac{|B|}{C} = -\frac{2}{9}\frac{B}{C}, \quad B < 0.$$ [53]

The entropy per molecule, s, can be obtained from $\mathfrak{F}^0(T)$ by differentiation:

$$s = -\frac{1}{N}\left(\frac{\partial \mathfrak{F}^0(T)}{\partial T}\right) = s_0 - \frac{3a}{4N}\langle S \rangle^2,$$ [54]

where $s_0 \equiv -(\partial f_0[T]/\partial T)/N$, and the term containing the temperature derivative of $\langle S \rangle$ is zero, because $\partial \mathfrak{F}^0(T)/\partial \langle S \rangle \equiv 0$. Substitution of $\langle S \rangle$, Eq. [51], into Eq. [54] gives the entropy per molecule of the high temperature phase as s_0 and the transition entropy per molecule Δs as

$$\Delta s = -\frac{1}{27}\frac{aB^2}{NC^2} .$$ [55]

Eqs. [52], [53] and [55] are usually used as the means by which the phenomenological parameters a, B, and C for any particular liquid crystal are determined from the experimental values of $\langle S \rangle_c$, Δs, and $(T_c - T_c^*)$. (The value of T_c^* can be obtained by light scattering experiments measuring order parameter fluctuations. See Section 5.) In Table 1 we give the values of a, B, C and $(T_c - T_c^*)$ for MBBA.

In Fig. 4 we note that at $T = T_c$ there is a free energy barrier between $\langle S \rangle = 0$ and $\langle S \rangle = \langle S \rangle_c$. It is interesting to calculate the height h of this barrier in terms of the phenomenological parameters. The peak of the barrier occurs at $S = S_h$ for which

$$(\partial \mathfrak{F}/\partial S)_{\substack{S=S_h \\ T=T_c}} = 0 \quad \text{and} \quad (\partial^2 \mathfrak{F}/\partial S^2)_{\substack{S=S_h \\ T=T_c}} < 0.$$

Simultaneous solution of these equations requires that $S_h = -B/9C$, which, when substituted into Eq. [45], gives

$$h = \frac{1}{11664} \frac{B^4}{C^3}.$$ [56]

The value of h for MBBA is given in Table 1. Eq. [56] illustrates the

Table 1—Values of the Various Phenomenological Parameters for MBBA[‡]

Parameters	Values	Units
a	0.042	J/cm^3 °K
$-B$	0.64	J/cm^3
C	0.35	J/cm^3
$T_c - T_c^*$	1	°K
L_1	6.1 × 10^{-7}	dyne
L_2	—	
h	3.35 × 10^{-4}	J/cm^3

[‡] Data obtained from refs [14] and [15].

fact that the order of the phase transition in the Landau–de Gennes theory is determined by the value of B. If B is nonzero, then $h > 0$, and we have a first order phase transition with $<S>_c$, Δs nonzero and $T_c > T_c^*$. For $B = 0$, $h = 0$ and the transition is second order with $<S>_c = \Delta s = 0$, $T_c = T_c^*$, and $<S> \sim (T_c - T)^{1/2}$ for $T < T_c$. For liquid-crystal phase transitions the values of B are nonzero and negative (see Section 2). However, its value is small compared to that for other first-order phase transitions, as evidenced by the relatively small Δs, and that is why liquid-crystal phase transitions are sometimes described as "nearly second order."

4. Fluctuation Phenomena

In this section we use the Landau–de Gennes theory to discuss thermal fluctuations in the isotropic phase of liquid crystals. For a physical system in thermal equilibrium, the instantaneous value of the order parameter will almost always be equal or close to its mean value (or equivalently, the equilibrium value). However, deviations from the mean value of the order parameter do occur, and the problem is to calculate the magnitude and the statistical distribution of these deviations, or fluctuations. We distinguish between two types of fluctuations: (1) homophase fluctuations, which occur within the range of stability of a single phase and are completely described by the rms deviation of the order parameter from its equilibrium value, and (2) heterophase fluctuations, which occur between two phases and are related to the problems of metastability and supercooling.[17] The qualitative difference between homophase and heterophase fluctuations is apparent in Fig. 4. Near $T = T_c$ there is a free energy barrier of height h (see Section 3), between the minima at $S = 0$ and $S = <S>_c$. All fluctuations occurring around $S = 0$, and to the left of the barrier are called homophase fluctuations. Fluctuations that carry the system from one minimum, over the barrier, to the other minimum are called heterophase fluctuations since the two minima correspond to different phases.

4.1 Homophase Fluctuations in the Isotropic Phase

Since homophase fluctuations in the isotropic phase involve only states close to $\mathbf{Q}(\vec{r}) = 0$, we can neglect the cubic and quartic terms in the expansion of $\mathfrak{F}_L$. If we also set $H = E = 0$, Eq. [43] reduces to

$$\mathfrak{F} = f_0[T] + \frac{1}{2}a(T - T_c^*)Q_{ij}(\vec{r})Q_{ji}(\vec{r}) +$$
$$\frac{1}{2}L_1 \partial_i Q_{jk}(\vec{r}) \partial_i Q_{jk}(\vec{r}) + \frac{1}{2}L_2 \partial_i Q_{ij}(\vec{r}) \partial_k Q_{kj}(\vec{r})$$
$$\pm 2q_0 L_1 \epsilon_{ijk} Q_{il}(\vec{r}) \partial_k Q_{jl}(\vec{r}). \qquad [57]$$

As shown in Section 2.3, $\mathbf{Q}(\vec{r})$ can be expressed in terms of $S(\vec{r})$ and $\hat{n}(\vec{r})$. If we were to make this substitution in Eq. [57], the resulting expression would contain several interaction terms coupling $S(\vec{r})$, $\hat{n}(\vec{r})$, and their spatial derivatives. Thus, in general, the problem of calculating the fluctuation spectra of $S(\vec{r})$ and $\hat{n}(\vec{r})$ can be quite complicated because of this coupling. In the latter part of this section we will show one way to handle the coupling between $S(\vec{r})$ and $\hat{n}(\vec{r})$, but first we want to consider a simple system that illustrates both the physics of homophase fluctuation phenomena and the mathematics involved in manipulating the formalism developed in previous sections.

Simple Illustrative Example of Homophase Fluctuations

The system we wish to consider is one in which we can neglect the spatial variation of $\hat{n}(\vec{r})$ and treat only the fluctuations of $S(\vec{r})$. Setting all spatial derivatives of $\hat{n}(\vec{r})$ to zero and limiting our discussion for the moment to nematic liquid crystals ($q_0 = 0$), Eq. [57] becomes

$$\mathfrak{F} = f_0[T] + \frac{3}{4}a(T - T_c^*)S^2(\vec{r}) +$$
$$\frac{3}{4}\left(L_1 + \frac{1}{6}L_2\right)[\vec{\nabla}S(\vec{r})]^2 + \frac{3}{8}L_2[\hat{n}\cdot\vec{\nabla}S(\vec{r})]^2, \qquad [58]$$

where $\hat{n}$ no longer depends upon $\vec{r}$. The problem before us is to calculate the mean square deviation of $S(\vec{r})$ from its equilibrium value $S(\vec{r}) = 0$. As a first step, we note that the partition function for the system is given by

$$Z(T) = \int \boldsymbol{D}S(\vec{r}) \exp[-\beta \int_{\substack{\text{volume} \\ \text{of the} \\ \text{sample}}} d^3\vec{r}\, \mathfrak{F}]. \qquad [59]$$

The difficulty in using Eq. [59] for actual calculations lies in the functional integral $\int \boldsymbol{D}S(\vec{r})$ and we therefore digress briefly to consider functional integration.

Functional Integration

It is customary to proceed by Fourier analyzing $S(\vec{r})$ into its Fourier components

$$S(\vec{r}) = \sum_{\vec{q}} S(\vec{q})\, e^{i\vec{q}\cdot\vec{r}}, \qquad [60]$$

where $\vec{q}$ is a particular wavevector, and $S(\vec{q})$ is usually a complex number representing the amplitude of the wave, $\exp(i\vec{q}\cdot\vec{r})$. $S(\vec{q})$ can be obtained from $S(\vec{r})$ by the inverse transformation

$$S(\vec{q}) = \frac{1}{V}\int d^3\vec{r}\, S(\vec{r})\, e^{-i\vec{q}\cdot\vec{r}}. \qquad [61]$$

Since $S(\vec{r})$ is real, Eq. [61] implies that

$$S^*(\vec{q}) = \frac{1}{V}\int d^3\vec{r}\, S(\vec{r})\, e^{i\vec{q}\cdot\vec{r}} = S(-\vec{q}), \qquad [62]$$

where $S^*(\vec{q})$ is the complex conjugate of $S(\vec{q})$. Since there is a one-to-one correspondence between any function $S(\vec{r})$ and its set of Fourier coefficients $\{S(\vec{q})\}$, the functional integral is equivalent to integrating over all possible sets of Fourier coefficients $\{S(\vec{q})\}$. This in turn is equivalent to a multi-dimensional integral in which each Fourier coefficient $S(q)$ is integrated over its possible values, with the constraint $S^*(\hat{q}) = S(-\hat{q})$ which implies that only half the $S(\hat{q})$'s in any set $\{S(q)\}$ can vary independently. We will label the set of all independent $S(\hat{q})$'s by $\{S(\hat{q})\}'$. $\{S(\hat{q})\}'$ can be specified by the condition that if $S(\hat{q})$ is a member of $\{S(\hat{q})\}'$, then $S(-\hat{q})$ is not a member of $\{S(\hat{q})\}'$.

The functional integral can then be written as

$$\int DS(\vec{r}) \equiv J[S(\vec{r}); S(\vec{q})] \prod_{\vec{q}}{}' \left\{ \int_0^{2\pi} d\varphi_{S(\hat{q})} \int |S(\vec{q})| \, d|S(\vec{q})| \right\}, \quad [63]$$

where $\prod'$ denotes the product only over those $\hat{q}$'s for which none is the negative of any other, $\varphi_{S(\hat{q})}$ is the phase angle of the complex number $S(\hat{q})$, $|S(\hat{q})| \equiv [S^*(\hat{q})S(\hat{q})]^{1/2}$ is the magnitude of $S(\hat{q})$, and $J[S(\vec{r}); S(\hat{q})]$ is the Jacobian of the transformation. In order to evaluate the Jacobian we make use of the fact (Section 1) that $DS(\vec{r})$ can be written as $\prod_{\alpha=1}^{M} dS(\vec{r}_\alpha)$, where $S(\vec{r}_\alpha)$ is the value of S in region α. Since each of the M regions has volume ΔV, we have from Eqs. [60] and [61]

$$S(\vec{r}_\alpha) = \sum_{\beta=1}^{M} S(\vec{q}_\beta) e^{i\vec{q}_\beta \cdot \vec{r}_\alpha}, \quad [60a]$$

and

$$S(\vec{q}_\beta) = \frac{\Delta V}{V} \sum_{\alpha=1}^{M} S(\vec{r}_\alpha) e^{-i\vec{q}_\beta \cdot \vec{r}_\alpha}. \quad [61a]$$

It is clear from Eq. [60a] and the definition of a Jacobian that $J[S(\vec{r}); S(\hat{q})]$ is simply the determinant of the $M \times M$ matrix whose $\alpha\beta$ − element is $\exp(i\vec{q}_\beta \cdot \vec{r}_\alpha)$. It is shown in Appendix B that this determinant is equal to $(V/\Delta V)^{M/2}$. Substituting this factor into Eq. [63] yields

$$\int DS(\vec{r}) \equiv \prod_{\vec{q}}{}' \frac{V}{\Delta V} \left\{ \int_0^{2\pi} d\varphi_{S(\vec{q})} \int |S(\vec{q})| \, d|S(\vec{q})| \right\}, \quad [63a]$$

where we have used the fact that the product $\prod'$ contains $M/2$ factors. We can now proceed with our evaluation of the partition function, Eq. [59], for our idealized example system.

We first integrate Eq. [58] term by term since we will need the volume integral of $\mathcal{F}$ in evaluating $Z(T)$. The volume integrals of $S^2(\vec{r})$, $[\vec{\nabla} S(\vec{r})]^2$, and $[\hat{n}\cdot\vec{\nabla} S(\vec{r})]^2$ can be expressed in terms of $S(\vec{q})$ and $\vec{q}$ in the form

$$\int_V d^3\vec{r}\, S^2(\vec{r}) = V\sum_{\vec{q}} |S(\vec{q})|^2 = 2V\sum_{\vec{q}}{}' |S(\vec{q})|^2, \quad [64a]$$

$$\int_V d^3\vec{r}\, [\vec{\nabla} S(\vec{r})]^2 = V\sum_{\vec{q}} q^2 |S(\vec{q})|^2 = 2V\sum_{\vec{q}}{}' q^2 |S(\vec{q})|^2, \quad [64b]$$

and

$$\int_V d^3\vec{r}\, [\hat{n}\cdot\vec{\nabla} S(\vec{r})]^2 = V\sum_{\vec{q}} (\hat{n}\cdot\vec{q})^2 |S(\vec{q})|^2$$

$$= 2V\sum_{\vec{q}}{}' (\hat{n}\cdot\vec{q})^2 |S(\vec{q})|^2, \quad [64c]$$

where V is the volume of the sample and $\sum'_{\vec{q}}$ denotes summing over the $\vec{q}$'s for which none is the negative of any other. In arriving at the above expressions we have used the identity

$$V\delta(\vec{q} + \vec{q}') = \int d^3\vec{r}\, \exp[i(\vec{q} + \vec{q}')\cdot\vec{r}]. \quad [65]$$

From Eqs. [58], [64a], [64b], and [64c] we obtain

$$\int_V d^3\vec{r}\, \mathcal{F} = Vf_0[T] +$$

$$2V\sum_{\vec{q}}{}' \left[\frac{3}{4}a(T - T_c^*) + \frac{3}{4}\left(L_1 + \frac{1}{6}L_2\right)q^2 + \frac{3}{8}L_2(\hat{n}\cdot\vec{q})^2\right] |S(\vec{q})|^2$$

$$= Vf_0[T] + \frac{3}{2}aV(T - T_c^*)\sum_{\vec{q}}{}' [1 + \xi^2(T, \psi_{\vec{q}})q^2] |S(\vec{q})|^2 \quad [66]$$

where

$$\xi(T, \psi_{\vec{q}}) = \frac{\xi_0(\psi_{\vec{q}})(T_c^*)^{1/2}}{(T - T_c^*)^{1/2}}, \quad [67]$$

with

$$\xi_0(\psi_{\vec{q}}) = \left[\frac{L_1 + \frac{1}{6}L_2(1 + 3\cos^2\psi_{\vec{q}})}{aT_c^*}\right]^{1/2}, \quad [68]$$

and $\psi_{\hat{q}}$ denotes the angle between $\hat{n}$ and $\hat{q}$. The parameter $\xi(T,\psi_{\hat{q}})$ has units of length and is referred to as a "correlation length" for reasons that will become apparent later. If we substitute typical values[14,15] for L_1, L_2, a, and T_c^* in Eq. [68] we find $\xi_0 \approx 10\text{Å}$. For simplicity we will write $\xi(T,\psi_{\hat{q}})$ and $\xi_0(\psi_{\hat{q}})$ as ξ and ξ_0, respectively. Substituting Eq. [66] into Eq. [59], we have for the partition function,

$$Z(T) = \exp\{-\beta V f_0[T]\} \times$$
$$\int DS(\vec{r}) \prod_{\vec{q}}{}' \exp\left\{-\frac{3}{2}\beta V a(T - T_c^*)(1 + \xi^2 q^2)|S(\vec{q})|^2\right\}, \quad [69]$$

which, according to Eq. [63], can be written

$$Z(T) = \exp\{-\beta V f_0[T]\} \prod_{\vec{q}}{}' \left\{\frac{\pi V}{\Delta V} \times \right.$$
$$\left.\int d|S(\vec{q})|^2 \exp\left[-\frac{3}{2}\beta V a(T - T_c^*)(1 + \xi^2 q^2)|S(\vec{q})|^2\right]\right\}, \quad [70]$$

where the angular integral has been explicitly evaluated. If we label the factors in the product in Eq. [70] by $Z(\hat{q},T)$, Eq. [70] reduces to

$$Z(T) = \exp\{-\beta V f_0[T]\} \prod_{\vec{q}}{}' Z(\vec{q}, T). \quad [71]$$

The total free-energy density of the system is calculated from Eq. [71] in Appendix B. For the moment we note that Eq. [71] can be used to calculate the thermal averages of $S(\hat{q})$ and $|S(\hat{q})|^2$. Since the partition function is the product of wavevector-dependent factors $Z(\hat{q},T)$, the thermal average of any quantity $X[S(\hat{q})]$ associated with wavevector $\hat{q}$ is given by

$$\langle X[S(\vec{q})]\rangle =$$
$$\frac{\frac{1}{2}\int_0^{2\pi} d\varphi_{S(\vec{q})} \int_0^\infty d|S(\vec{q})|^2 X[S(q)] \exp\left\{-\frac{3}{2}\beta V a(T - T_c^*)(1 + \xi^2 q^2)|S(\vec{q})|^2\right\}}{\pi \int_0^\infty d|S(\vec{q})|^2 \exp\left\{-\frac{3}{2}\beta V a(T - T_c^*)(1 + \xi^2 q^2)|S(\vec{q})|^2\right\}},$$
$$[72]$$

where, in place of the maximum possible value of $|S(\hat{q})|^2$, the upper limit of the $d|S(\hat{q})|^2$ integral has been extended to ∞. This is possible because the factor V in the exponent makes the integrand sharply peaked at $|S(\hat{q})|^2 = 0$, and the error introduced by extending the limit of integration is therefore negligible. We can now calculate the thermal average of $S(\hat{q})$ and $|S(\hat{q})|^2$ using Eq. [72]. Expressing $S(\hat{q})$ in the form $|S(q)| \exp(i\varphi_{S(\hat{q})})$ and substituting it for $X[S(\hat{q})]$ in Eq. [72] we find

$$\langle S(\vec{q}) \rangle = 0,$$

and

$$\langle S^2(\vec{q}) \rangle = \langle S(\vec{q}) S(\vec{q}') \rangle = \langle S(\vec{q}) \rangle \langle S(\vec{q}') \rangle = 0, \quad \vec{q} \neq -\vec{q}' \quad [73]$$

Similar calculation yields

$$\langle |S(\vec{q})|^2 \rangle = \frac{2}{3} \frac{k_B T}{Va(T - T_c^*)(1 + \xi^2 q^2)}$$

$$= \frac{2}{3} \frac{k_B(T/T_c^*)}{Va\left[\left(\dfrac{T}{T_c^*} - 1\right) + \xi_0^2 q^2\right]}, \quad [74]$$

where the last expression is obtained by the substitution of Eq. [67] for ξ. $\langle|S(\hat{q})|^2\rangle$ is the square of the amplitude of fluctuation in $S(\hat{r})$ with wavevector $\hat{q}$, and Eq. [74] is simply an expression of the equipartition theorem. It is evident from Eq. [66] that in thermal equilibrium each of the independent modes in the set $\{S(\hat{q})\}'$ contributes a term

$$\frac{3}{2} Va(T - T_c^*)(1 + \xi^2 q^2)\langle |S(\vec{q})|^2 \rangle$$

to the Landau free energy. Since $S(\hat{q})$ is a complex number with two degrees of freedom (real and imaginary parts), the equipartition theorem requires that

$$\frac{3}{2} Va(T - T_c^*)(1 + \xi^2 q^2)\langle |S(\vec{q})|^2 \rangle = k_B T, \quad [75]$$

which is identical to Eq. [74]. The equipartition theorem therefore offers a simple method for calculating the thermal averages of the square of the amplitudes for independent modes.* Three points concerning Eq. [74] warrant further discussion.

First, it appears that all fluctuations vanish in the thermodynamic limit $V \to \infty$. However, this is not the case. Although the fluctuation amplitude for each mode decreases as V increases, the density of modes increases such that the sum of all fluctuation amplitudes remains constant. This is simply demonstrated by calculating the average fluctuation amplitude in real space:

$$\left\langle \frac{1}{V} \int d^3\vec{r}\, S^2(\vec{r}) \right\rangle = \frac{1}{V} \left\langle \int d^3\vec{r}\, S^2(\vec{r}) \right\rangle = \sum_{\vec{q}} \langle |S(\vec{q})|^2 \rangle,$$

where we have used Eq. [60]. Making the replacement

$$\sum_{\vec{q}} \longrightarrow \frac{V}{(2\pi)^3} \int d^3\vec{q},$$

and substituting Eq. [74] for $\langle |S(\hat{q})|^2 \rangle$, we get

$$\left\langle \frac{1}{V} \int d^3\vec{r}\, S^2(\vec{r}) \right\rangle =$$

$$\frac{V}{(2\pi)^3} \int_0^{q_{max}} 4\pi q^2 dq\, \frac{2}{3} \frac{k_B T}{Va(T - T_c^*)(1 + \xi^2 q^2)}$$

$$= \frac{k_B T}{3\pi^2 \xi_0^3 a T_c^*} \left[\xi_0 q_{max} - \sqrt{\frac{T}{T_c^*} - 1}\, \tan^{-1}\left(\frac{\xi_0 q_{max}}{\sqrt{\frac{T}{T_c^*} - 1}} \right) \right],$$

[76]

where $q_{max} \equiv 2\pi/(\text{smallest length for which the theory is valid}) \simeq$

---

* If, instead of considering only the independent modes, all possible modes were counted, then in thermal equilibrium the contribution of each mode to the Landau free energy would be $(3/4)Va(T - T_c^*)(1 + \xi^2 q^2) \langle |S(\hat{q})|^2 \rangle$. However, each $S(\hat{q})$ now has only one degree of freedom due to the reality restriction $S^*(\hat{q}) = S(-\hat{q})$ and therefore can only have $\frac{1}{2} k_B T$ of energy. The resulting equation

$$\frac{3}{4} Va(T - T_c^*)(1 + \xi^2 q^2) \langle |S(\vec{q})|^2 \rangle = \frac{1}{2} k_B T$$

is identical to Eq. [75]. Hence, the same expression for $\langle |S(\hat{q})|^2 \rangle$ is obtained independent of how the modes are counted.

$2\pi/(\Delta V)^{1/3}$, ΔV being the volume of an elemental region as defined in Section 2. Eq. [76] expresses the physically reasonable result that the average fluctuation amplitude in real space is independent of V.

The second point we wish to stress concerns the dependence of $\langle|S(\vec{q})|^2\rangle$ on q and T, shown in Fig. 6. For a given temperature $T >$

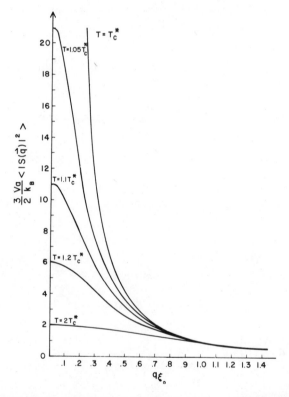

Fig. 6—The square of the fluctuation amplitude plotted as a function of wavevector for different temperatures $T \geq T_c{}^*$. Note that for fixed T the fluctuation amplitude decreases monotonically with increasing q and that for fixed q it increases as T decreases toward $T_c{}^*$.

$T_c{}^*$, $\langle|S(\vec{q})|^2\rangle$ decreases with increasing q, and, for fixed q, $\langle|S(\vec{q})|^2\rangle$ increases as $T \rightarrow T_c{}^*$ from above. We can qualitatively understand this behavior from the following mechanical analogy. Consider the string of coupled pendula shown in Fig. 7. Let the gravitational field play the role of the term $\frac{3}{4} a(T - T_c{}^*)S^2(\vec{r})$ in $\mathfrak{F}_L$, and the coupling springs play the role of the terms proportional to the spatial derivatives of $S(\vec{r})$. It is obvious that there are two kinds of oscillations; one is the oscillation in the gravitational field and the other is the normal mode oscillation of a string of mass points connected by springs. If q denotes the wavenumber associated with the normal

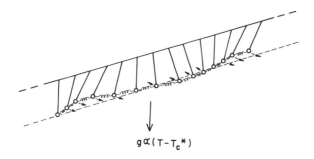

Fig. 7—Mechanical analog illustrating the qualitative features of Fig. 6. The springs play the role of the $[\vec{\nabla} S(\vec{r})]^2$ terms and the gravitational field plays the role of the $(T - T_c^*)S^2(\vec{r})$ term in the Landau free energy density function.

mode oscillation of the elastic linear chain, then for $q = 0$ the springs are not distorted and we have uniform in-phase oscillation of all the pendula in the gravitational field. For $q \neq 0$ each mass point experiences the force from the springs (caused by the relative motion of neighboring mass points) as well as the force due to the gravitational field. The square of the amplitude for the mode with wavevector q is analogous to $|S(\hat{q})|^2$. Suppose now that the system of pendula is in thermal equilibrium at temperature T with each mode having energy $k_B T/2$. For the $q = 0$ mode the entire $k_B T/2$ of energy goes into the oscillation in the gravitational field. However, for a $q \neq 0$ mode part of the $k_B T/2$ of energy is spent in distorting the springs, the remainder going into oscillation in the gravitational field. Thus the amplitude of oscillation for any $q \neq 0$ mode is smaller than that for the $q = 0$ mode. Since the energy stored in the springs increases with q, it follows that, for a fixed temperature, the amplitude of oscillation (and by analogy $<|S(\hat{q})|^2>$) should decrease with increasing q. What happens as $T \to T_c^*$? This corresponds in our analogy to weakening the gravitational field, with the result that, for a fixed energy input and fixed q, the amplitude of oscillation increases. This increase in fluctuation amplitude as T decreases toward T_c^* is an indication of the growing instability in the high-temperature phase. In fact, at $T = T_c^*$ the fluctuation amplitude $<|S(\hat{q})|^2>$ diverges for $\hat{q} = 0$. T_c^* is thus the temperature below which the isotropic phase is absolutely unstable.

The third point concerns the physical significance of ξ. Consider the correlation function $\langle S(\vec{r}_1)S(\vec{r}_2)\rangle$. It is obvious that if $L_1 = L_2 = 0$, a disturbance at $\vec{r}_1$ would have no effect on any other spatial point $\vec{r}_2 \neq \vec{r}_1$. The relative phase of fluctuations at two different points would be completely random. Therefore, it is expected that

$$\langle S(\vec{r}_1)S(\vec{r}_2)\rangle = \langle S(\vec{r}_1)\rangle\langle S(\vec{r}_2)\rangle = 0.$$

The interesting case arises when $L_1, L_2 \neq 0$. Then a thermal fluctuation at $\vec{r}_1$ can cause a certain amount of in-phase fluctuation at $\vec{r}_2 \neq \vec{r}_1$ and as a result $\langle S(\vec{r}_1)S(\vec{r}_2)\rangle$ should be nonzero. In general, we can write

$$\langle S(\vec{r}_1)S(\vec{r}_2)\rangle = \sum_{\vec{q}}\sum_{\vec{q}'} \langle S(\vec{q})S(\vec{q}')\rangle \exp\{i(\vec{q}\cdot\vec{r}_1 + \vec{q}'\cdot\vec{r}_2)\}. \quad [77]$$

$\langle S(\vec{q})S(\vec{q}')\rangle$ can have the following values:

$$\langle S(\vec{q})S(\vec{q}')\rangle = \begin{cases} \langle S(\vec{q})\rangle\langle S(\vec{q}')\rangle = 0, & \vec{q}' \neq -\vec{q} \\ \langle |S(\vec{q})|^2\rangle, & \vec{q}' = -\vec{q} \end{cases} \quad [78]$$

Therefore,

$$\langle S(\vec{r}_1)S(\vec{r}_2)\rangle = \sum_{\vec{q}} \langle |S(\vec{q})|^2\rangle \exp\{i\vec{q}\cdot(\vec{r}_1 - \vec{r}_2)\}$$

$$= \sum_{\vec{q}} \frac{2}{3}\frac{k_BT}{Va(T-T_c^*)(1+\xi^2q^2)} \exp\{i\vec{q}\cdot(\vec{r}_1-\vec{r}_2)\}$$

$$\simeq \frac{k_BT}{6\pi^2 a(T-T_c^*)} \int_{-1}^{1} d(\cos\theta) \int_0^\infty q^2 dq \frac{\exp\{iq|\vec{r}_1-\vec{r}_2|\cos\theta\}}{1+\xi^2q^2}$$

$$= \frac{k_BT}{6\pi\xi_0(aT_c^*)^{1/2}} \frac{\exp\{-|\vec{r}_1-\vec{r}_2|/\xi\}}{|\vec{r}_1-\vec{r}_2|}. \quad [79]$$

In converting the sum over all $\vec{q}$ vectors to an integral, we have made the approximation of replacing q_{max}, the upper limit of the integral over dq, by ∞, valid for $q_{max}|\vec{r}_1 - \vec{r}_2| \gg 1$. From Eq. [79] it can easily be checked that if $L_1, L_2 \to 0$, then $\xi \to 0$, and $\langle S(\vec{r}_1)S(\vec{r}_2)\rangle = 0$ for $\vec{r}_1 \neq \vec{r}_2$. For $L_1, L_2 \neq 0$, Eq. [79] states that a thermal disturbance at $\vec{r}_1$ decays exponentially to the equilibrium condition of its surroundings in a characteristic distance ξ. Since ξ is the distance over which fluctuations occur in phase, it is called the "correlation length." ξ diverges as $T \to T_c^*$, and for materials with T_c sufficiently close to T_c^* (such as liquid crystals) this behavior of ξ gives rise to various observable phenomena, such as the increase in the light scattering cross section and in the Cotton-Mouton coefficient when T_c is approached (See Sections 5 and 6).

The General Case of Homophase Fluctuations

We will now investigate in more detail what happens when the restriction placed on $\hat{n}(\vec{r})$ in the previous subsection is relaxed. It was pointed out earlier that in such a situation the fluctuations of $S(\vec{r})$ cannot be calculated independently from the fluctuations of $\hat{n}(\vec{r})$, due to interaction terms between them in Eq. [57]. The problem of coupling between different fluctuation modes can be avoided if one considers the thermal fluctuation amplitudes of the tensor order parameter components, $Q_{ij}(\vec{r})$, rather than the fluctuation amplitudes of $S(\vec{r})$ and $\hat{n}(\vec{r})$.

Since $\mathbf{Q}(\vec{r})$ is a symmetric, traceless tensor, only five of its nine components are independent, viz. $Q_{11}(\vec{r}) - Q_{22}(\vec{r})$, $Q_{33}(\vec{r})$, $Q_{12}(\vec{r})$, $Q_{13}(\vec{r})$, and $Q_{23}(\vec{r})$. As before we Fourier analyze each independent component,

$$Q_{ij}(\vec{r}) = \sum_{\vec{q}} Q_{ij}(\vec{q}) \exp\{i\vec{q} \cdot \vec{r}\} \qquad [80]$$

with $Q_{ij}(\vec{q})$ defined as

$$Q_{ij}(\vec{q}) = \frac{1}{V}\int d^3\vec{r}\, Q_{ij}(\vec{r}) \exp\{-i\vec{q}\cdot\vec{r}\} = Q_{ij}{}^*(-\vec{q}). \qquad [81]$$

The volume integral of $\mathfrak{F}$, (Eq. [57]), can then be expressed as

$$\int_V \mathfrak{F}\, d^3\vec{r} = Vf_0[T] + V\sum_{\vec{q}} \mathfrak{F}(\vec{q}). \qquad [82]$$

For a wavevector $\vec{q}$ pointing along the 3-axis (z-axis), $\mathfrak{F}(\vec{q})$ has the form

$$\mathcal{F}(\vec{q}) = a(T - T_c^*)\left[\frac{3}{4}(1 + \xi_\parallel^2 q^2)|Q_{33}(\vec{q})|^2 + \right.$$
$$\frac{1}{4}(1 + \xi_1^2 q^2)|Q_{11}(\vec{q}) - Q_{22}(\vec{q})|^2 + (1 + \xi_1^2 q^2)|Q_{12}(\vec{q})|^2 +$$
$$\left. (1 + \xi_\perp^2 q^2)(|Q_{13}(\vec{q})|^2 + |Q_{23}(\vec{q})|^2)\right] \pm$$
$$2qq_0 L_1 i \left[Q_{12}(\vec{q})(Q_{11}^*(\vec{q}) - Q_{22}^*(\vec{q})) + \right.$$
$$\left. Q_{12}^*(\vec{q})(Q_{22}(\vec{q}) - Q_{11}(\vec{q})) + Q_{23}(\vec{q})Q_{13}^*(\vec{q}) - Q_{13}(\vec{q})Q_{23}^*(\vec{q})\right],$$
[83]

where $\xi_\parallel(\xi_\perp)$ is the value of ξ when $\psi_{\hat{q}}$ is zero ($\pi/2$) and ξ_1 is the value of ξ when $L_2 = 0$. It is clear from Eq. [83] that for $q_0 = 0$, i.e., for nematics, the last term of Eq. [83] vanishes and $Q_{33}(\hat{q})$, $Q_{11}(\hat{q}) - Q_{22}(\hat{q})$, $Q_{12}(\hat{q})$, $Q_{13}(\hat{q})$, $Q_{23}(\hat{q})$ are independent quantities. Since there are no interaction terms containing both $\hat{q}$ and $\hat{q}'$ ($\hat{q} \neq \hat{q}'$) in Eq. [82], we are free to choose the spatial axes for each $\hat{q}$. Thus, Eqs. [82] and [83] represent a general solution, valid for any $\hat{q}$, provided we remember that the 3-axis is parallel to $\hat{q}$. The thermal average of the square of the fluctuation amplitude for each independent mode can be calculated in exactly the same manner as was that for $S(\hat{r})$ in Section 4.1.1. As before, the results can be obtained directly from the equipartition theorem. Therefore, we have, for $q_0 = 0$,[7]

$$\langle|Q_{33}(\vec{q})|^2\rangle = \frac{2}{3}\frac{k_B T}{Va(T - T_c^*)(1 + \xi_\parallel^2 q^2)}, \quad [84a]$$

$$\langle|Q_{11}(\vec{q}) - Q_{22}(\vec{q})|^2\rangle = \frac{2k_B T}{Va(T - T_c^*)(1 + \xi_1^2 q^2)}, \quad [84b]$$

$$\langle|Q_{12}(\vec{q})|^2\rangle = \frac{k_B T}{2Va(T - T_c^*)(1 + \xi_1^2 q^2)}, \quad [84c]$$

and

$$\langle|Q_{13}(\vec{q})|^2\rangle = \langle|Q_{23}(\vec{q})|^2\rangle = \frac{k_B T}{2Va(T - T_c^*)(1 + \xi_\perp^2 q^2)}. \quad [84d]$$

When $q_0 \neq 0$ the situation is quite different, and some of the modes which are independent of each other for $q_0 = 0$ become mixed.

LANDAU–deGENNES THEORY

However, if we again take the 3-axis parallel to $\hat{q}$, Eq. [83] can be diagonalized to give

$$\mathcal{F}(\vec{q}) = a(T - T_c^*)\left[\frac{3}{4}(1 + \xi_\parallel^2 q^2)|Q_{33}(\vec{q})|^2 + \frac{1}{4}(1 + \xi_\perp^2 q^2)|Q_{11}(\vec{q}) - Q_{22}(\vec{q}) \pm \frac{8q_0 q \xi_1^2 i}{(1 + \xi_1^2 q^2)}Q_{12}(\vec{q})|^2 + \left(1 + \xi_1^2 q^2 - \frac{16q_0^2 q^2 \xi_1^4}{(1 + \xi_1^2 q^2)}\right)|Q_{12}(\vec{q})|^2 + (1 + \xi_\perp^2 q^2)|Q_{13}(\vec{q}) \pm \frac{2q_0 q \xi_1^2 i}{(1 + \xi_\perp^2 q^2)}Q_{23}(\vec{q})|^2 + \left(1 + \xi_\perp^2 q^2 - \frac{4q_0^2 q^2 \xi_1^4}{(1 + \xi_\perp^2 q^2)}\right)|Q_{23}(\vec{q})|^2\right]. \quad [85]$$

It can be seen that the independent modes for cholesterics are different from those for nematics. If we denote the quantity $[Q_{11}(\hat{q}) - Q_{22}(\hat{q}) \pm (8q_0 q\ \xi_1^2 i/(1 + \xi_1^2 q^2))Q_{12}(\hat{q})]$ by $Q_a(\hat{q})$ and $[Q_{13}(\hat{q}) \pm (2q_0 q\ \xi_1^2 i/(1 + \xi_\perp^2 q^2))Q_{23}(\hat{q})]$ by $Q_b(\hat{q})$, then the thermal averages of the fluctuation amplitudes of the independent modes are[7]

$$\langle |Q_{33}(\vec{q})|^2\rangle = \frac{2}{3}\frac{k_B T}{Va(T - T_c^*)(1 + \xi_\parallel^2 q^2)}, \quad [86a]$$

$$\langle |Q_a(\vec{q})|^2\rangle = \frac{2k_B T}{Va(T - T_c^*)(1 + \xi_1^2 q^2)}, \quad [86b]$$

$$\langle |Q_{12}(\vec{q})|^2\rangle = \frac{k_B T(1 + \xi_1^2 q^2)}{2Va(T - T_c^*)[(1 + \xi_1^2 q^2)^2 - 16q_0^2 q^2 \xi_1^4]}, \quad [86c]$$

$$\langle |Q_b(\vec{q})|^2\rangle = \frac{k_B T}{2Va(T - T_c^*)(1 + \xi_\perp^2 q^2)}, \quad [86d]$$

and

$$\langle |Q_{23}(\vec{q})|^2\rangle = \frac{k_B T(1 + \xi_\perp^2 q^2)}{2Va(T - T_c^*)[(1 + \xi_\perp^2 q^2)^2 - 4q_0^2 q^2 \xi_1^4]} \quad [86e]$$

The thermal averages $\langle|Q_{11}(\hat{q})|^2\rangle$, $\langle|Q_{22}(\hat{q})|^2\rangle$, and $\langle|Q_{13}(\hat{q})|^2\rangle$ can also be calculated in terms of the thermal averages of the independent modes given in Eq. [86]. Remembering that $Q_{11}(\hat{q}) + Q_{22}(\hat{q}) + Q_{33}(\hat{q}) = 0$, we obtain

$$\langle|Q_{22}(\vec{q})|^2\rangle = \langle|Q_{11}(\vec{q})|^2\rangle$$

$$= \left\langle \left| \frac{1}{2}\left[Q_a(\vec{q}) \mp \frac{8q_0 q \xi_1^2 i}{(1+\xi_1^2 q^2)} Q_{12}(\vec{q}) - Q_{33}(\vec{q}) \right] \right|^2 \right\rangle$$

$$= \frac{1}{4}\left[\langle|Q_a(\vec{q})|^2\rangle + \frac{64 q_0^2 q^2 \xi_1^4}{(1+\xi_1^2 q^2)^2}\langle|Q_{12}(\vec{q})|^2\rangle + \langle|Q_{33}(\vec{q})|^2\rangle \right]$$

$$= \langle|Q_{12}(\vec{q})|^2\rangle + \frac{1}{4}\langle|Q_{33}(\vec{q})|^2\rangle, \qquad [86f]$$

$$\langle|Q_{23}(\vec{q})|^2\rangle = \langle|Q_{13}(\vec{q})|^2\rangle$$

$$= \left\langle \left| Q_b(\vec{q}) \mp \frac{2q_0 q \xi_1^2 i}{(1+\xi_\perp^2 q^2)} Q_{23}(\vec{q}) \right|^2 \right\rangle$$

$$= \langle|Q_b(\vec{q})|^2\rangle + \frac{4 q_0^2 q^2 \xi_1^4}{(1+\xi_\perp^2 q^2)^2} \langle|Q_{23}(\vec{q})|^2\rangle. \qquad [86g]$$

In deriving Eqs. [86f] and [86g] we have used the fact that the thermal averages of the cross terms between independent modes, such as $\langle Q_a(\hat{q})Q_{12}^*(\hat{q})\rangle$, all vanish.

Due to the fact that the order parameter **Q** is related to macroscopic observable quantities such as the susceptibility tensor χ (Eq. [32]) and the dielectric tensor ϵ, fluctuations in the components of **Q** are directly manifested as fluctuations in ϵ and in χ and are therefore experimentally measurable. In Section 5 we will show how the fluctuation amplitude, $\langle|Q_{ij}(\hat{q})|^2\rangle$, can be used to calculate cross section of light scattering by fluctuations in the isotropic phase of liquid crystals.

4.2 Heterophase Fluctuations

In this section we will investigate fluctuations that result in the formation of small spatial regions of the low temperature, liquid-crystal phase in the isotropic phase, as the temperature T approaches T_c from above. These so-called "heterophase" fluctuations produce a sudden change in the value of the order parameter, in a local spatial region, from $S = 0$ to a value appropriate to the liquid crystal phase at temperature T. Referring to Fig. 4, such fluctuations carry local regions from the free energy minimum at $S = 0$, over the barrier to the free energy minimum near $S = <S>_c$. The small regions of liquid crystal phase so produced are sometimes called "embryos"; their stability and statistical distribution for $T \sim T_c$ are the subject of the following discussion.

Let $\bar{\mathcal{F}}_i$ denote the value of the Landau free energy density of the spatially uniform isotropic phase and $\bar{\mathcal{F}}_a$ denote the analogous quantity for the anisotropic, liquid crystal phase. $\bar{\mathcal{F}}_i$ and $\bar{\mathcal{F}}_a$ correspond to the values of the two minima in Fig. 4 for T close to T_c, and their difference, $\Delta \bar{\mathcal{F}} = \bar{\mathcal{F}}_a - \bar{\mathcal{F}}_i$, is positive for $T > T_c$, zero for $T = T_c$, and negative for $T < T_c$. For the moment let us set $L_1 = L_2 = 0$ in the Landau free energy expression, which implies perfectly sharp boundaries between the embryos and their surroundings; that is, every point in the system sits in one or the other of the two minima. In this case, the probability of heterophase fluctuations can be easily calculated. If we let P_a denote the probability for a spatial region of volume V_{em} to spontaneously fluctuate from the isotropic phase to the anisotropic phase and P_i be the probability that the same spatial volume would remain in the isotropic phase, then in thermal equilibrium

$$\frac{P_a}{P_i} = \exp[-\beta \Delta \bar{\mathcal{F}} V_{em}]. \qquad [90]$$

Eq. [90] is correct, independent of the barrier height h between the two minima of the Landau free energy density, so long as we neglect the energy contributions due to the gradient terms. In this case the barrier height influences only the rate at which the thermal equilibrium distribution is reached, but not the distribution itself. According to Eq. [90] one should expect to observe large heterophase fluctuations near any first order phase transition since $\Delta \bar{\mathcal{F}} \approx 0$ for T close to T_c. This, of course, is not the case and the reason lies precisely in our neglect of the gradient terms in the Landau free energy density expansion.

If we let $L_1, L_2 \neq 0$, it is clear that formation of an embryo with an infinitely sharp boundary is energetically unfavorable. Therefore, there must necessarily be a boundary layer within which the value of the order parameter varies continuously from that inside the embryo to that of the surrounding isotropic fluid. The tendency of the gradient terms to maximize the spatial extent of the boundary layer is opposed by the accompanying increase in energy associated with values of the order parameter between the two minima, and the boundary layer that obtains represents a compromise between these two forces. In reality, then, we see that the barrier height plays a vital role not only in the rate of approaching the thermal distribution of embryos but also in the final distribution itself. That is, in addition to the free energy $\Delta \bar{\mathcal{F}}\, V_{em}$, creation of an embryo also requires a certain amount of surface energy μ per unit area.[17] Later in this section we will estimate the magnitude of μ for liquid crystals, but for the present we will calculate the total work Φ required to produce a spherical embryo of radius ρ.

From the above discussion we have

$$\Phi(\rho) = \frac{4}{3}\pi\rho^3 \Delta\bar{\mathcal{F}} + 4\pi\rho^2 \mu, \qquad [91]$$

so that the probability for thermally generating an embryo of radius ρ is $\sim \exp[-\beta\Phi(\rho)]$. Fig. 8 shows the behavior of $\Phi(\rho)$ for three different cases: $\Delta\bar{\mathcal{F}} > 0$, $\Delta\bar{\mathcal{F}} = 0$, and $\Delta\bar{\mathcal{F}} < 0$. For $\Delta\bar{\mathcal{F}} \geq 0$, $T \geq T_c$, $\Phi(\rho)$ is a monotonically increasing function of ρ, and embryos of any radius, once generated, inevitably shrink and disappear. It is also clear that if the value of the surface energy μ is large, then even at $T = T_c$, where $\Delta\bar{\mathcal{F}} = 0$, the probability for the occurrence of heterophase fluctuations is small. Since the magnitude of the surface energy μ is directly related to the barrier height between the two free energy minima (higher barrier $\Rightarrow$ larger value of μ), it follows that in physical systems where the barrier is high, such as in the liquid-solid phase transitions, fluctuations are usually small and unobservable.

The interesting case arises when $\Delta\bar{\mathcal{F}} < 0$, i.e., for $T < T_c$. In this case $S = 0$ is a relative minimum, and it is possible to have a supercooled metastable state. In Fig. 8 it is shown that for $\Delta\bar{\mathcal{F}} < 0$, $\Phi(\rho)$ has a maximum at

$$\rho^* = \frac{2\mu}{|\Delta\bar{\mathcal{F}}|} = -\frac{2\mu}{\Delta\bar{\mathcal{F}}} \qquad [92]$$

due to the competition between the volume free energy, which is negative, and the surface energy, which is positive. For embryos of radius $\rho < \rho^*$ the surface energy dominates, and the embryos tend to shrink. However, for those embryos whose radius is greater than ρ^* the volumes free energy dominates, and the embryos tend to grow until the whole sample transforms into the low temperature phase. It follows that if one could avoid producing embryos with radius $\rho \geq \rho^*$, then supercooling of the sample would occur.

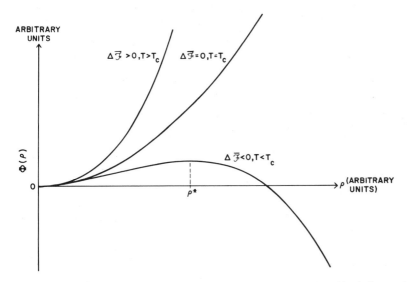

Fig. 8—Plot of the work required to generate an embryo as a function of its radius ρ at different temperatures. For $T \geq T_c$ the work required is an increasing function of ρ so that an embryo of any radius, once generated, must ultimately shrink and disappear. For $T < T_c$, embryos whose radius is greater than a critical radius ρ^* can grow indefinitely until the entire sample has transformed to the ordered phase. Note that if the formation of embryos of radius $\rho > \rho^*$ can be prevented, the sample can be supercooled.

Since the phenomenon of supercooling depends critically on the absence of embryos with radius $\rho > \rho^*$, it is of interest to estimate the relative probability Λ of thermally generating an embryo with radius ρ^* by heterophase fluctuations. We have, for T near T_c,

$$\Lambda \equiv \frac{\text{Probability of generating an embryo with radius } \rho^*}{\text{Probability of the system remaining in isotropic state}}$$

$$\simeq \exp[-\beta\Phi(\rho^*)] = \exp\left[-\frac{1}{3k_BT_c}\frac{16\pi\mu^3}{|\Delta\overline{\mathcal{F}}|^2}\right]. \quad [93]$$

Since $\Delta\overline{\mathcal{F}}$ changes sign at $T = T_c$, we approximate it by the form

$$\Delta \bar{\mathcal{F}} = \text{const.} \ (T - T_c)$$

for T close to T_c. The constant must be equal to $-N\Delta s$, the transition entropy per unit volume, since, by definition, $-\partial \Delta \bar{\mathcal{F}}/\partial T|_{T=T_c} = N\Delta s$. Thus,

$$\Delta \bar{\mathcal{F}} = -N\Delta s(T - T_c), \quad \Delta s < 0, \qquad [94]$$

which can be substituted into Eq. [93] to give[18]

$$\Lambda \simeq \exp\left[-\frac{16\pi\mu^3}{3k_B T_c^3 \left(\dfrac{T}{T_c} - 1\right)^2 N^2(\Delta s)^2}\right] \qquad [95]$$

From Eq. [95] it is evident that if the dimensionless ratio $\mu^3/(k_B T_c^3 N^2(\Delta s)^2)$ is large, Λ is small and supercooling should be easily observable.

We can estimate the value of μ within the framework of Landau–de Gennes theory. For this purpose we will use Eq. [42] with $\bar{H}$, $\bar{E}$, q_0, and L_2 set equal to zero and we will neglect any spatial variation in $\hat{n}(\vec{r})$. Eq. [42] then becomes

$$\mathcal{F} = f_0[T] + \frac{3}{4}a(T - T_c^*)S^2(\vec{r}) + \frac{1}{4}BS^3(\vec{r}) +$$
$$\frac{9}{16}CS^4(\vec{r}) + \frac{3}{4}L_1[\vec{\nabla}S(\vec{r})]^2. \qquad [96]$$

We will consider a semi-infinite planar sample in which $S(\vec{r})$ varies only as a function of z with boundary values $S(-\infty) = <S>_c$ and $S(\infty) = 0$. The origin $z = 0$ is defined arbitrarily by the condition $S(z = 0) = <S>_c/2$. From arguments presented previously we expect to find a transition layer in which the value of S varies continuously from $<S>_c$ to 0. The free energy per unit area of the transition layer is given by

$$\frac{F\{S(z)\}}{A} = \int_{-\infty}^{\infty} dz\{\mathcal{F} - f_0[T]\}$$
$$= \int_{-\infty}^{\infty} dz\left\{\frac{3}{4}a(T - T_c^*)S^2(z) + \frac{1}{4}BS^3(z) + \frac{9}{16}CS^4(z) + \frac{3}{4}L_1\left(\frac{dS(z)}{dz}\right)^2\right\} \qquad [97]$$

where the dependence of F on $S(z)$ is shown explicitly. The variation of $S(z)$ with z is obtained by minimizing the free energy F. The resulting $S(z)$ must satisfy the Euler-Lagrange equation

$$\frac{d}{dS(z)} \left[\frac{3}{4} a(T - T_c^*) S^2(z) + \frac{1}{4} B S^3(z) + \frac{9}{16} C S^4(z) \right]$$
$$= \frac{3}{2} L_1 \frac{d^2 S(z)}{dz^2} \qquad [98]$$

Multiplying both sides of Eq. [98] by $dS(z)/dz$ and integrating once with respect to z gives

$$\text{const.} + \frac{3}{4} a(T - T_c^*) S^2(z) + \frac{1}{4} B S^3(z) + \frac{9}{16} C S^4(z)$$
$$= \frac{3}{4} L_1 \left(\frac{dS(z)}{dz} \right)^2. \qquad [99]$$

The constant in Eq. [99] must be zero since $S(\infty) = 0$ and $[dS(z)/dz|_{z=\infty}] = 0$. Therefore, using the fact that $z = 0$ is defined as the point at which $S = \langle S \rangle_c/2$, we have

$$\int_{\langle S \rangle_c/2}^{S(z)} \left[3(T/T_c^* - 1) S'^2 + \frac{B}{aT_c^*} S'^3 + \frac{9C}{4aT_c^*} S'^4 \right]^{1/2} dS'$$
$$= - \left(\frac{aT_c^*}{3L_1} \right)^{1/2} z$$

or
$$\frac{z}{\xi_1} = G[S(z)] - G[\langle S \rangle_c/2], \qquad [100]$$

where

$$G(S) = \ln \left\{ \frac{-\sqrt{3[(T/T_c^*) - 1]} + (B/aT_c^*)S + (9C/4aT_c^*)S^2}{S} \right.$$
$$\left. - \frac{\sqrt{3[(T/T_c^*) - 1]}}{S} - \frac{(B/aT_c^*)}{2\sqrt{3[(T/T_c^*) - 1]}} \right\}. \qquad [101]$$

For $T = T_c$, Eq. [100] can be simplified to the form

$$S(z) = \frac{\langle S \rangle_c}{\exp\left(\dfrac{z}{\xi_1}\right) + 1} \qquad [100a]$$

with the aid of Eqs. [52] and [53]. Equations [100], [100a], and [101] give the spatial variation of S as a function of z. Using Eq. [100a] and $\langle S \rangle_c = 0.33$, we obtain $S(z)$ as plotted in Fig. 9 for $T = T_c$. The fact

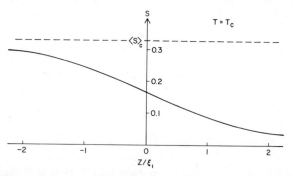

Fig. 9—Variation of the order parameter S as a function of distance in the boundary layer separating the ordered phase, $S = \langle S \rangle_c$, from the disordered phase, $S = 0$, at $T = T_c$.

that $S(z)$ decays most rapidly for $S \simeq 0.2$ is understandable because $S = 0.17$ is the position of the barrier peak and therefore such values of S should occupy the smallest portion of the boundary layer in order to minimize the total free energy.

We can also calculate the surface energy[7] per unit area μ using Eq. [99]. Substituting Eq. [99] into Eq. [97], we get

$$\mu = \int_{-\infty}^{\infty} dz \frac{3}{2} L_1 \left(\frac{dS(z)}{dz}\right)^2 = \frac{3}{2} L_1 \int_{-\infty}^{\infty} dz \frac{dS(z)}{dz}\left(\frac{dS(z)}{dz}\right)$$

$$= \frac{3}{2} L_1 \int_{\langle S \rangle_c}^{0} dS(z) \left(\frac{dS(z)}{dz}\right) . \qquad [102]$$

Using Eq. [99] for $dS(z)/dz$ (with the minus sign for $z > 0$), Eq. [102] becomes

$$\mu = \frac{(3L_1 aT_c{}^*)^{1/2}}{2} \times$$

$$\int_0^{\langle S \rangle_c} dS' S' \left[3\left(\frac{T}{T_c{}^*} - 1\right) + \frac{B}{aT_c{}^*} S' + \frac{9C}{4aT_c{}^*} S'^2 \right]^{1/2}.$$

[103]

Setting $T = T_c$, $L_1 = 6 \times 10^{-7}$ dyne, and using the values of a, B, C, T_c, $\langle S \rangle_c$, and $T_c{}^*$ for MBBA, we find $\mu \simeq 0.02$ erg/cm². This is in good agreement with the measured value[19] of $\mu \simeq 0.023$ erg/cm². Using this value of μ, $\Delta s \simeq 0.16\, k_B$, and $N \simeq 2.3 \times 10^{21}$ cm⁻³, we obtain the dimensionless ratio $\mu^3/(k_B T_c{}^3 N^2 (\Delta s)^2) \simeq 0.7 \times 10^{-6}$ for MBBA. Putting this value into Eq. [95] yields the result that Λ is of order e^{-1} ($\Phi(\rho^*) = kT_c$) when $T \simeq T_c - 1°K$, which means that MBBA can at most be supercooled to about $1°K$ below T_c.† Recalling from Section 2.3 that $T_c{}^*$ was interpreted as that temperature below which supercooling is impossible, we conclude from the above calculation that for MBBA, $T_c{}^* \simeq T_c - 1°K$. This is in very good agreement with experimental value[14] of $T_c - T_c{}^* \simeq 1°K$.

5. Observation of Fluctuations Using Light Scattering

Fluctuations in the order parameter are reflected in various physical properties of a liquid crystal material. In this section we will focus on the elastic (Rayleigh) scattering of light by such fluctuations in the isotropic phase of nematic and cholesteric materials near T_c.

We begin by expressing the dielectric constant at an arbitrary point $\vec{r}$ in terms of the order parameter

$$\boldsymbol{\epsilon}(\vec{r}) = \frac{2}{3}(\Delta\epsilon)_{\max} \mathbf{Q}(\vec{r}) + \bar{\epsilon}\mathbf{I}, \qquad [104]$$

where $\bar{\epsilon}$ is the dielectric constant of the isotropic liquid and $\mathbf{I}$ is the unit tensor of the second rank. Since the polarizability $\boldsymbol{\alpha}(\vec{r}) \equiv [\boldsymbol{\epsilon}(\vec{r}) - \mathbf{I}]/4\pi$, we can write

$$\boldsymbol{\alpha}(\vec{r}) = \frac{1}{6\pi}(\Delta\epsilon)_{\max} \mathbf{Q}(\vec{r}) + \frac{\bar{\epsilon} - 1}{4\pi}\mathbf{I}, \qquad [105]$$

which, upon Fourier transformation, becomes

---

† At this temperature the value of ρ^* calculated from Eqs. [92] and [94] is about 80 Å.

$$\alpha(\vec{q}) = \frac{1}{6\pi}(\Delta\epsilon)_{max}\mathbf{Q}(\vec{q}) + \frac{\bar{\epsilon} - 1}{4\pi}\delta(\vec{q})\mathbf{I}. \qquad [106]$$

Consider an incident light wave (Fig. 10) of the form

$$\vec{E}_{in} = \hat{e}_{in}E_0 \exp[i(\vec{k}_{in}\cdot\vec{r} - \omega t)], \qquad [107]$$

with frequency ω and wavevector $|\vec{k}_{in}| = \omega/c$. $\hat{e}_{in}$ in Eq. [107] is a unit

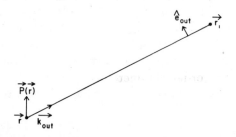

Fig. 10—Graphic illustration of the vector identity, Eq. [110].

polarization vector pointing along the electric field direction of the incident light. This wave will induce a dipole moment per unit volume $\vec{P}(\vec{r})$ in the sample

$$\vec{P}(\vec{r}) = \alpha(\vec{r})\cdot\vec{E}_{in} = \alpha(\vec{r})\cdot\hat{e}_{in}E_0 \exp[i(\vec{k}_{in}\cdot\vec{r} - \omega t)]. \qquad [108]$$

At a point $\vec{r}_1$ far from the scattering volume the radiation from a volume $d^3\vec{r}$ around $\vec{r}$ due to the induced oscillating dipole moment $\vec{P}(\vec{r})$ is[20]

$$d\vec{E}_{out} = -\frac{\omega^2 d^3\vec{r}}{c^2|\vec{r}_1 - \vec{r}|} \frac{[(\vec{P}(\vec{r}) \times \vec{k}_{out}) \times \vec{k}_{out}]}{|\vec{k}_{out}|^2} \exp[i\vec{k}_{out}\cdot(\vec{r}_1 - \vec{r})], \qquad [109]$$

where c is the speed of light and $\vec{k}_{out}$ is a vector of magnitude ω/c pointing in the direction of $(\vec{r}_1 - \vec{r})$. Letting $\hat{e}_{out}$ be the unit vector pointing along the direction of $d\vec{E}_{out}$, we see from Fig. 10 that

$$-\frac{[(\vec{P}(\vec{r}) \times \vec{k}_{out}) \times \vec{k}_{out}]}{|\vec{k}_{out}|^2} = [\hat{e}_{out}\cdot\vec{P}(\vec{r})]\hat{e}_{out}. \qquad [110]$$

Substitution of Eqs. [108] and [110] into Eq. [109] yields

$$d\vec{E}_{out} = \hat{e}_{out}\frac{\omega^2 E_0}{c^2 R_0}\exp[i(\vec{k}_{out}\cdot\vec{r}_1 - \omega t)](\hat{e}_{out}\cdot\boldsymbol{\alpha}(\vec{r})\cdot\hat{e}_{in})\times$$
$$\exp[i\vec{q}\cdot\vec{r}]d^3\vec{r}, \quad [111]$$

where $R_0 \equiv |\vec{r}_1 - \vec{r}|$ and $\vec{q} \equiv \vec{k}_{out} - \vec{k}_{in}$. Integrating over the volume of the sample with R_0 treated as a constant ($\vec{r}_1$ is very far from the sample), we obtain an expression for the total scattered electric field at $\vec{r}_1$

$$\vec{E}_{out} = \hat{e}_{out}\frac{\omega^2 E_0 V}{c^2 R_0}\exp[i(\vec{k}_{out}\cdot\vec{r}_1 - \omega t)](\hat{e}_{out}\cdot\boldsymbol{\alpha}(\vec{q})\cdot\hat{e}_{in}), \quad [112]$$

and for the differential Rayleigh scattering cross section $d\sigma_R$

$$d\sigma_R = \frac{|\vec{E}_{out}|^2}{|E_0|^2}R_0^2 d\Omega = \frac{\omega^4 V^2}{c^4}|\hat{e}_{out}\cdot\boldsymbol{\alpha}(\vec{q})\cdot\hat{e}_{in}|^2 d\Omega, \quad [113]$$

where $d\Omega$ is the differential solid angle in the direction of $\vec{k}_{out}$. Substituting Eq. [106] into Eq. [113], we find the light scattering power, or Rayleigh ratio, to be

$$R \equiv \frac{1}{V}\frac{d\sigma_R}{d\Omega} = \frac{\omega^4(\Delta\epsilon)^2_{max}V}{c^4(6\pi)^2}|\hat{e}_{out}\cdot\mathbf{Q}(\vec{q})\cdot\hat{e}_{in}|^2, \quad [114]$$

for $\vec{q} \neq 0$. In the isotropic phase of a nematic or cholesteric liquid crystal the scattering results from thermal fluctuations, and the experimentally measured Rayleigh ratio is therefore given by

$$R = \frac{\omega^4(\Delta\epsilon)^2_{max}V}{c^4(6\pi)^2}\langle|\hat{e}_{out}\cdot\mathbf{Q}(\vec{q})\cdot\hat{e}_{in}|^2\rangle. \quad [115]$$

In Fig. 11 we show a scattering geometry in which the wavevector of the incoming light makes an angle v with that of the outgoing light and $\hat{e}_{out}$ is perpendicular to $\hat{e}_{in}$. If $\hat{e}_1$, $\hat{e}_2$, and $\hat{e}_3$ are unit vectors along the x,y, and z axes, respectively, with $\vec{q} \parallel z$, then

$$\hat{e}_{in} = \hat{e}_2,$$

and

$$\hat{e}_{out} = \cos\frac{\upsilon}{2}\hat{e}_3 - \sin\frac{\upsilon}{2}\hat{e}_1. \qquad [116]$$

It follows that

$$\langle|\hat{e}_{out} \cdot \mathbf{Q}(\vec{q}) \cdot \hat{e}_{in}|^2\rangle = \left\langle\left|\cos\frac{\upsilon}{2}Q_{32}(\vec{q}) - \sin\frac{\upsilon}{2}Q_{12}(\vec{q})\right|^2\right\rangle. \qquad [117]$$

Since $Q_{23}(\hat{q})$ and $Q_{12}(\hat{q})$ are independent, the cross terms in Eq. [117] average to zero, leaving

$$\langle|\hat{e}_{out} \cdot \mathbf{Q}(\vec{q}) \cdot \hat{e}_{in}|^2\rangle = \cos^2\frac{\upsilon}{2}\langle|Q_{32}(\vec{q})|^2\rangle + \sin^2\frac{\upsilon}{2}\langle|Q_{12}(\vec{q})|^2\rangle. \qquad [118]$$

Substituting Eq. [118] into Eq. [115] and using the results of Eq. [86] we have, finally, that the Rayleigh ratio for the scattering geometry of Fig. 11 is

$$R = \frac{\omega^4(\Delta\epsilon)_{max}^2 k_B T}{c^4(6\pi)^2 a(T - T_c^*)}\left[\cos^2\frac{\upsilon}{2}\frac{(1 + \xi_\perp^2 q^2)}{[(1 + \xi_\perp^2 q^2)^2 - 4q_0^2\xi_1^4 q^2]} + \sin^2\frac{\upsilon}{2}\frac{(1 + \xi_1^2 q^2)}{[(1 + \xi_1^2 q^2)^2 - 16q_0^2\xi_1^4 q^2]}\right], \qquad [119]$$

where $|\vec{q}| = (2\omega/c)\sin\upsilon/2$.

For nematic materials $q_0 = 0$ and Eq. [119] shows that the scattering intensity decreases with increasing q. For cholesteric materials, on the other hand, $q_0 \neq 0$ and the second term of Eq. [119] can be peaked for a value of q such that

$$q = \frac{(4q_0\xi_1 - 1)^{1/2}}{\xi_1} > 0, \qquad [120]$$

provided $q_0\xi_1 > \frac{1}{4}$. Physically this means that if the correlation length ξ_1 is of order $\lambda_0/10$ then fluctuations in cholesteric order will produce incomplete helices with $\sim\frac{1}{10}$ of a pitch, which couple strongly to the incident and scattered light waves when the transfer wavevector $\vec{q}$ satisfies Eq. [120]. As $T \to T_c^*$ from above, ξ_1 increases and the peak in the second term of Eq. [119] diverges at $q_0\xi_1 = \frac{1}{2}$ which corresponds to the temperature $T = T_c^* + 4q_0^2 L_1/a$ (Recall that $\xi_1 = [L_1/a(T - T_c^*)]^{1/2}$).

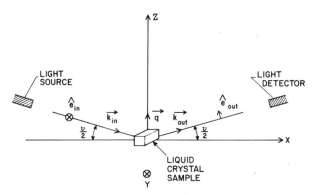

Fig. 11—Illustration of the geometry used in the light scattering experiment discussed in Section 5. Note that $\vec{q}$ is chosen to be along the z-direction to conform with the convention used in the text.

The Rayleigh ratio has been measured in the isotropic phase of both nematic[14,21] and cholesteric[22] liquid crystals. Agreement between these measurements and the predictions of the Landau–de Gennes theory is excellent in both cases.

6. Magnetic Birefringence and the Paranematic Susceptibility

Consider the effect of applying a static or slowly varying magnetic field to a sample of nematic liquid crystal in its isotropic phase. Assuming that $\hat{n} \parallel \vec{H}$, we can write the magnetic contribution to the free energy density as (see Eq. [38])

$$\mathcal{F}_M = -\frac{1}{3}(\Delta\chi)_{\max} S(\vec{r}) H^2. \qquad [121]$$

The equilibrium state in the presence of the magnetic field will have finite order since $\mathcal{F}_M$ is negative; the average value of this order $\bar{S}(H)$ can be calculated straightforwardly. We have

$$\bar{S}(H) \equiv \left\langle \frac{1}{V} \int_V d^3\vec{r}\, S(\vec{r}) \right\rangle_H \equiv \langle S(\vec{q} = 0) \rangle_H =$$

$$\frac{\int DS(\vec{r}) \left[\frac{1}{V}\int d^3\vec{r} S(\vec{r})\right] \exp\left\{-\beta \int d^3\vec{r}\left(\mathcal{F}_L - \frac{(\Delta \chi)_{\max}}{3} S(\vec{r}) H^2\right)\right\}}{\int DS(\vec{r}) \exp\left\{-\beta \int d^3\vec{r}\left(\mathcal{F}_L - \frac{(\Delta \chi)_{\max}}{3} S(\vec{r}) H^2\right)\right\}}, \quad [122]$$

where $\langle \ \rangle_H$ denotes the thermal average with the magnetic field H present.

$\bar{S}(H)$ depends on H quadratically and, furthermore, since $\bar{S}(H)$ is expected to be small and $\bar{S}(H = 0) = 0$, we can make the following expansion

$$\bar{S}(H) = \eta H^2 + \ldots$$

where

$$\eta = \left.\frac{\partial \bar{S}(H)}{\partial(H^2)}\right|_{H^2=0}$$

is the paranematic susceptibility. Substituting for $\bar{S}(H)$ from Eq. [122] and evaluating the resulting expression in the limit $H^2 \to 0$, we find

$$\eta = \frac{1}{k_B T} \frac{(\Delta \chi)_{\max}}{3} V \left\{\langle |S(\vec{q}=0)|^2\rangle_{H^2=0} - \langle S(\vec{q}=0)\rangle^2_{H^2=0}\right\}$$

$$= \frac{(\Delta \chi)_{\max} V}{3 k_B T} \langle |S(\vec{q}=0)|^2 \rangle, \quad [123]$$

In the approximation $B = C = 0$, we can substitute Eq. [74] into Eq. [123] and get

$$\eta = \frac{2(\Delta \chi)_{\max}}{9a(T - T_c^*)} \quad [124]$$

Taking $(\Delta \chi)_{\max} \simeq 10^{-6}$ cgs units, $a \simeq 4.2 \times 10^5$ erg/°K cm^3, $T = T_c$, and $T_c - T_c^* = 1$°K, we find $\eta \simeq 5 \times 10^{-13}$ cm^3/erg. For a field $H \simeq 10$ KOe this results in an equilibrium order $\bar{S}(10 \text{ KOe}) \simeq 5 \times 10^{-5}$.

The paranematic susceptibility is directly related to the phenomenon of magnetic birefringence. Any anisotropic property is proportional to the induced order. Specifically we can write

$$\Delta\epsilon \equiv \epsilon_\| - \epsilon_\perp = M_0 \langle S(q=0) \rangle_H \equiv M_0 \bar{S}(H) \qquad [125]$$

where M_o is a constant and $\epsilon_\|(\epsilon_\perp)$ is the long wavelength ($q = 0$) dielectric constant parallel (perpendicular) to $\vec{H}$. However, $\Delta\epsilon = (\epsilon_\|^{1/2} - \epsilon_\perp^{1/2})(\epsilon_\|^{1/2} + \epsilon_\perp^{1/2}) \simeq 2n\Delta n$, n being the refractive index, and using our previous result for $\bar{S}(H)$, Eq. [125] becomes

$$n\Delta n = \frac{M_0(\Delta\chi)_{max}}{9a(T - T_c^*)} H^2. \qquad [126]$$

Rearranging Eq. [126] gives

$$\frac{\Delta n}{H^2} = \frac{M_0(\Delta\chi)_{max}}{9an(T - T_c^*)}$$

$$= \frac{M_0 \eta}{2n} \qquad [127]$$

The quantity $\Delta n/H^2$ is called the Cotton-Mouton coefficient which is seen to diverge at $T = T_c^*$ and to fall off as $(T - T_c^*)^{-1}$ above T_c^*. Measurements of magnetic birefringence have been made for MBBA,[14] and the inverse Cotton-Mouton coefficient was plotted as a function of temperature. The experimental behavior is in complete agreement with Eq. [127]. Analysis of the data yields $T_c^* \simeq T_c - 1°K$.

Acknowledgment

We express our appreciation to Dr. R. Cohen for helpful discussions and to Dr. P. J. Wojtowicz for assistance in some numerical calculations and careful reading of the manuscript.

Appendix A

In this appendix we illustrate some implications of Eq. [15]. Consider a hypothetical system in which the function $\gamma[\vec{\nabla}\sigma(\vec{r}), T]$ is given by

$$\exp\{-\beta\gamma[\vec{\nabla}\sigma(\vec{r}), T]\} = \delta[\vec{\nabla}\sigma(\vec{r})]. \qquad [128]$$

The interaction energy defined by Eq. [128] reflects a rigid coupling between neighboring spatial regions. Substituting Eq. [128] into Eq. [15], we see that the only form of $\sigma(\vec{r})$ that can contribute to the integral is $\sigma(\vec{r})$ = constant, since any spatial variation of $\sigma(\vec{r})$ would make the integrand vanish due to the particular form of $\gamma[\vec{\nabla}\sigma(\vec{r}),T]$. The delta function thus simplifies the functional integral $\int D\sigma(\vec{r})$, which can be written as $\int d\sigma(1) \ldots \int d\sigma(M)$, (see Section 1) to a one-dimensional integral $\int d\sigma$. Eq. [15] therefore becomes

$$Z(T) = \int d\sigma \, \exp\{-\beta V f[\sigma, T]\}. \qquad [129]$$

At any particular temperature, $f[\sigma,T]$ has a minimum. If $\sigma_0(T)$ is the value of σ which minimizes $f[\sigma,T]$ at temperature T, then the integrand in Eq. [129] has a peak at $\sigma = \sigma_0(T)$. In the spirit of Landau's approach, let us approximate $Z(T)$, Eq. [129], by using a Taylor series approximation for $f[\sigma,T]$ around its minimum:

$$f[\sigma, T] \simeq f_0[T] + \frac{1}{2} A(T)[\sigma - \sigma_0(T)]^2, \; A(T) > 0 \qquad [130]$$

where $f_0[T] \equiv f[\sigma_0(T),T]$. Then

$$Z(T) \simeq \exp\{-\beta V f_0[T]\} \int d\sigma \, \exp\{-\beta V A(T)[\sigma - \sigma_0(T)]^2/2\}. \qquad [131]$$

Note that the integrand in Eq. [131] is a Gaussian centered at $\sigma_0(T)$ with half width $[\beta V A(T)]^{-1/2}$. As $V \to \infty$, the width of the peak approaches zero and the approximation, Eq. [131], becomes exact. It also follows that in the limit $V \to \infty$ the thermal distribution function, $\exp\{-\beta V f[\sigma,T]\}/Z(T)$, approaches a delta function peaked at $\sigma_0(T)$. This observation is the basis of the well-known result that in the thermodynamic limit the thermal average of σ, denoted by $\langle\sigma\rangle$, is equal to the value of $\sigma_0(T)$ that minimizes the Landau free energy function $f[\sigma,T]$, that is, the equilibrium value.

For a sufficiently large volume V the integral in Eq. [131] can be evaluated to give

$$Z(T) \simeq \left(\frac{2\pi}{\beta V A(T)}\right)^{1/2} \exp\{-\beta V f_0[T]\}. \qquad [132]$$

From Eq. [132] the free energy density in the thermodynamic limit $V \to \infty$ is evaluated to be

$$\lim_{V \to \infty} \frac{F(T)}{V} = f_0[T] + \lim_{V \to \infty} \frac{k_B T}{2V} \left[\ln \frac{2\pi}{\beta VA(T)} \right] = f_0[T]. \quad [133]$$

Eq. [133] states the result that in a spatially uniform system, where the partition function can be expressed by Eq. [129], the free energy density at temperature T is equal to the value of the minimum of the Landau free energy density function $f[\sigma,T]$ at that temperature.

The magnitude of the fluctuations in this hypothetical system can also be evaluated. If the thermal average of the quantity σ is defined by

$$\langle \sigma^\nu \rangle = \frac{1}{Z(T)} \int d\sigma \, \sigma^\nu \exp\{-\beta V f[\sigma, T]\}, \quad [134]$$

then the root mean square fluctuation of the order parameter is

$$[\langle (\sigma - \langle \sigma \rangle)^2 \rangle]^{1/2} = [\langle \sigma^2 \rangle - \langle \sigma \rangle^2]^{1/2}. \quad [135]$$

Carrying out the calculation with the approximation of Eq. [130] and V large, we obtain

$$[\langle \sigma^2 \rangle - \langle \sigma \rangle^2]^{1/2} = [\pi/\beta VA(T)]^{1/4}, \quad [136]$$

which vanishes in the limit of $V \to \infty$ unless, of course, $A(T) \to 0$. This result, which is the same as the corresponding result of the mean field theory, is purely an artifact of having required the whole system to fluctuate in phase by the specific choice of $\gamma[\vec{\nabla}\sigma(\vec{r}),T]$ in Eq. [128].

To see the effect of specifying a different $\gamma[\vec{\nabla}\sigma(\vec{r}),T]$, let us set

$$\gamma[\vec{\nabla}\sigma(\vec{r}), T] = 0 \quad [137]$$

and calculate the free energy density. In this case Eq. [15] becomes

$$Z(T) = \int D\sigma(\vec{r}) \exp\{-\beta(\int d^3\vec{r} f[\sigma(\vec{r}), T])\}$$
$$= \int d\sigma(1) \ldots \int d\sigma(M) \exp\left\{-\beta \Delta V \sum_{\alpha=1}^{M} f[\sigma(\alpha), T]\right\}$$
$$= \left[\int d\sigma \exp\{-\beta \Delta V f[\sigma, T]\}\right]^M, \quad [138]$$

where ΔV is the elementary volume as defined before, $M = V/\Delta V$, and V is the volume of the whole system. Using the same approximation for $f[\sigma,T]$ as that given by Eq. [130], we get

$$Z(T) \simeq \left(\frac{2\pi}{\beta \Delta V A(T)} \right)^{\frac{M}{2}} \exp\{-\beta V f_0[T]\}. \qquad [139]$$

The free energy density is then

$$\lim_{V \to \infty} \frac{F(T)}{V} = f_0[T] + \lim_{V \to \infty} \frac{k_B T M}{2V} \left[\ln \frac{2\vec{\pi}}{\beta \Delta V A(T)} \right]. \qquad [140]$$

As $V \to \infty$, we have $M \to \infty$ such that $V/M = \Delta V$ = constant. Therefore,

$$\lim_{V \to \infty} \frac{F(T)}{V} = f_0[T] + \frac{k_B T}{2 \Delta V} \left[\ln \frac{2\pi k_B T}{\Delta V A(T)} \right]. \qquad [141]$$

Eq. [141] differs from Eq. [133] by an additional term which can be attributed to spatial fluctuations of the order parameter. In contrast to the previous case where the whole system fluctuates in phase, fluctuations in a system where $\gamma[\vec{\nabla}\sigma(\vec{r}),T] = 0$ are completely uncorrelated from one spatial region to the next.

In real systems, coupling between a small part of the sample with the rest is neither zero nor rigid but can be described as elastic. This implies that for physical systems the order parameter fluctuations at two different spatial points are partially correlated, the degree of correlation being a decreasing function of the separation between the two points. These expectations are made explicit by calculations in Section 4.

Appendix B

In this appendix we wish to obtain the Jacobian factor for Eq. [63] and thence to evaluate the free energy density given by Eq. [70].

From Eq. [60a] it is clear that the Jacobian is given by $\|\exp(i\hat{q}_\beta \cdot \vec{r}_\alpha)\|$, the determinant of the $M \times M$ matrix with elements $\exp(i\hat{q}_\beta \cdot \vec{r}_\alpha)$ in column α and row β. The inverse for the matrix for which the $\alpha\beta$ − element is $[\exp(i\hat{q}_\beta \cdot \vec{r}_\alpha)]$ can be obtained from Eqs. [60a] and [61a];

LANDAU–deGENNES THEORY

$$S(\vec{r}_\alpha) = \sum_\beta S(\vec{q}_\beta) \exp\{i\vec{q}_\beta \cdot \vec{r}_\alpha\}$$

$$= \frac{\Delta V}{V} \sum_\beta \sum_\gamma \exp\{-i\vec{q}_\beta \cdot \vec{r}_\gamma\} \exp\{-i\vec{q}_\beta \cdot \vec{r}_\alpha\} S(\vec{r}_\gamma) \quad [142]$$

which implies

$$\frac{\Delta V}{V} \sum_\beta \exp\{-i\vec{q}_\beta \cdot \vec{r}_\gamma\} \exp\{i\vec{q}_\beta \cdot \vec{r}_\alpha\} = \delta_{\gamma\alpha} = \begin{cases} 1 & \gamma = \alpha \\ 0 & \gamma \neq \alpha \end{cases} \quad [143]$$

Therefore, the inverse matrix has element

$$\frac{\Delta V}{V} e^{-i\vec{q}_\beta \cdot \vec{r}_\alpha}$$

in row α and column β. From matrix algebra and Eq. [143] we get

$$\|\delta_{\alpha\gamma}\| = 1 = \|\exp(i\vec{q}_\beta \cdot \vec{r}_\alpha)\| \cdot \| \cdot \frac{\Delta V}{V} \exp(-i\vec{q}_\beta \cdot \vec{r}_\alpha)\|$$

$$= \|\exp(i\vec{q}_\beta \cdot \vec{r}_\alpha)\| \cdot \|\frac{\Delta V}{V}\| \cdot \|\exp(i\vec{q}_\beta \cdot \vec{r}_\alpha)\|^*$$

$$= \left(\frac{\Delta V}{V}\right)^M \|\exp(i\vec{q}_\beta \cdot \vec{r}_\alpha)\| \cdot \|\exp(i\vec{q}_\beta \cdot \vec{r}_\alpha)\|^*. \quad [144]$$

Therefore, apart from a phase factor, which we set equal to zero, the Jacobian is given by $(V/\Delta V)^{M/2}$.

From Eq. [70] the integral is easily evaluated to give

$$Z(\vec{q}, T) = \frac{2\pi}{3} \frac{k_B T}{\Delta V a (T - T_c^*)(1 + \xi^2 q^2)}, \quad [145]$$

where the upper limit of integration for $d|S(\hat{q})|^2$ is set equal to ∞. The total free energy of the system is given by

$$F(T) = -k_B T \ln Z(T) = V f_0[T] -$$

$$k_B T \sum_{\vec{q}}{}' \ln\left[\frac{2\pi}{3} \frac{k_B T}{\Delta V a (T - T_c^*)(1 + \xi^2 q^2)}\right] = V f_0[T] -$$

$$\frac{k_B T}{2} \frac{V}{(2\pi)^3} \int_0^{q_{max}} 4\pi q^2 dq \, \ln\left[\frac{2\pi}{3} \frac{k_B T}{\Delta V a (T - T_c^*)(1 + \xi^2 q^2)}\right] \quad [146]$$

where the summation over the independent set of $\hat{q}$ vectors is extended to every $\vec{q}$ with the final sum multiplied by ½. The integral can be evaluated to give a free-energy density expression

$$\frac{F(T)}{V} = f_0[T] - \frac{k_B T q^3_{max}}{12\pi^2} \left\{ \ln\left[\frac{1}{12\pi^2} \frac{k_B T q^3_{max}}{a(T - T_c^*)(1 + \xi^2 q^2_{max})}\right] + \frac{2}{3} - \frac{2}{\xi^2 q^2_{max}} + \frac{2\tan^{-1}\xi q_{max}}{\xi^3 q^3_{max}} \right\} \quad [147]$$

where we have set $\Delta V \equiv (2\pi)^3/q^3_{max}$. Eq. [147] can be easily reduced to the two limits, Eqs. [133] and [141], discussed in Appendix A. On the one hand, the rigid coupling limit can be obtained by reducing the degrees of freedom of the system from M independently varying components to a single independently varying component. This can be accomplished by letting $q_{max} \to 0$ in Eq. [147], which is equivalent to requiring $\Delta V \to V$ and $M \to 1$. The zero coupling limit, on the other hand, can simply be obtained by letting $\xi \to 0$ in Eq. [147].

References:

[1] P. J. Wojtowicz, "Introduction to the Molecular Theory of Nematic Liquid Crystals," Chapter 3.
[2] P. Sheng, "Hard Rod Model of the Nematic-Isotropic Phase Transition," Chapter 5.
[3] E. B. Priestley, "Nematic Order: The Long Range Orientational Distribution Function," Chapter 6.
[4] L. D. Landau, "On the Theory of Phase Transitions, Part I and Part II," *Collected Papers of L. D. Landau*, Edited by D. ter Haar, Gordon and Breach, Science Publishers, N. Y., 2nd Edition, p. 193–216 (1967).
[5] The values of the various critical exponents predicted by the Landau theory has been shown to disagree with experimental findings. See H. E. Stanley, *Introduction to Phase Transitions and Critical Phenomena*, Oxford University Press, N. Y. (1971).
[6] C. Kittel, *Introduction to Solid State Physics*, 4th Ed., John Wiley and Sons, N. Y., p. 477.
[7] P. G. de Gennes, "Short Range Order Effect in the Isotropic Phase of Nematics and Cholesterics," *Mol. Cryst. Liq. Cryst.*, **12**, p. 193 (1971).
[8] R. P. Feynman and A. R. Hibbs, *Quantum Mechanics and Path Integrals*, McGraw-Hill Book Co., N. Y. (1965).
[9] L. D. Landau and E. M. Lifshitz, *Statistical Physics*, 2nd Edit. Addison-Wesley, Reading, Mass. p. 425 (1969).
[10] J. A. Gonzalo, "Critical Behavior of Ferroelectric Triglycine Sulfate," *Phys. Rev.*, **144**, p. 662 (1966).
[11] V. L. Ginzburg, "On a Macroscopic Theory of Superconductivity for All Temperatures," *Soviet Phys. Doklady*, **1**, p. 541 (1956–57).
[12] P. Sheng, "Introduction to the Elastic Continuum Theory of Liquid Crystals," Chapter 8.
[13] F. C. Frank, "On the Theory of Liquid Crystal," *Faraday Soc. Disc.*, **25**, p. 19 (1958).
[14] T. W. Stinson and J. D. Litster, "Pretransitional Phenomena in the Isotropic Phase of a Nematic Liquid Crystal," *Phys. Rev. Lett.*, **25**, p. 503 (1970).
[15] T. W. Stinson and J. D. Litster, "Correlation Range of Fluctuations of Short-Range Order in the Isotropic Phase of a Liquid Crystal," *Phys. Rev. Lett.*, **30**, p. 688 (1973).
[16] The cholesteric liquid crystal state consists of either all left-handed helices or all right-handed helices. The physical difference between these two states is manifested in their optical activities. See, for example, H. de Vries, "Rotatory Power and Other Optical Properties of Certain Liquid Crystals," *Acta. Cryst.*, **4**, p. 219 (1951).

[17] J. Frenkel, *Kinetic Theory of Liquids,* Oxford University Press, London, Chapter VII (1946).
[18] Ref. 9, p. 473.
[19] D. Langevin and M. A. Bouchiat, "Molecular Order and Surface Tension for the Nematic-Isotropic Interface of MBBA, Deduced from Light Reflectivity and Light Scattering Measurements," *Mol. Cryst. Liq. Cryst.,* **22,** p. 317 (1973).
[20] L. D. Landau and E. M. Lifshitz, *The Classical Theory of Fields,* Addison-Wesley, Reading, Mass., p. 200 (1962).
[21] T. W. Stinson, J. D. Litster and N. A. Clark, "Static and Dynamic Behavior Near the Order-Disorder Transition of Nematic Liquid Crystals," *J. Phys. (Paris),* Suppl. **33,** Cl-69 (1972).
[22] C. C. Yang, "Light Scattering Study of the Dynamical Behavior of Ordering Just Above the Phase Transition to a Cholesteric Liquid Crystal," *Phys. Rev. Lett.,* **28,** p. 955 (1972).

Introduction to the Optical Properties of Cholesteric and Chiral Nematic Liquid Crystals

E. B. Priestley

RCA Laboratories, Princeton, N. J. 08540

1. Introduction

The results derived here, and the related discussion, apply both to cholesteric and to chiral nematic liquid crystals; however, in the interest of brevity, we refer specifically only to cholesteric liquid crystals.

The helical arrangement of the molecules in a cholesteric phase has been described in an earlier chapter.[1] On a sufficiently microscopic scale one cannot distinguish between cholesteric and nematic ordering. However, as we consider larger and larger volumes of the two types of material, a difference in the molecular ordering begins to become apparent; we observe that the cholesteric director $\hat{n}$ follows a helix

$$\begin{aligned}(\hat{n})_x &= \cos(q_0 z + \varphi) \\ (\hat{n})_y &= \sin(q_0 z + \varphi) \\ (\hat{n})_z &= 0\end{aligned} \biggr\}, \qquad [1]$$

as shown in Fig. 1, whereas this secondary, helical structure is absent in the nematic phase. In general, both the direction of the helix axis z in space and the magnitude of the constant φ are arbitrary. It is evident from Fig. 1 that the structure of a cholesteric liquid crystal is pe-

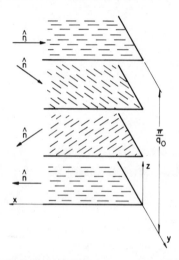

Fig. 1—A schematic representation of the helical arrangement of the constituent molecules in a cholesteric liquid crystal.

riodic with a spatial period

$$L = \pi / |q_0|. \qquad [2]$$

In a right-handed coordinate system, Eq. [1] describes a right-handed helix for positive q_0 and a left-handed helix for negative q_0. Thus the sign of q_0 determines the sense of the helix and its magnitude determines the spatial periodicity.

Several unique optical properties arise from this spatially periodic, helical structure of cholesteric liquid crystals.

(1) Bragg reflection of light beams is observed. For light incident parallel to the helix axis z, only the lowest-order reflection is al-

lowed; for oblique incidence the higher-order reflections also become allowed.
(2) The Bragg reflected light is circularly polarized if the incident wave propagates parallel to the z-axis and elliptically polarized for oblique incidence.
(3) Only the component of optical polarization for which the instantaneous spatial electric field pattern matches the spiraling cholesteric director is strongly reflected. The other component is transmitted with no significant reflection loss.
(4) Very strong rotatory power is observed. Rotations of tens of revolutions per millimeter are typical, compared to the fraction of a revolution per millimeter characteristic of isotropic, optically active liquids.

A detailed treatment of the optical properties of cholesteric liquid crystals for obliquely incident light beams involves extensive numerical calculations[2] and is outside the scope of this chapter. However, much of the physics underlying the observed optical behavior of these spiral structures can be understood by considering the more restricted case in which the wave vector of the incident light is everywhere normal to the local director, i.e., $\mathbf{k} \| z$. The development presented below parallels closely that of de Vries;[3] however, the approach is somewhat different.[4]

2. Maxwell's Equations

As noted above we treat only waves propagating along the helix axis z; $\mathbf{D}$ and $\mathbf{E}$ are therefore confined to the xy-plane and are related by a two-dimensional, second-rank dielectric tensor $\overset{\leftrightarrow}{\epsilon}$. In addition to this restriction, we neglect (1) the weak intrinsic optical activity of the constituent molecules which persists even in the isotropic phase, (2) energy dissipation by absorption, and (3) magnetic permeability ($\mu = 1$). Finally, we assume all waves to be of the form

$$\mathrm{Re}\{f(z)\exp(-i\omega t)\} \qquad [3]$$

(Re $\equiv$ real part) so that $\nabla^2 \to \partial^2/\partial z^2$ and $\partial^2/\partial t^2 \to -\omega^2$. With these assumptions, Maxwell's equations reduce to

$$\frac{\partial^2 \mathbf{E}}{\partial z^2} + \left(\frac{\omega}{c}\right)^2 \overset{\leftrightarrow}{\epsilon} \mathbf{E} = 0. \qquad [4]$$

One could at this point proceed to solve Eq. [4] with $\overset{\leftrightarrow}{\epsilon}$ suitably expressed as a periodic function of z in the fixed laboratory frame of reference. Then, by Floquet's theorem,[6] we know that there exist solutions to Eq. [4] such that

$$E(z + L) = K E(z) \qquad [5]$$

where K is a constant that may be complex. However, it is somewhat simpler to solve Eq. [4] if we first transform it to a coordinate system that rotates with the cholesteric helix. In the rotating frame $\bar{\epsilon}$ has a simple diagonal form for all values of z. In making this transformation we will utilize the Pauli matrices and we therefore digress briefly to consider the properties of these matrices.

Pauli Matrices

$$\sigma_1 = \begin{pmatrix} 0 & 1 \\ 1 & 0 \end{pmatrix}, \; \sigma_2 = \begin{pmatrix} 0 & -i \\ i & 0 \end{pmatrix}, \; \sigma_3 = \begin{pmatrix} 1 & 0 \\ 0 & -1 \end{pmatrix}$$

are known as the Pauli matrices. Their principal properties, which can be deduced from their explicit form, are summarized by

$$\sigma_j \sigma_k = i \epsilon_{jkl} \sigma_l + \delta_{jk} \sigma_0 \qquad [6]$$

where j, k and l can independently take on the values 1, 2 or 3. ϵ_{jkl} in Eq. [6] is the Levi–Cevita antisymmetric symbol,[6] which behaves as follows:

$$\epsilon_{jkl} = \begin{cases} 1 & \text{if } jkl = 123, 231, 312 \\ -1 & \text{if } jkl = 213, 321, 132 \\ 0 & \text{otherwise.} \end{cases}$$

σ_0 is the 2 × 2 unit matrix. Inspection of Eq. [6] reveals that the Pauli matrices anticommute, i.e., $\sigma_j \sigma_k + \sigma_k \sigma_j = 0, j \neq k$.

In transforming Eq. [4] to the rotating frame of reference we make use of the exponential operator, $\exp(-i \sigma_j \theta)$. Its properties can best be seen from a series expansion of the exponential, viz,

$$\exp(-i\sigma_j \theta) = \sigma_0 - i\sigma_j \theta + \frac{1}{2!}(i\sigma_j \theta)^2 - \frac{1}{3!}(i\sigma_j \theta)^3 + \frac{1}{4!}(i\sigma_j \theta)^4 + \ldots \qquad [7]$$

Regrouping the terms in this expansion, and bearing in mind the properties of the Pauli matrices summarized in Eq. [6], we obtain

$$\exp(-i\sigma_j\theta) = \sigma_0\left(1 - \frac{\theta^2}{2!} + \frac{\theta^4}{4!} - + \ldots\right)$$
$$- i\sigma_j\left(\theta - \frac{\theta^3}{3!} + \frac{\theta^5}{5!} - + \ldots\right), \qquad [8]$$

where the two power series in parentheses will be recognized as expansions of $\cos\theta$ and $\sin\theta$. Thus

$$\exp(-i\sigma_j\theta) = \sigma_0\cos\theta - i\sigma_j\sin\theta. \qquad [9]$$

The reader can check that for $j = 2$, Eq. [9] becomes

$$\exp(-i\sigma_2\theta) = \begin{pmatrix} \cos\theta & -\sin\theta \\ \sin\theta & \cos\theta \end{pmatrix}, \qquad [10]$$

which is the rotation matrix for a vector in a plane. The utility of Eq. [10] in our present problem results from the ease and compactness with which the coordinate transformation can be made using the exponential operator notation. Notice that the z-dependence is contained in the relationship

$$\theta = 2\pi z/P, \qquad [11]$$

where $P = 2L$ is the pitch of the cholesteric structure.

As mentioned previously, $\overleftrightarrow{\epsilon}$ can be written in diagonal form in the rotating coordinate system. Letting $\epsilon_\parallel$ and $\epsilon_\perp$ represent the dielectric constant parallel and perpendicular to the local director $\hat{n}$, respectively, it follows that

$$\overleftrightarrow{\epsilon} = \begin{pmatrix} \epsilon_0 + \epsilon_1 & 0 \\ 0 & \epsilon_0 - \epsilon_1 \end{pmatrix},$$

which, in terms of σ_0 and σ_3, is simply

$$\overleftrightarrow{\epsilon} = \epsilon_0\sigma_0 + \epsilon_1\sigma_3. \qquad [12]$$

$\epsilon_0 = (\epsilon_\parallel + \epsilon_\perp)/2$ is the mean dielectric constant, and $\epsilon_1 = (\epsilon_\parallel - \epsilon_\perp)/2$ is a measure of the dielectric anisotropy, in the xy plane.

The wave equation in the rotating frame of reference is (see Appendix A)

$$\left(\frac{\partial^2 E'}{\partial z^2}\right) - \left(\frac{4\pi i \sigma_2}{P}\right)\left(\frac{\partial E'}{\partial z}\right) + \left[\left(\frac{\omega}{c}\right)^2(\epsilon_0 \sigma_0 + \epsilon_1 \sigma_3) - \left(\frac{4\pi^2 \sigma_0}{P^2}\right)\right] E' = 0 \quad [13]$$

where the x and y axes of the rotating coordinate system have been fixed parallel and perpendicular to $\hat{n}$, respectively. We try as a solution to Eq. [13]

$$E' = \text{Re}\left\{\hat{u} E_0 \exp\left[-i\left(\omega t - \frac{2\pi m \epsilon_0^{1/2} z}{\lambda}\right)\right]\right\}, \quad [14]$$

where

$$\hat{u} = \begin{pmatrix} \mu_1 \\ \mu_2 \end{pmatrix} \quad [15]$$

is a (complex) two-component vector that describes the state of polarization of the wave in the rotating frame, E_0 is a real constant that gives the amplitude of the wave, and λ is the wavelength in *vacuo* of the wave. The product $m \epsilon_0^{1/2}$ plays the role of the refractive index of the cholesteric material; however it is not strictly correct to think of it as such. As we shall see later, the m values are complicated functions of the pitch and dielectric anisotropy of the medium and of the wavelength of the electromagnetic wave. ϵ_0 in Eq. [12] is an average (optical frequency) dielectric constant defined above. Eq. [13] then reduces to

$$\left[-\frac{4\pi^2 c^2}{\omega^2 \epsilon_0}\left(\frac{\sigma_0}{P^2} - \frac{2\sigma_2 m \epsilon_0^{1/2}}{P\lambda} + \frac{m^2 \epsilon_0 \sigma_0}{\lambda^2}\right) + \sigma_0 + \left(\frac{\epsilon_1}{\epsilon_0}\right)\sigma_3\right]\hat{\mu} = 0. \quad [16]$$

Notice that $4\pi^2 c^2/\omega^2$ is just the square of the vacuum wavelength λ. Defining $\alpha = \epsilon_1/\epsilon_0$ and $\lambda' = \lambda/\epsilon_0^{1/2} P$ simplifies Eq. [16] to

$$[(1 - (\lambda')^2 - m^2)\sigma_0 + \alpha \sigma_3 + 2m\lambda' \sigma_2]\hat{\mu} = 0, \quad [17]$$

which can be written explicitly as

$$\begin{pmatrix} 1 - (\lambda')^2 - m^2 + \alpha & -2im\lambda' \\ 2im\lambda' & 1 - (\lambda')^2 - m^2 - \alpha \end{pmatrix}\begin{pmatrix} \mu_1 \\ \mu_2 \end{pmatrix} = 0. \quad [18]$$

The two simultaneous equations represented by Eq. [18] have a nontrivial solution only if

$$\begin{vmatrix} 1 - (\lambda')^2 - m^2 + \alpha & -2im\lambda' \\ 2im\lambda' & 1 - (\lambda')^2 - m^2 - \alpha \end{vmatrix} = 0. \qquad [19]$$

The resulting fourth-order equation in m

$$m^4 - 2[1 + (\lambda')^2]m^2 + [1 - (\lambda')^2]^2 - \alpha^2 = 0 \qquad [20]$$

has solutions

$$m^2 = 1 + (\lambda')^2 \pm [4(\lambda')^2 + \alpha^2]^{1/2}. \qquad [21]$$

It will be useful to have more explicit expressions for the roots given by Eq. [21] in two limiting cases. In both limits we will expand Eq. [21] and keep only the lowest-order terms.

(a) The $4(\lambda')^2/\alpha^2 \ll 1$ Limit

In this limit the pitch is large compared to the wavelength of the light. The square root in Eq. [21] can be expanded to give

$$m^2 = 1 + (\lambda')^2 \pm \left(\alpha + \frac{2(\lambda')^2}{\alpha} + \ldots \right) \qquad [22]$$

whence

$$m_1 = -m_3 = (1 - \alpha)^{1/2} + \frac{(\lambda')^2(\alpha - 2)}{2\alpha(1 - \alpha)^{1/2}} + \ldots \qquad [23]$$

and

$$m_2 = -m_4 = (1 + \alpha)^{1/2} + \frac{(\lambda')^2(\alpha + 2)}{2\alpha(1 + \alpha)^{1/2}} + \ldots \qquad [24]$$

The positive and negative roots are associated with waves traveling through the cholesteric medium in the positive and negative z directions, respectively.

(b) The $4(\lambda')^2/\alpha^2 \gg 1$ Limit

In this case the pitch is small compared to the wavelength of the light and the appropriate expansion of Eq. [21] is

$$m^2 = 1 + (\lambda')^2 \pm \left(2\lambda' + \frac{\alpha^2}{4\lambda'} + \ldots \right), \qquad [25]$$

from which it follows that

$$m_1 = -m_3 = 1 - \lambda' - \frac{\alpha^2}{8\lambda'(1 - \lambda')} + \ldots \qquad [26]$$

and
$$m_2 = -m_4 = 1 + \lambda' + \frac{\alpha^2}{8\lambda'(1 + \lambda')} + \ldots \quad [27]$$

Again the plus and minus signs correspond to waves traveling in opposite directions along the helix axis.

The results we have obtained above are identical to those given by de Vries.[3]

Now that we have explicit expressions for the roots m, it is simple to determine the polarization vectors $\hat{\mu}$ of the corresponding modes. We begin by rewriting Eq. [18] schematically as

$$\begin{pmatrix} a + \alpha & -ib \\ ib & a - \alpha \end{pmatrix} \begin{pmatrix} \mu_1 \\ \mu_2 \end{pmatrix} = 0, \quad [28]$$

where $a = 1-(\lambda')^2-m^2$ and $b = 2m\lambda'$. Expanding Eq. [28], we have

$$(a + \alpha)\mu_1 - ib\mu_2 = 0, \quad [29]$$

and

$$ib\mu_1 + (a - \alpha)\mu_2 = 0. \quad [30]$$

Eqs. [29] and [30] require

$$\mu_2\mu_2^* = \left(\frac{a + \alpha}{a - \alpha}\right)\mu_1\mu_1^*, \quad [31]$$

which, when combined with the normalization condition,

$$\mu_1\mu_1^* + \mu_2\mu_2^* = 1,$$

leads to

$$\mu_1\mu_1^* = \left(\frac{a - \alpha}{2a}\right), \quad [32]$$

and hence

$$\mu_2\mu_2^* = \left(\frac{a + \alpha}{2a}\right). \quad [33]$$

The relative phase of μ_1 and μ_2 is fixed by Eq. [29], viz

$$\frac{\mu_2}{\mu_1} = -i\left(\frac{a + \alpha}{b}\right). \quad [34]$$

Substituting for a and b in Eqs. [32] through [34] yields the following

OPTICAL PROPERTIES

expressions which, together with the appropriate m values, determine the polarization vectors $\hat{\mu}$:

$$\mu_1\mu_1^* = \frac{1 - (\lambda')^2 - m^2 - \alpha}{2[1 - (\lambda')^2 - m^2]}, \qquad [35]$$

$$\mu_2\mu_2^* = \frac{1 - (\lambda')^2 - m^2 + \alpha}{2[1 - (\lambda')^2 - m^2]}, \qquad [36]$$

$$\frac{\mu_2}{\mu_1} = -i\left(\frac{1 - (\lambda')^2 - m^2 + \alpha}{2m\lambda'}\right). \qquad [37]$$

For the case $4(\lambda')^2/\alpha^2 \ll 1$, we have

$$\hat{\mu}(m_1) = \hat{\mu}(m_3) \approx \begin{pmatrix} 0 \\ 1 \end{pmatrix} \qquad [38]$$

$$\hat{\mu}(m_2) = \hat{\mu}(m_4) \approx \begin{pmatrix} 1 \\ 0 \end{pmatrix}. \qquad [39]$$

These vectors describe linearly polarized light. Thus, the modes are essentially linearly polarized near the origin of the λ' axis in Fig. 2. In the other extreme, $4(\lambda')^2/\alpha^2 \gg 1$, the eigenvectors are

$$\hat{\mu}(m_1) = \hat{\mu}(m_4) \approx \frac{1}{\sqrt{2}}\begin{pmatrix} 1 \\ -i \end{pmatrix} \qquad [40]$$

$$\hat{\mu}(m_2) = \hat{\mu}(m_3) \approx \frac{1}{\sqrt{2}}\begin{pmatrix} 1 \\ i \end{pmatrix}, \qquad [41]$$

which describe circularly polarized light. Thus the modes are essentially circularly polarized for large values of λ'.

3. Discussion

Eq. [21] has been evaluated numerically with α arbitrarily set equal to 0.1. The results for m_1 and m_2 are plotted in Fig. 2. The qualitative features of Fig. 2 are apparent from the approximate expressions for the roots, Eqs. [23], [24], [26], and [27]. For example, in the limit of very small λ', Eqs. [23] and [24] reduce to

$$m_1 = (1 - \alpha)^{1/2} \text{ and } m_2 = (1 + \alpha)^{1/2}.$$

These equations also show that the initial dependence on λ' is quadratic. For large values of λ', on the other hand, Eqs. [26] and [27] become linear in λ', as observed in Fig. 2.

$$m_1 = 1 - \lambda' \text{ and } m_2 = 1 + \lambda'.$$

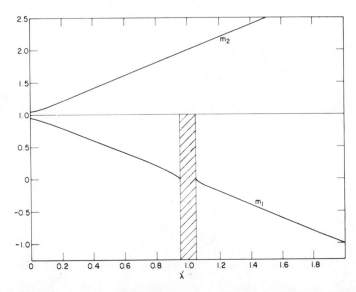

Fig. 2—Solutions given by Eq. [21], with $\alpha = 0.1$ as a function of reduced wavelength λ' for the two waves propagating in the positive z direction. In the shaded region the m_1 wave is strongly reflected.

We have seen by Eqs. [38] and [39] that the normal waves are linearly polarized in the local frame of reference for $\lambda' \to 0$. As the local frame rotates ($\theta = 2\pi z/P$), so do the polarization vectors of the normal waves. This is the "waveguide" regime discussed by de Gennes[7] and is also the regime in which twisted nematic field-effect devices[8] operate (values typical of such devices are $P \sim 50$ μm, $\epsilon_0^{1/2} \sim 1.65$ and $\lambda \sim 0.5$ μm, leading to a value of $\lambda' \sim 0.006$).

For values of λ' near unity, i.e., within the shaded region of Fig. 2, m_1 and m_3 are imaginary and the corresponding waves are nonpropagating. It is apparent from Eq. [21] that this reflection band extends over the range of λ'

$$(1 - \alpha)^{1/2} < \lambda' < (1 + \alpha)^{1/2}.$$

The m_2 and m_4 waves are unaffected and are observed to propagate freely for all values of λ'. Considering now only the waves traveling in the plus z- direction, we see from Eq. [41] that the m_2 wave is left circularly polarized and that its instantaneous electric field pattern is of opposite sense to the (right-handed) cholesteric helix (see Appendix B). The strongly reflected m_1 wave, on the other hand, is right circularly polarized and has an instantaneous electric field pattern that is superposable on the cholesteric helix. Thus, for $\lambda' \sim 1$, a right-handed cholesteric liquid crystal reflects right circularly polarized light and transmits left circularly polarized light. The reverse is true for a left-handed cholesteric material.

For larger values of λ', the normal waves are nearly circularly polarized (see Fig. 3), have opposite signs of rotation, and propagate

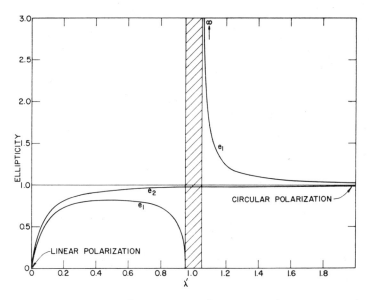

Fig. 3—The ellipticities $e_1 = |\mu_1(m_1)/\mu_2(m_1)|$ and $e_2 = -|\mu_2(m_2)/\mu_1(m_2)|$ as a function of reduced wavelength λ' for the two waves propagating in the positive z direction.

with different phase velocities. A superposition of two waves of opposite circular polarization can be thought of as a linearly polarized wave whose plane of polarization lies along the bisector of the instantaneous angle between the two rotating vectors. Since the two circularly polarized waves travel at different velocities in the cholesteric medium, there is a net rotation of the plane of polarization when a linearly polarized wave is passed through a slab of the cholesteric material. We can estimate the magnitude of this rotation in a straightforward manner using the results derived above.

Again we consider only the m_1 and m_2 waves and we restrict ourselves to the $4(\lambda')^2/\alpha^2 \gg 1$ limit. We have seen that in this limit the m_1 wave has the form

$$\mathbf{E}_1' = \operatorname{Re}\left\{\frac{1}{\sqrt{2}}\begin{pmatrix}1\\-i\end{pmatrix}E_0 \exp\left[-i\left(\omega t - \frac{2\pi m_1 \epsilon_0^{1/2} z}{\lambda}\right)\right]\right\} \quad [42]$$

in the rotating coordinate system, with components

$$(E_1')_x = \frac{E_0}{\sqrt{2}} \cos\left(\omega t - \frac{2\pi m_1 \epsilon_0^{1/2} z}{\lambda}\right) \quad [43]$$

and

$$(E_1')_y = -\frac{E_0}{\sqrt{2}} \sin\left(\omega t - \frac{2\pi m_1 \epsilon_0^{1/2} z}{\lambda}\right). \quad [44]$$

Thus, by Eq. [52], we can write $(E_1)_x$ in the laboratory frame of reference as

$$(E_1)_x = (E_1')_x \cos\left(\frac{2\pi z}{P}\right) - (E_1')_y \sin\left(\frac{2\pi z}{P}\right), \quad [45]$$

which, upon substitution of Eqs. [43] and [44], reduces immediately to

$$(E_1)_x = \frac{E_0}{\sqrt{2}} \cos\left[\frac{2\pi z}{P} - \left(\omega t - \frac{2\pi m_1 \epsilon_0^{1/2} z}{\lambda}\right)\right] \quad [46]$$

From Eq. [46] we see that the phase angle θ_1 of the m_1 wave is

$$\theta_1 = \frac{2\pi z}{P}\left(1 + \frac{m_1}{\lambda'}\right) - \omega t \quad [47]$$

in the laboratory frame of reference. Similarly we find for the m_2 wave

$$\theta_2 = \frac{2\pi z}{P}\left(1 - \frac{m_2}{\lambda'}\right) + \omega t. \quad [48]$$

The angular position of the resultant linear polarization vector is determined by

$$\psi = \frac{1}{2}(\theta_1 + \theta_2), \qquad [49]$$

which is the bisector of the instantaneous angle between the two rotating electric vectors. Combining Eqs. [47], [48] and [49], and substituting the results of Eqs. [26] and [27], we find

$$\frac{d\psi}{dz} = -\frac{2\pi}{P} \frac{\alpha^2}{8(\lambda')^2[1-(\lambda')^2]} \qquad [50]$$

for the rotation per unit length. Taking $\alpha = 0.1$, $p = 0.5$ μm, $\epsilon_0^{1/2} = 1.65$ and $\lambda = 1$ μm, $d\psi/dz$ is calculated to be approximately 36 revolutions cm^{-1}.

Finally, we plot in Fig. 3 the ellipticities $e_1 = |\mu_1(m_1)/\mu_2(m_1)|$ and $e_2 = -|\mu_2(m_2)/\mu_1(m_2)|$ of the waves corresponding to m_1 and m_2 as a function of reduced wavelength λ'. These have been defined so that linear polarization is associated with a value of zero; one could equally well have chosen to plot the reciprocal relations in which case linear polarization would correspond to infinite ellipticity. With either definition, the ellipticity of circularly polarized waves is unity. It is apparent that the ellipticity of the m_1 wave is anomalous in the region of the reflection band whereas the m_2 wave is "well behaved" for all values of λ'.

4. Conclusion

The optical properties of cholesteric liquid crystals have been examined in the restricted case of waves propagating parallel to the helix axis. Solutions to the wave equation were obtained in order to determine the polarization states of the normal waves in the medium. It was found that the waves corresponding to two of the four solutions are strongly reflected when the wavelength becomes comparable to the helix pitch. Closer examination revealed that the instantaneous electric field pattern of these waves is superposable on the cholesteric helix. That is, the instantaneous electric field pattern of the reflected waves is a right-handed helix that, when the wavelength is correct, matches exactly the right-handed cholesteric helix. An expression was derived for the optical rotatory power (rotation per unit length), that correctly accounts for the sign and magnitude of observed optical rotations in cholesteric liquid crystals.

Acknowledgment

I am indebted to P. Sheng for helpful discussions.

Appendix A

In this appendix we transform the wave equation to a coordinate system that rotates with the cholesteric helix described by Eq. [1]. Let q_0 be positive so the resulting helix is right handed. Then, an arbitrary electric field vector $\mathbf{E}$ with components E_x and E_y in the laboratory frame of reference has components $E_x{'}$ and $E_y{'}$

$$\left.\begin{array}{l} E_x{'} = E_x\cos\theta + E_y\sin\theta \\ E_y{'} = -E_x\sin\theta + E_y\cos\theta \end{array}\right\} \qquad [51]$$

in the $z = P\theta/2\pi$ plane of the rotating coordinate system. Thus, by Eq. [9], we can write

$$\left.\begin{array}{l} \mathbf{E} = \exp(-i\sigma_2\theta)\mathbf{E'} \\ \mathbf{D} = \exp(-i\sigma_2\theta)\mathbf{D'} \end{array}\right\}. \qquad [52]$$

and

Also, in the rotating frame of reference, we know that

$$\mathbf{D'} = \overleftrightarrow{\epsilon}_{\text{local}}\mathbf{E'}, \qquad [53]$$

where $\overleftrightarrow{\epsilon}_{\text{local}}$ has the simple diagonal form given by Eq. [12]. Hence the wave equation

$$\frac{\partial^2 \mathbf{E}}{\partial z^2} + \left(\frac{\omega}{c}\right)^2 \mathbf{D} = 0 \qquad [54]$$

in the laboratory frame of reference becomes

$$\frac{\partial^2 \mathbf{E'}}{\partial z^2} - \left(\frac{4\pi i\sigma_2}{P}\right)\frac{\partial \mathbf{E'}}{\partial z} + \left[\left(\frac{\omega}{c}\right)^2(\epsilon_0\sigma_0 + \epsilon_1\sigma_3) - \frac{4\pi^2\sigma_0}{P^2}\right]\mathbf{E'} = 0 \qquad [55]$$

in the rotating frame of reference.

Appendix B

Here we consider the spatial and temporal variation of the electric vector for a circularly polarized light wave. We begin with a plane wave propagating in the positive z direction

$$\mathbf{E}(t,z) = \text{Re}\left\{\hat{\mu}E_0\exp\left[-i\omega\left(t - \frac{z}{c}\right)\right]\right\} \qquad [56]$$

where Re ≡ real part, $\hat{\mu}$ is a vector describing the polarization state of the wave, and E_0 is the amplitude of the wave. Eq. [56] can be rewritten

$$\mathbf{E}(t,z) = \text{Re}\left\{\hat{\mu} E_0 \left[\cos\left[\omega\left(t - \frac{z}{c}\right)\right] - i\,\sin\left[\omega\left(t - \frac{z}{c}\right)\right]\right]\right\}. \quad [57]$$

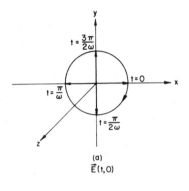

(a)
$\vec{E}(t,0)$

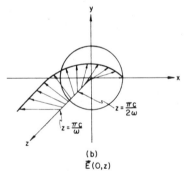

(b)
$\vec{E}(0,z)$

Fig. 4—Behavior of the electric vector for a right circularly polarized light wave: (a) rotation of the electric vector as a function of time t at a fixed position on the z axis; (b) position of the electric vector as a function of z at a fixed instant in time.

Choosing

$$\hat{\mu} = \frac{1}{\sqrt{2}}\begin{pmatrix} 1 \\ -i \end{pmatrix}, \quad [58]$$

it is apparent that

$$E_x(t,z) = \frac{E_0}{\sqrt{2}} \cos\left[\omega\left(t - \frac{z}{c}\right)\right] \quad [59]$$

and

$$E_y(t,z) = -\frac{E_0}{\sqrt{2}} \sin\left[\omega\left(t - \frac{z}{c}\right)\right].$$ [60]

Consider first the temporal development of the wave at some fixed value of z, say $z = 0$ for convenience. Examination of Eqs. [59] and [60] shows that at $t = 0$, the electric vector lies along the x axis while at a later time $t = \pi/2\omega$ it is along the negative y axis and at a still later time $t = \pi/\omega$ it is along the negative x axis, and so on. Thus, as viewed from a point on the positive z axis, looking back toward the origin, the electric vector in the $z = 0$ plane rotates in a clockwise sense from left to right, and by convention this is called a right circularly polarized wave (see Fig. 4a).

Next we consider the spatial distribution of the electric vector along the z axis at one instant of time, e.g., at $t = 0$. We see from Eqs. [59] and [60] that at $z = 0$ the electric vector points along the x axis, as before. As we move along the positive z axis we find the electric vector rotates first to the positive y direction at $z = \pi c/2\omega$ and then to the negative x direction at $z = \pi c/\omega$, etc. Thus the instantaneous electric field pattern traces out a right-handed spiral (see Fig. 4b). All directions are reversed for left circularly polarized light, which is described by the polarization vector

$$\hat{\mu} = \frac{1}{\sqrt{2}}\begin{pmatrix}1\\i\end{pmatrix}.$$ [61]

These properties of circularly polarized light are important in understanding the optical properties of cholesteric liquid crystals.

References

[1] E. B. Priestley, "Liquid Crystal Mesophases," Chapter 1.

[2] See, for example, D. W. Berreman and T. J. Scheffer, "Reflection and Transmission by Single-Domain Cholesteric Liquid Crystal Films: Theory and Verification," *Mol. Cryst. and Liquid Cryst.*, **11**, p. 395 (1970); R. Dreher and G. Meier, "Optical Properties of Cholesteric Liquid Crystals," *Phys. Rev.*, **A8**, p. 1616 (1973).

[3] This problem has been studied in detail by H. De Vries, "Rotatory Power and Other Optical Properties of Certain Liquid Crystals," *Acta Cryst.*, **4**, p. 219 (1951).

[4] The approach presented in this chapter follows that contained in some unpublished lecture notes of P. S. Pershan on the optical properties of cholesteric liquid crystals.

[5] See, for example, A. Messiah, *Quantum Mechanics*, Vol. II, pp. 544–549, John Wiley & Sons, Inc. (1966).

[6] J. Mathews and R. L. Walker, *Mathematical Methods of Physics*, W. A. Benjamin Inc., N. Y. (1965).

[7] P. G. de Gennes, *The Physics of Liquid Crystals*, p. 228, Oxford University Press, Oxford (1974).

[8] L. A. Goodman, "Liquid-Crystal Displays—Electro-optic Effects and Addressing Techniques," Chapter 14.

12

Liquid-Crystal Displays—
Packaging and Surface Treatments

L. A. Goodman

RCA Laboratories, Princeton, N. J. 08540

1. Introduction

For the proper operation of liquid-crystal display devices, appropriate cell construction and surface treatment methods are necessary. In this chapter, first, some of the general packaging techniques that are being used today are outlined. Second, the common conductive coatings are listed and standard preparation methods given. Third, the diverse surface treatment methods are summarized and, finally, some relationships between cell construction and device operation are discussed.

2. Packaging

The standard sandwich cell configuration for liquid-crystal displays is schematically illustrated in Fig. 1. Normally, low cost soda-lime soft glass is used. However, more expensive borosilicate and fused silica substrates can also be utilized. The spacing between top and bottom glass plates varies from 5 to 50 μm with nominal values being in the 10 to 20 μm range. The spacer composition is restricted by possible chemical reactions with the liquid crystal. Glass frits and relatively inert organic materials such as Teflon and Mylar can be used as the spacer materials.

The sealing materials are limited both by the need for compatibility with the liquid crystal and by the need for a fairly hermetic seal

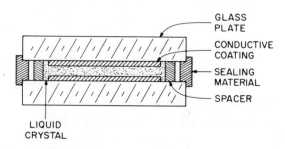

Fig. 1—Side view of schematic representation of a liquid-crystal cell.

since both moisture and oxygen can react with many mesomorphic systems in a deleterious fashion. Other criteria for selecting sealing materials are thermal expansion match with the glass plates, bonding strength, sealing temperature, and manufacturing cost. Glass frits, solder glasses, and polymeric materials are suitable sealing materials.

3. Electrodes

At least one of the conductive coatings in a liquid-crystal display must be transparent. The most common transparent conductive coating material is a mixture of indium oxide and tin oxide. This material can be prepared by several different techniques which include (1) rf sputtering of indium–tin oxide powder targets,[1] (2) dc sputtering of highly conducting indium–tin oxide powder targets,[2] (3) dc reactive sputtering of an indium–tin alloy,[3] (4) thermal evaporation,[4] and (5) high-temperature (400°C) bake-out of spun-on solutions.[5] The sheet resistance of as-deposited indium–tin oxide coatings can vary from 2 ohms/square up to many kilohms per square.[2] After bake-out in air at temperatures in excess of 500°C, the resistance increases by a factor of 3 to 10.[1,2] The sheet resistance for liquid-crystal displays is nominally in the 100 to 500 ohms/square range. The optical transmission of these films is excellent, and some typical results are given in Fig. 2.

Another compound commonly used for conductive coatings is antimony-doped tin oxide, which is usually deposited by a pyrolitic spray process at temperatures between 600 and 700°C. The chemical durability, abrasion resistance, electrical conductivity, and typical transmission are all very adequate. Surface resistances as low as 40 ohms/square can be achieved with a transmission in excess of 75–80%.

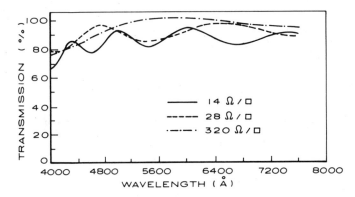

Fig. 2—Transmission versus wavelength for In_2O_3-SnO_2 films deposited in argon to different thicknesses.[1]

The chemically deposited tin oxide suffers from two important disadvantages as compared to the vacuum-deposited indium–tin oxide. At the high temperatures necessary for the spray deposition process, soft glass loses its flatness. The different approaches previously listed for creating the indium–tin oxide films can all be performed at sufficiently low temperature that the soft glass substrates do not warp. In liquid-crystal displays, where a relatively narrow spacing between opposite electrodes is required, the loss in flatness caused by the spray process is a distinct drawback.

Antimony-doped tin oxide can also be deposited by sputtering, but the sheet resistance is not as low as with sputtered indium–tin oxide. In addition, for the tin oxide concentrations normally used, the indium–tin oxide films can be readily etched in hydrochloric acid.[3,6] Antimony-doped tin oxide films are not readily soluble in acids or bases, but they can be etched by a procedure using zinc dust and hydrochloric acid.[7]

Reflective electrodes are readily obtained by the evaporation of metals—for example, aluminum or chromium. Other coatings such as layers of insulating dielectrics are also used to obtain specular surfaces. These can be deposited by both electron-beam and thermal-evaporation methods.[4]

4. Surface Orientation

A unit vector called the director denotes the average orientation of the long molecular axes in any local region of the fluid. For the purpose of maximizing the contrast ratio of the display, it is desirable that the orientation of the director be the same throughout the fluid whether the applied voltage is on or off. Two important examples of maximum ordering are shown in Fig. 3. For perpendicular (or homeo-

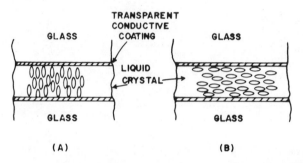

Fig. 3—Side view (A) of homeotropic and (B) of homogeneous orientations.

tropic) alignment, all of the long molecular axes are perpendicular to the cell walls. Looking down through the cell, the fluid appears to be isotropic, and this state is optically clear. When the mesomorphic medium possesses uniform parallel (or homogeneous) orientation, the director is parallel to the cell walls over dimensions of a millimeter or more and points in only one direction. As observed in a top view, the fluid has the optical properties of a uniaxial single crystal with the director describing the main axis (see Fig. 4A). The "head" and the "tail" of the director can be interchanged without changing the observable fluid properties.

There are two other examples of fluid orientation that are closely related to uniform parallel alignment. In the first case, the director also lies parallel to the cell surfaces. However, the director orientation in the plane parallel to the cell walls is not uniform; rather, it changes randomly over dimensions on the order of micrometers. This orientation is known as random parallel alignment (see Fig. 4B).

The second related example is the planar or Grandjean texture of the cholesteric mesophase. In the planar state, the main helix axis is perpendicular to the electrode surfaces of the cell. Consequently, the director is always oriented parallel to the surface with the orientation of the director varying in helical fashion with linear distance along the helix axis (see Figure 4C).

Four states of bulk orientation have been described so far. Because of the long-range ordering forces that operate in liquid crystals, the preceding bulk orientations can be produced by the proper treatment

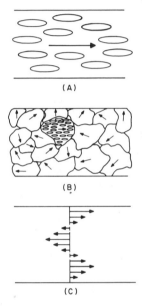

Fig. 4—Top view (A) of homogeneous orientation and (B) of random parallel alignment (the arrows represent the directors) (C) Side view of cholesteric planar texture. Length of director arrows illustrates the amount of twist of each layer.

of the surface region between the liquid crystal and the cell walls. The mechanisms of surface alignment have been poorly understood, but recent investigations seem to be leading toward a better comprehension of them.

In his work on optical textures in mesomorphic materials, Friedel presented an interesting discussion of both the uniform parallel and perpendicular states.[8] However, the techniques he describes for achieving these two states are rather cumbersome. Since that time, a number of investigators have described different methods for obtaining perpendicular alignment. These have included chemical etching,[9,10] coating with lecithin,[11] and physical adsorption of organic surfactant additives such as the polyamide resin Versamid[12] or impurities in the fluid[13] that are the thermal decomposition products of the mesomorphic material.

Petrie et al[14] indicated that additives of certain surface-active molecules such as tertiary amines, quaternary ammonium salts, and pyridinium salts produced perpendicular alignment. Specifically, Haller and Huggins[15] reported that the quaternary ammonium salt, hexadecyltrimethylammoniumbromide (HTAB), caused perpendicular alignment.

Zocher[16] stated that uniform parallel alignment could be obtained if the cell surfaces were first unidirectionally rubbed. Various materials such as paper, tissue, and cotton wool were used to achieve uniform parallel alignment. Chatelain[17] hypothesized that the parallel orientation resulted from the forces generated by the presence of an adsorbed layer of fatty contaminants on the cell surface, and the directionality was achieved by the unidirectional rubbing. However, he could not eliminate the possibility that mechanical deformation of surface might have induced the observed alignment.

Several recent articles support the theory that physicochemical forces, e.g., van der Waals, hydrogen bonding, and dipolar forces, are dominant. Proust et al[18] have obtained both uniform parallel and perpendicular alignment by depositing monolayers of hexadecyltrimethylammoniumbromide (HTAB) from aqueous solution onto glass slides prior to the insertion of the liquid crystal between the slides. They stated that perpendicular alignment was obtained when the bromide compound was densely packed in the monolayer; consequently, the molecules were oriented normal to the glass slide. Uniform parallel alignment was observed when the monolayer was less densely packed and the molecules were oriented parallel to the surface. They achieved the directionality necessary for uniform parallel orientation by withdrawing the glass slides from the bromide-containing solution in one direction.

Kahn[19] has also demonstrated the importance of physicochemical forces. He has prepared highly stable surface aligning conditions by the utilization of silane coupling agents that were chemically bonded to the glass surface. The preparation of the organosilane layer was affected by the nature of the metal oxide surface, the degree of surface hydration, and the pH of the solutions used for deposition. The alkoxysilane monomers of the general type $RSiX_3$ were found to be quite useful for aligning liquid crystals. R is an organofunctional orienting group and X designates a hydrolyzable group attached to the silicon. Schematic illustrations of the cured coatings and the chemical formulas for the silane agents are given in Fig. 5. The DMOAP coating resulted in perpendicular alignment of the liquid crystal; parallel alignment of the mesomorphic molecules was produced by the cured MAP layer.

Berreman[20,21] calculated the difference in elastic strain energy that occurs when a nematic fluid lies parallel to grooves in a surface rather than perpendicular to them. As a consequence of his calculations and experimental observations made by himself and Dryer,[11] Berreman has suggested that the anisotropy in elastic strain energy of the liquid crystal is sufficient to induce the alignment of the fluid director parallel to the grooves in the surface. Furthermore, he concluded that

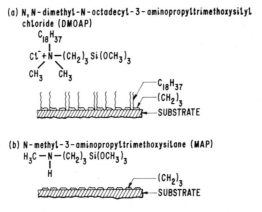

Fig. 5—Chemical formulae for two silane coupling agents and their geometric relationship to a substrate when cured.[19]

the elastic energy considerations explain the tendency of some nematic molecules to align perpendicular to a surface that is rough in two dimensions even if physicochemical forces would normally cause the molecules to lie parallel to a flat surface of the same solid.

Creagh and Kmetz[22,23] have recently suggested an explanation of surface orientation forces that is a synthesis of both the physicochemical and geometric factor hypotheses. Using chemically cleaned tin oxide coated glass substrates, they investigated a number of different surface aligning conditions. Thorough cleaning with chromic acid was required because they found that, in the absence of the cleaning, trace remnants of carbon on the glass surface caused nematic fluids such as MBBA and PAA (p-azoxyanisole) to align with the long axes of the molecules parallel to the surface. With the appropriate cleaning, these materials aligned in the perpendicular state.

The data presented in Fig. 6 summarizes the results of their tests on surface alignment. Cells were prepared with different aligning

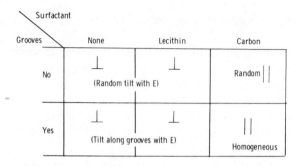

Fig. 6—Matrix of alignment results for various grooving and surfactant conditions.[23]

agents and either with or without grooves. The grooves were obtained by rubbing the surfaces with a diamond paste. Fig. 7 is a typical electron micrograph of a rubbed surface. The tabulated results show that the evaporated carbon layer gave parallel orientation with the grooves providing the unidirectionality necessary for uniform parallel alignment.

With both lecithin-coated and chemically cleaned surfaces, rub-

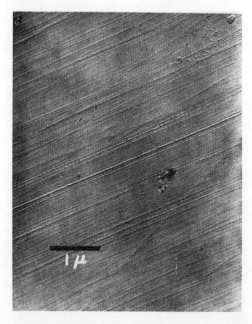

Fig. 7—Scanning electron micrograph of a substrate grooved with 1-μm diamond paste and then cleaned.[23]

bing did not result in uniform parallel alignment. In both situations, perpendicular orientation occurred for no applied voltage. When a voltage was applied to the fluid, either one of two configurations appeared. With no grooving, the nematic fluid adopted the random parallel condition, whereas when the surfaces were rubbed with the diamond paste, the nematic director adopted a uniform orientation in the direction of rubbing.

As a result of their experimental observations, Creagh and Kmetz claimed that the determination of whether a liquid crystal adopts the perpendicular or parallel orientation can be made on the basis of the relative surface energy of the substrate and the surface tension of the fluid. They asserted that when the surface energy of the substrate is low compared to that of the liquid crystal the fluid does not wet the surface, and intermolecular forces within the fluid produce the perpendicular texture. However, if the relative surface energy of the solid is high, the fluid wets the substrate, and the long axes of the molecules align parallel to the surface. They also quoted the results of Proust et al[18] in support of their theory. Apparently, the surface energy is lower when the hexadecyltrimethylammoniumbromide molecules are densely packed on the surface than when they are diffusely packed.

Their observation that the nematic director was normal to thoroughly cleaned surfaces is somewhat at variance with the above conclusions, since one would expect the solid surface to possess a high surface free energy, but they have given an explanation. The apparent discrepancy was presumably caused by the presence of a thin layer of water on the surface. They cited the results of Shafrin and Zisman,[24] namely, that at normal relative humidity, the critical surface tension of glass at 20°C is about 30 dynes/cm for nonhydrophilic liquids. This surface water can only be removed by prolonged heating. In further support of their theory, Creagh and Kmetz were able to achieve random parallel alignment on flame-fired platinum substrates. Fig. 8 is a summary of their results on the effect of surface energy on orientation.

In the model of Creagh and Kmetz, the main effect of grooving appears to have been the provision of a preferred direction for the molecules in accord with calculations made by Berreman[20,21] and Grabmeier et al.[25,26] The physicochemical forces determined whether the molecules were parallel or perpendicular to the substrate surface. They concluded that Chatelain's hypothesis was correct—that the conventional rubbing technique provided uniform parallel alignment by grooving an organic surface layer produced by the rubbing medium, be it cloth or paper.

Kahn, Taylor, and Schonhorn[27] have expanded upon the ideas discussed by Creagh and Kmetz.[22,23] In essence, Kahn et al agreed with the concepts propounded by Creagh and Kmetz with regard to the relative importance of the physicochemical and elastic forces in the

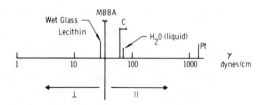

Fig. 8—Alignment of MBBA on substrates with different surface energies.[23]

fluid. However, they did not describe the physicochemical interaction between the liquid and the solid in terms of the liquid surface tension, γ_L, and the surface tension of the solid, γ_S. Rather they characterized the solid by its critical surface tension, γ_c, and the liquid by its surface tension, γ_L. The critical surface tension of a solid is an empirically determined quantity first proposed by Zisman and associates[28] to classify low-energy solid surfaces.

As explained by Adamson,[29] γ_c is not a fundamental property of the solid surface alone, but also depends upon the nature of fluid in contact with the solid. Apparently, it is only a fixed quantity for a given homologous series of organic liquids on the solid. γ_c values can vary somewhat from one homologous series of liquids to another.

Kahn et al,[27] also noted that the γ_c of a solid, in particular one of high surface energy, can be drastically lowered by the presence of a layer of liquid molecules that has been preferentially adsorbed at the solid–liquid interface. If the molecules in the monolayer have both polar and nonpolar ends, the polar ends will often attach themselves to the high-energy surface, with the nonpolar portion facing into the liquid. As far as the bulk liquid is concerned, the nonpolar groups present a lower surface energy to the fluid than the free solid surface, and consequently the effective γ_c is lowered. The adsorbed monolayer can consist of intentionally added impurities in the fluid[12–14] or unintentional impurities. In addition, for certain impurity-free liquids, the molecular structure may be such that the fluid cannot spread on its own monolayer. These are known as "autophobic" liquids.[28]

Proust and Ter-Minassian-Saraga[30] have measured the contact angle ϕ and calculated the work of adhesion, $W_A = \gamma_L (1 + \cos \phi)$, of MBBA to glass surfaces coated with a monolayer of HTAB. As previously described,[18] the MBBA had either uniform parallel or perpendicular orientation depending upon the surface density of the HTAB. They have found that the contact angle associated with parallel alignment was 32° instead of 0°. Also, in spite of the fact that the alignment changed sharply from parallel to perpendicular due to a small increase in the surface density of HTAB, the contact angle only increased slightly—to 37°. The work of adhesion gradually decreased during the transition from parallel to perpendicular orientation. The general criterion enunciated by Creagh and Kmetz[22,23] that parallel alignment is caused by stronger solid–liquid forces than those present with perpendicular alignment was verified, but parallel alignment was not associated with wetting of the solid by the surface.

Haller[31] has measured the contact angle at the solid–liquid interface for three different liquid crystals in contact with coated solid surfaces and has also observed the alignment properties of the various liquid crystals in sandwich cells with the specially treated surfaces. The surface energy of the glass plates was varied by treatment with either an organic monolayer or multi-molecular layer. Of the three liquid crystals, only MBBA exhibited perpendicular orientation on glass surfaces coated with octadecyltrichlorosilane, HTAB, or barium stearate. All three liquid crystals, MBBA, BECS (4-n-butyl-4-ethoxy-2-chlorostilbene) and LiCristal IV (isomeric mixture of 4-methoxy-4'-butylazoxybenzenes) had surface tensions far in excess of the γ_c's of the coatings that were measured with unnamed isotropic test liquids. It should be noted that the three liquid crystals had different central linkages, and it is possible that this variation in chemical structure accounted for the diverse results. Haller concluded that the proposed correlation between wetting and alignment properties was too general, and specific details of the interaction forces were necessary for a precise prediction of the alignment properties.

A technique for producing uniform parallel alignment that does not involve organic coatings has been described by Janning.[32] The surface preparation consisted of evaporating materials such as gold, aluminum, platinum, or silicon monoxide onto the substrate at an angle of 85° to the substrate normal. The films were 100 Å or less thick. Cells containing MBBA and using glass plates coated in this manner exhibited uniform parallel alignment.

Recently Guyon, Pieranski, and Boix[33] investigated the dependence of liquid-crystal orientation on the angle, θ, between the evaporation direction and the normal to the substrate. Their principal results were:

(1) $0 < \theta < 45°$. No preferred direction for the alignment of MBBA was observed.
(2) $45° < \theta < 80°$. Uniform parallel alignment was obtained with the preferred direction being perpendicular to the plane containing the direction of evaporation and the normal to the substrate.
(3) $80° < \theta < 90°$. The fluid director possessed a preferred direction of orientation in the plane of evaporation; however, the direction did not lie parallel to the substrate, but instead it was tilted out of the substrate plane at an angle between 20° and 30.°

The authors hypothesized that the evaporated coatings were deposited with a sawtooth surface profile whose shape depended on the oblique angle of incidence. For $45° < \theta < 80,°$ it was felt that the long axes of the fluid lay parallel to the long axes of the sawtooth grooves. They also argued for $80° < \theta < 90°$ that the liquid crystal director was pointed into the teeth of sawtooth. Most of their data was obtained with SiO, but they stated that similar data was achieved with C and Au.

Complete microscopic confirmation of their model has not yet been obtained, although using high magnification with an electron microscope, Dixon, Brody, and Hester[34] have observed fine structure in 85° evaporated SiO_x films with the structure oriented in the plane of evaporation.

Guyon et al[33] did not state the exact nature of the liquid-crystal orientation with films evaporated at $\theta = 0°$. Meyerhofer[35] has measured random parallel alignment for $\theta = 0°$ evaporations. A reasonable conclusion that can be drawn from the data is that, when in contact with mesomorphic fluids, freshly evaporated SiO_x films produce parallel alignment in liquid crystals through physicochemical forces. At the present time, it is still not clear whether the directionality induced in the liquid crystal by obliquely evaporated SiO_x film is caused by the sawtooth model or some other anisotropic property of the SiO_x.

5. Influence of Packaging on Surface Orientation

In the section on packaging we have indicated that some liquid crystals are adversely affected by coming in contact with moisture; consequently, for these materials a good water-tight seal is necessary. The moisture can increase the fluid conductivity, produce a lowering of the mesomorphic–isotropic transition, or can participate in the production of impurities in the fluid that are adsorbed onto the surface and that can modify the alignment of the liquid crystal through a change of the forces at the solid–liquid interface.

Not only may the choice of sealing technique be important for controllable liquid-crystal orientation, but the effect of the interaction at the glass–liquid crystal interface must be considered in the selection of the type of glass to be used in the liquid crystal cells. Workers at E. Merck Co.[36] stated that it was easier to achieve perpendicular alignment on borosilicate glass than soda-lime glass when both types of glass pieces had been heated above 100°C. They claimed that the alkali impurities in the soda-lime glass caused a disorientation of the liquid crystal in some unknown manner. They were able to prevent the alkali-induced misorientation by rinsing the heated plates in chromosulfuric acid or water. They suggested that overcoating the glass surface with inorganic thin films such as MgF_2 and SiO_2 should alleviate the condition. Also, they found that the addition of certain surfactants such as lecithin or tetraalkyl ammonium salts to the liquid crystal produces good perpendicular alignment even between two soft glass plates that had been heated above 100°C.

We have found[37] that even with as much as 0.1% of HTAB added to MBBA, misorientation gradually occurred on solid glass surfaces that had been heated to 400°–500°C. When the cells were first made, good perpendicular alignment was achieved. However, over a period of time of up to several months at 25–30°C, or within a few hours at 85°C, the fluid slowly lost its perpendicular orientation. Cells made with fused silica plates did not exhibit this degradation. The data obtained from an ion-scattering analysis of the surfaces of the glass plates fired at high temperature showed a strong excess of cations, in particular, alkali ions. These results strongly support the original suggestion of alkali-induced misalignment.

6. Summary

A short outline of the standard conductive coatings and packaging methods has been presented.

Most of the chapter has been devoted to a discussion of the different surface treatments used to obtain controllable molecular alignment in liquid-crystal cells. The methods reviewed for obtaining parallel or perpendicular alignment were the use of surfactant additives, organic coatings on the solid surface, rubbing of the solid surface, and evaporation of inorganic materials. The data in the literature strongly suggests that the strength of the physicochemical forces at the liquid-crystal–solid interface is the most important element in determining whether parallel or perpendicular orientation will occur. However, the details of the interaction are sufficiently complicated that it is very difficult at the present time to make totally valid predictions about orientation for a specific solid surface and a particular liquid crystal. The directionality necessary for uniform parallel alignment can be provided by physical grooving of the solid surface, unidirectional withdrawal of the glass substrates from a coating solution, or by using an oblique angle of incidence during the evaporation of inorganic materials onto the glass surfaces.

Finally, two examples have been presented which show the relationship between packaging techniques and liquid-crystal molecular orientation.

References

[1] J. L. Vossen, "RF Sputtered Transparent Conductors II: The System In_2O_3-SnO_2," *RCA Rev.*, **32**, p. 289, June 1971.

[2] D. B. Fraser and H. D. Cook, "Highly Conductive Transparent Films of Sputtered $In_{2-x}Sn_xO_{3-y}$," *J. Electrochem. Soc.*, **119**, p. 1368 (1972).

[3] F. H. Gillery, "Transparent Conductive Coatings of Indium Oxide," *Information Display*, **9**, p. 17 (1972).

[4] R. Clary, Optical Coating Lab., Inc., Santa Rosa, California; private communication.

[5] Emulsitone Solution No. 673, Emulsitone Co., Millburn, N.J.

[6] Bulletin on Indium Oxide Conductive Coatings, Optical Coating Laboratory, Inc., Santa Rosa, Calif.

[7] Bulletin entitled "Nesa and Nesatron Glass," PPG Industries, Industrial Glass Products, Pittsburgh, Penna.

[8] G. Friedel, "The Mesomorphic States of Matter," *Ann. Physique*, **18**, p. 273 (1922).

[9] H. Zocher, *Z. Phys. Chem.*, **132**, 285 (1928).

[10] M. F. Schiekel and K. Fahrenschon, "Deformation of Nematic Liquid Crystals with Vertical Orientation in Electrical Fields," *Appl. Phys. Lett.*, **19**, p. 391 (1971).

[11] J. F. Dryer, "Epitaxy of Nematic Liquid Crystals," p. 1113 in *Liquid Crystals 3*, G. H. Brown and M. M. Labes, eds., Gordon and Breach, London (1973).

[12] W. Haas, J. Adams, and J. Flannery, "New Electro-Optic Effect in a Room-Temperature Nematic Liquid Crystal," *Phys. Rev. Lett.*, **25**, p. 1326 (1970).

[13] T. Uchida, H. Watanabe, and M. Wada, "Molecular Arrangement of Nematic Liquid Crystals," *Jap. J. Appl. Phys.*, **11**, p. 1559, (1972).

[14] S. E. Petrie, H. K. Bucher, R. T. Klingbiel, and P. I. Rose, "Aspects of Physical Properties and Applications of Liquid Crystals," Organic Chemical Bulletin 45, No. 2 (1973), Eastman Kodak Co., Rochester, N.Y.

[15] I. Haller and H. A. Huggins, Additive for Liquid Crystal Material, U.S. Patent 3,656,834, April 18, 1972.

[16] H. Zocher and K. Coper, *Z. Phys. Chem.*, **132**, p. 195 (1928).

[17] P. Chatelain, *Bull. Soc. Franc. Miner. Christ.* **66**, p. 105 (1943).

[18] J. E. Proust, L. Ter-Minassian-Saraga, and E. Guyon, "Orientation of a Nematic Liquid Crystal By Suitable Boundary Conditions," *Sol. St. Commun.*, **11**, p. 1227 (1972).

[19] F. J. Kahn, "Orientation of Liquid Crystals by Surface Coupling Agents," *App. Phys. Lett.*, **22**, p. 386 (1973).

[20] D. W. Berreman, "Solid Surface Shapes and the Alignment of an Adjacent Nematic Liquid Crystal," *Phys. Rev. Lett.*, **28**, p. 1683 (1972).

[21] D. W. Berreman, "Alignment of Liquid Crystals by Grooved Surfaces," *Mol. Cryst. and Liq. Cryst.*, **23**, p; 215 (1974).

[22] L. T. Creagh and A. R. Kmetz, "Performance Advantages of Liquid Crystal Displays with Surfactant-Produced Homogeneous Alignment," Digest of 1972 Soc. for Information Display International Symp., San Francisco, Calif., p. 90.

[23] L. T. Creagh and A. R. Kmetz, "Mechanism of Surface Alignment in Nematic Liquid Crystals," *Mol. Cryst. and Liq. Cryst.*, **24,** p. 59 (1973).

[24] E. G. Shafrin and W. A. Zisman, "Effect of Adsorbed Water on the Spreading of Organic Liquids on Soda-Lime Glass," *J. Amer. Ceramic Soc.,* **50,** p. 478 (1967).

[25] J. G. Grabmeier, W. F. Greubel, H. H. Kruger, and U. W. Wolff, "Homogeneous Orientation of Liquid Crystal Layers," 4th International Liquid Crystal Conf., Kent, Ohio, Aug. 1972, Paper No. 103.

[26] U. W. Wolff, W. F. Greubel, and H. H. Kruger, "The Homogeneous Alignment of Liquid Crystal Layers," *Mol. Cryst. and Liq. Cryst.*, **23,** p. 187 (1973).

[27] F. J. Kahn, G. N. Taylor, and H. Schonhorn, "Surface-Produced Alignment of Liquid Crystals," *Proc. IEEE,* **61,** p. 823 (1973).

[28] W. A. Zisman, "Relation of the Equilibrium Contact Angle to Liquid and Solid Constitution," *Adv. Chem. Ser.,* **43,** p. 1 (1964).

[29] A. W. Adamson, *Physical Chemistry of Solids,* 2nd ed. New York: Interscience, 1967, Chap. VII.

[30] J. E. Proust and L. Ter-Minassian-Saraga, "Notes des Membres et Correspondants et Notes Présentees ou Transmises Par Leurs Soins," *C. R. Acad. Sci.,* **276C,** p. 1731 (1973).

[31] I. Haller, "Alignment and Wetting Properties of Nematic Liquids," *Appl. Phys. Lett.,* **24,** p. 349 (1974).

[32] J. L. Janning, "Thin-Film Surface Orientation for Liquid Crystals," *Appl. Phys. Lett.,* **21,** p. 173 (1972).

[33] E. Guyon, P. Pieranski, and M. Boix, "On Different Boundary Conditions of Nematic Films Deposited on Obliquely Evaporated Plates," *Letters in Appl. and Eng. Science,* **1,** p. 19 (1973).

[34] G. D. Dixon, T. P. Brody, and W. A. Hester, "Alignment Mechanism in Twisted Nematic Layers," *Appl. Phys. Lett.,* **24,** p. 47 (1974).

[35] D. Meyerhofer, Private Communication.

[36] Current Information on Liquid Crystals No. 4 (1973), E. Merck Co., Darmstadt, Federal Republic of Germany.

[37] L. Goodman and F. DiGeronimo, "Nematic Liquid Crystal Misalignment Induced by Excess Alkali Impurities in Soft Glass," 5th International Liquid Crystal Conf., Stockholm, Sweden, June 1974.

13

Pressure Effects in Sealed Liquid-Crystal Cells

Richard Williams

RCA Laboratories, Princeton, N. J. 08540

1. Introduction

Liquid crystal cells are hermetically sealed glass containers completely filled with liquid. Two plane-parallel plates are sealed all around the edges to a frit glass spacer. The cell is then filled with liquid through two holes and sealed off with plugs of fusible metal. This construction gives rise to some internal pressure effects, because the thermal expansion coefficient of the liquid is about 100 times that of the glass. If the cell is filled and sealed off at room temperature, the liquid will exert a pressure at all higher temperatures. At lower temperatures it will be under tension. The pressure will deform the cell, making the walls bow out. This makes the volume enclosed by the cell a little larger and reduces the pressure but does not eliminate it completely. Some pressure or tension will always remain for temperatures different from the filling temperature. Repeated expansion and contraction may lead to loss of hermeticity or other cell failure. In what follows, the magnitude of the effect is calculated and the important factors are analyzed.

Fig. 1 shows the effect schematically and gives the notation used for the cell dimensions. The thickness of the layer of the liquid crystal is z, the length and width of the cell are b and a, and t is the thickness of the glass plates.

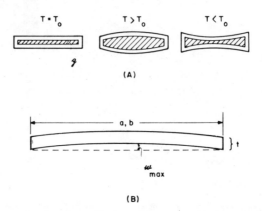

Fig. 1—(A) Expansion and contraction effects as the filled cell is maintained at a temperature T higher or lower than the filling temperature T_0. **(B)** Deformation of one of the plates: a, b are the width and length and t is the thickness of the glass. The maximum displacement w_{max} is at the center of the plate.

If the temperature is raised after filling, the liquid expands, and the cell walls bow out to accommodate the change in volume. We can neglect the thermal expansion of the glass and the changes of the liquid volume due to changes in pressure, since both these effects are small compared to the volume change of the liquid due to thermal expansion. The cell walls are held at the edges and the pressure of the liquid exerts a uniform force per unit area over the surface.

Consider the case where the cell is filled and sealed at temperature T_0 and later warmed to temperature T. As the walls bow out, the edges stay fixed. The maximum displacement, w_{max}, of the plate from its unstressed position will be at the center (Fig. 1). We need to determine the volume change Δv due to this deformation of the cell walls. The volume v_0 of the original undeformed cell is abz. Simple geometric considerations show that when $w_{max} \ll a$ the increase in volume due to the bowing out of the walls is

$$\Delta v = ab \frac{w_{max}}{2} \qquad [1]$$

Δv is also equal to the increase in volume of the liquid due to thermal expansion. Using the thermal expansion coefficient α_L of the liquid, we can express this as

$$\Delta v = abz\alpha_L(T - T_0). \qquad [2]$$

Since Eqs. [1] and [2] must be equal,

$$w_{max} = 2z\alpha_L(T - T_0). \qquad [3]$$

To get the pressure p required to give a deformation w_{max} is a standard problem in the strength of materials. It involves the cell dimensions and the mechanical properties of glass. For our particular case of a plate of uniform thickness held at the edges,[1] the solution is

$$w_{max} = \frac{Cpa^4}{Et^3}, \qquad [4]$$

where E is Young's modulus and, for typical glasses, has the value 6×10^{11} dynes/cm^2. C is a tabulated function that depends on the ratio a/b, i.e., the ratio of cell width to cell length. t is the thickness of the glass. From Eqs. [3] and [4] we get

$$p = \frac{2\alpha_L z E t^3 (T - T_0)}{Ca^4}. \qquad [5]$$

For MBBA (N-(p'-methoxybenzylidene)-p-n butylaniline), α_L has the value of 0.85×10^{-3} over the range of interest.[2] (The volume change at the nematic-isotropic transition is small compared to the thermal expansion and will be neglected.)

2. Effect of Temperature Change

The magnitude of the pressure developed by thermal expansion is shown in Fig. 2 for a cell 1 cm wide, 2 cm long, made of glass 1.0 mm thick and filled with a layer of MBBA 12.5 μm thick (½ mil). For these dimensions the value of C is 0.11.

Three cases of filling and sealing are shown, corresponding to three different temperatures T_0: 0°, 25°, and 50°C. The pressure change Δp may be either positive or negative, depending on whether the ambient temperature is greater than or less than T_0. The final cell pressure difference may amount to about 1 atmosphere for an operating range of 100°C. The pressure inside the cell will fluctuate continuously as the ambient temperature changes. This will be a continuous test of the hermetic sealing plugs. Eq. [5] shows the factors in cell design that lead to high pressures. These are the thickness of the liquid

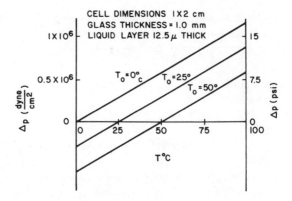

Fig. 2—Pressure changes Δp that result when a cell is filled at one temperature, T_0, and put in an ambient at another temperature, T.

layer, the thickness of the glass plates, and the overall cell size. The effect will be most serious for small cells, such as watch displays, and it can be alleviated by using thinner glass and thinner layers of liquid crystal.

3. Effect of Glass Thickness

Fig. 3 shows the effect of the thickness of the glass plates for a given temperature change with other conditions fixed. This is the t^3 depen-

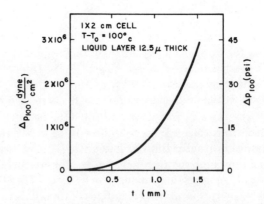

Fig. 3—Effect of the thickness t of the glass plates on the pressure change Δp_{100}, caused by heating the cell 100° above the filling temperature.

dence of Eq. [5] and emphasizes the merits of using thin glass, insofar as this is compatible with other cell requirements. In general, the pressure effects are less serious for larger cells and would be negligible for cells larger than 2 inches on a side.

4. The Case of a Rigid Container

In very small cells or in cells made of thick glass, there may be so little deformation of the cell walls that the liquid behaves as if it were in an ideally rigid container. Very high pressures develop and these can be estimated from readily available thermodynamic data. The quantity required is $(\partial p/\partial T)_v$, the pressure increase per degree of temperature rise when the volume is held constant. From general thermodynamic arguments this can be related to α_L and the compressibility β_L of the liquid.

$$\left(\frac{\partial p}{\partial T}\right)_v = -\left(\frac{\partial p}{\partial v}\right)_T \left(\frac{\partial v}{\partial T}\right)_p = \frac{\alpha_L}{\beta_L}. \qquad [6]$$

For most liquids the magnitude of α_L is around 1×10^{-3} deg^{-1} and the magnitude of β_L is around 1×10^{-4} atm^{-1}. The pressure rises by about 10 atmospheres per degree of temperature rise. This would be a severe design limitation. About the only way it might arise in practice would be if the lateral dimensions of the cell were small; if, for example the individual elements of a numeric display were made as separate closed-off cells.

An interesting possibility, especially important in the rigid container case, arises when the cell temperature is lowered after filling and sealing. The resulting negative pressure tends to make any dissolved gas come out of solution in the form of a bubble—a gas "embolism." This is very similar to what happens when a diver gets the bends. Dissolved gas, equilibrated in a liquid at a high pressure, comes out of solution when the pressure is lowered, with disastrous consequences. The same remedy may be useful in both cases—the use of a helium atmosphere. The solubility of helium in organic liquids is about one-tenth that of air. By handling and storing liquid-crystal materials under an atmosphere of helium, the gas available to produce an embolism would be reduced by a factor of ten.

In summary, the liquid-crystal cell, though ideally a closed system, is one that reacts significantly to changes in external conditions, and this is an important consideration in applications.

Acknowledgment

I am indebted to L. A. Goodman, D. Meyerhofer, E. B. Priestley, and P. J. Wojtowicz for valuable discussions of this problem.

References

[1] J. P. Den Hartog, *Advanced Strength of Materials*, pp. 132, McGraw-Hill Book Co., New York (1952).

[2] M. J. Press and A. S. Arrott, "Expansion Coefficient of Methoxybenzylidene Butylaniline through the Liquid-Crystal Phase Transition," *Phys. Rev.*, **A8,** p. 1459, Sept. 1973.

14

Liquid-Crystal Displays—Electro-Optic Effects and Addressing Techniques

L. A. Goodman

RCA Laboratories, Princeton, N. J. 08540

1. Introduction

Many of the physical properties of mesomorphic materials, such as birefringence, optical activity, viscosity, and thermal conductivity are sensitive to relatively weak external stimuli. Electric fields, magnetic fields, heat energy, and acoustical energy can all be used to induce optical effects. At the present time, most of the display-related research is centered on the application of electro-optic effects because of the relative ease and efficiency of excitation with an applied voltage as compared with other means of stimulation. Liquid-crystal electro-optic effects are important because they do not require the emission of light; instead they modify the passage of light through the liquid crystal either by light scattering, modulation of optical density, or color changes. The salient properties are low-voltage operation, very low power dissipation, size and format flexibility, and washout immunity in high-brightness ambients.

This chapter is divided into three major sections. The first describes the various liquid-crystal electro-optic phenomena; the second discusses important display-related parameters; and the third describes the operation of liquid-crystal devices in matrix-addressed and beam-scanned modes of operation.

2. Electro-optic Phenomena

Liquid-crystal electro-optic phenomena can be divided into two categories—those caused only by dielectric forces and those induced by the combination of dielectric and conduction forces. The two conduction-induced phenomena discussed later are dynamic scattering and the storage effect. Four of the dielectric phenomena, or field effects as they are sometimes known, are discussed first: (1) induced birefringence, (2) twisted nematic effect, (3) guest–host interaction, and (4) cholesteric–nematic transition.

In all of the present theories about the excitation of nematic or cholesteric liquids by an electric field, the mesomorphic material is treated as a continuous elastic anisotropic medium. The Oseen[1]–Frank[2] elastic theory is used to describe the interaction between the applied field and the fluid. The application of an electric field causes the liquid crystal to deform. For a material with a positive dielectric anisotropy, $\Delta\epsilon = \epsilon_\parallel - \epsilon_\perp > 0$, the director aligns in the direction of the field; if the dielectric anisotropy is negative, the director tends to align perpendicular to the applied field. The elastic forces attempt to restore the field-driven fluid to the initial orientation, which is determined by the surface alignment. The interplay between dielectric and elastic torques leads to the occurrence of the threshold voltage or field. In another chapter in this book,[3] the interaction between the liquid crystal and an applied field is discussed in detail. The interested reader can refer to that chapter and to papers published elsewhere[4-6] for the calculations of the field–liquid-crystal interaction.

2.1 Field-Induced Birefringence

The first electro-optic effect is field-induced birefringence or deformation of aligned phases.[7-10] Schematic representations of the fluid with zero applied volts and for a voltage exceeding the threshold voltage are presented in Fig. 1. With no applied voltage, the nematic liquid is in the perpendicular state. In the discussion of induced birefringence and the other effects considered, the assumption is made that the surface orientation of the molecules remains constant even when the field is applied, while the voltage-induced deformation increases toward the center of the cell. When the applied voltage exceeds the threshold voltage, the liquid crystal distorts if it has negative dielectric anisotropy. A maximum director rotation of 90° is possible. The threshold voltage is given in m.k.s. units by

$$V_{TH} = \pi \sqrt{\frac{K_{33}}{\epsilon_0 \Delta \epsilon}}, \quad [1]$$

where K_{33} is the bend elastic constant.[11,12]

The perpendicular texture is optically isotropic to light propagating perpendicular to the cell walls. Consequently, with crossed polarizer and analyzer, no light is transmitted through the analyzer. Dur-

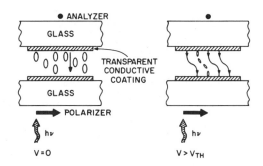

Fig. 1—Schematic illustration of the induced birefringence effect with and without an applied voltage. Thin arrows represent director orientation.

ing fluid deformation, the liquid crystal becomes birefringent to the transmitted light, and part of the light passes through the analyzer. The intensity of the emerging light is expressed by[9]

$$I = I_p \sin^2 2\phi \sin^2 \frac{\delta}{2} \quad [2]$$

where $\delta = [2\pi d \Delta n(V)/\lambda]$, I_p is the light transmitted through two parallel polarizers, ϕ is the angle between the input-light optical vector and the projection of the director on the plane parallel to the cell walls, d is the cell thickness, $\Delta n(V)$ is the voltage-induced change in birefringence, and λ is the wavelength of the light.

The transmitted light intensity is maximum when $\phi = 45°$. Normally, the angle ϕ is not well-defined because of the cylindrical symmetry that results when a perpendicularly aligned fluid is deformed by the electric field (see Fig. 2b). However, as described in the surface investigations of Creagh and Kmetz,[13] a preferential direction can be established in a plane parallel to the cell walls by grooving or rubbing the substrates. Samples prepared in this manner deform with a well-defined direction for the fluid director.[14] This preferred direction can be set at a 45° angle to the crossed polarizer and analyzer.

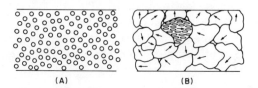

Fig. 2—Top views of perpendicularly oriented cells. (A) $V = 0$ and the long axes of the molecules are perpendicular to the electrodes. (B) The fluid is partially deformed with no prescribed direction.

The maximum value of $\Delta n(V)$ is the index of refraction anisotropy, $\Delta n = n_\parallel - n_\perp$. Typically, Δn is about 0.2 to 0.3 This anisotropy is so large that, for monochromatic radiation, the transmitted light intensity undergoes many maxima and minima as the voltage increases above threshold. With white light, variable colors can be observed as a function of voltage.

To first order, the frequency response of induced birefringence is constant in amplitude from low frequency to the molecular dispersion frequency in the dielectric constant where the dielectric anisotropy changes.[9] This property is typical of all the field effects.

For a nematic material with positive dielectric anisotropy, induced birefringence can also be observed. However, the liquid crystal must be in the uniform parallel orientation at zero volts.[15] Above the threshold voltage, the director aligns itself parallel to the applied field. With crossed polarizer and analyzer, the voltage dependence of the light intensity is reversed from that described previously for a fluid of negative dielectric anisotropy.[6,11]

For materials with positive dielectric anisotropy, the threshold voltage can be as low as 1.0 volt, whereas devices using negative dielectric anisotropy fluids typically possess threshold voltages in the 4–6 volt range. Since the elastic constants are relatively independent of material, the difference in threshold voltages is ascribed to the much larger magnitude of anisotropy normally found in fluids with positive anisotropy as compared to that occurring in materials with negative anisotropy.

2.2 Twisted Nematic Effect

The twisted nematic field effect is probably the most important of the field effects because of its combined properties of very low voltage threshold, low resistive power dissipation, and relatively wide viewing angle in the reflective mode.

The typical cell structure used in the twisted nematic device is shown in Fig. 3.[16] The molecules in each surface layer of the liquid crystal are uniformly aligned in one direction, but with a twist angle of 90° between the preferred direction for the two surfaces. With no

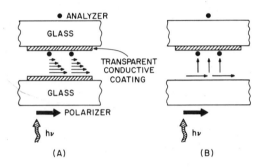

Fig. 3—Side view of twisted nematic effect for (A) $V = 0$ and (B) $V > V_{TH}$. The thin arrows represent the orientation of the nematic molecules.

applied voltage, the bulk fluid distorts so as to provide a gradual rotation of the molecular alignment from one cell wall to the other.[3] With a nematic fluid of positive dielectric anisotropy, voltages exceeding the threshold voltage cause the nematic director to become untwisted and to tend to align parallel to the applied field. The threshold voltage is

$$V_{TH}^2 = \frac{1}{\Delta\epsilon\epsilon_0} [K_{11}\pi^2 + (K_{33} - 2K_{22})\varphi_0^2], \qquad [3]$$

where K_{11} is the splay elastic constant, K_{22} is the twist elastic constant, and the twist angle φ_0 is equal to $\pi/2$. Eq. [3] is the corrected version of the expression derived by Schadt and Helfrich for the dielectric analog to the magnetic case originally solved by Leslie.[17] The theoretical dependence of the threshold voltage on the anisotropy has been verified.[18] Also, several mixtures with large dielectric anisotropy have been reported with threshold voltages less than 1.0 volt.[19-21]

The optical properties of the twisted nematic field effect are particularly interesting. Linearly polarized light propagating perpendicular to the cell is rotated by approximately 90° as it passes through the fluid when there is no applied voltage.[22,23] Maximum light transmission is obtained for the zero-field case by orienting the crossed polarizer and analyzer with the polarizer optic axis parallel to one of the preferred surface alignment directions in the cell. The transmitted light decreases when the applied voltage exceeds the threshold voltage, and the fluid starts to align in the perpendicular state. Fig. 4 presents the data obtained by Schadt and Helfrich[16] with parallel polarizer and analyzer, which can be used instead of crossed polarizer and analyzer. Extinction is obtained with no applied voltage, while light transmission occurs for voltages exceeding the threshold voltage. Hence, depending on the orientation of the polarizer and analyzer, either a black-on-white or white-on-black display can be obtained.

The formula presented in Eq. 3 indicates the threshold voltage at which the director starts to reorient. Gerritsma, DeJeu, and Van Zanten[24] have measured the magnetic threshold by both capacitive and optical techniques and found that the capacitive threshold is lower than the optical one. Van Doorn[25] has shown that this difference is to be expected, since the fluid starts to reorient by the tilting of the director toward the applied magnetic field before the twist has appreciably changed. Consequently the capacitive threshold, which occurs when the director starts to tilt toward the applied field, is lower than the optical threshold, which occurs when the twist becomes sufficiently nonuniform that the optical vector of the light does not "follow" the twist. A similar difference has been observed in twisted nematic devices excited with electric fields.[26,27] Berreman's[28] explanation of the static characteristics of electric-field-excited devices is similar to that of Van Doorn.[25]

The data presented in Fig. 4 and the optical results given by Gerritsma et al[24] were obtained with light normally incident upon the cell and coaxial with the detector. Recently, the angle and voltage dependence of the light transmission characteristics of twisted nematic devices prepared by the rubbing technique has been measured by several investigators. Kobayashi and Takeuchi[29] have obtained the data presented in Fig. 5. The transmitted light as a function of voltage is not symmetric with the viewing angle. The curves also demonstrate that the apparent optical threshold is a function of viewing angle. Both of these phenomena have been explained in terms of the relative ability of the incident light to "follow" the twist for different angles of incidence.[30] As a result, the true threshold is the capacitive threshold, and it is always lower than the angle-dependent optical threshold. Nonuniformities in cell appearance caused by reverse

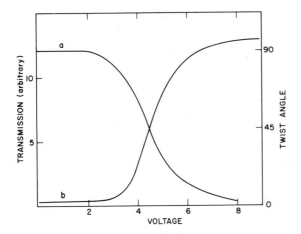

Fig. 4—Curve (a) shows rotation angle of linearly polarized light versus voltage for a nematic liquid crystal at room temperature and 1 kHz; curve (b) is transmission versus voltage with parallel polarizers (Ref. [16]).

tilt[3,31] and reverse twist[32,33] have been studied and can be eliminated by the addition of a cholesteric to the nematic and by proper surface preparation.[33,34]

With field-induced birefringence, variable colors are transmitted when the voltage exceeds the critical value. These angle-dependent color effects are unavoidable and constitute one of the main disadvantages of the induced birefringence effect. In the twisted nematic

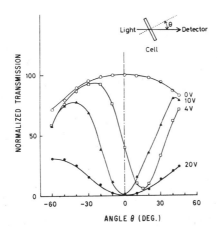

Fig. 5—Light transmission of a twisted nematic cell with crossed polarizers as a function of turning angle for various values of applied field at 5 kHz. Cell thickness is 30 μm (Ref. [29]).

phenomenon, relatively little birefringence occurs except at applied voltages just above the threshold voltage. For a twisted nematic device operating either with zero volts or with a voltage more than two or three times the threshold value, where most of the molecules are aligned perpendicular to the walls, the principal color phenomena are minor effects in the fieldless state associated with the inhomogenous thickness of the fluid.[35]

In reflective applications, the small angle visibility of twisted nematic devices about the normal to the cell can be much better than with scattering displays. For both scattering and nonscattering devices, a specularly reflecting mirror would normally be used behind the cell to obtain a bright, legible display. However, not only is the desired display information seen by the observer, but quite often unwanted specular glare can also be detected.

With nonscattering displays, such as the twisted nematic device, this glare can be prevented by the use of a diffuse reflector in place of the mirror reflector. However, there is a loss of brightness due to the depolarization of radiation reflected from the diffuse surface. The compromise is acceptable because the glare-free viewability is usually considered more important than high brightness.

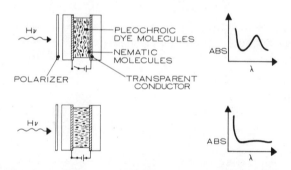

Fig. 6—Schematic diagrams of electronic color switching phenomenon.

2.3 Guest–Host Effect

A third phenomenon depending solely upon dielectric forces is the guest–host or electronic color-switching interaction[36,37] in which "guest" pleochroic dyes are incorporated within nematic "host" materials. The dyes have different absorption coefficients parallel and perpendicular to their optical axes. As illustrated in Fig. 6, the dye molecules can be oriented by the liquid crystal. With zero field, the liquid crystal is in the uniform parallel orientation and the dye molecules are aligned with long axes parallel to the optical vector of the linearly polarized light. In this configuration, the dye molecules have absorption bands in the visible. Above the threshold voltage, the nematic fluid of positive dielectric anisotropy tends to align parallel to the field. This is the condition for low dye absorption. Consequently, a color variation can be observed between the two states. Only one polarizer need be used with this effect. With optimum dye concentrations, optical density changes as large as 1.5 have been measured and a threshold voltage of approximately 2.0 volts has been observed.[37]

2.4 Cholesteric-to-Nematic Transition

The electric-field-induced cholesteric-to-nematic phase transition was observed by Wysocki et al.[38] The magnetic analog had been previously measured by Sackmann et al,[39] and the theoretical magnetic and electric field dependence has been calculated by deGennes[40] and Meyer.[41]

The phase transition, which is illustrated in Fig. 7, only occurs in an electric field with a cholesteric fluid of positive dielectric anisotro-

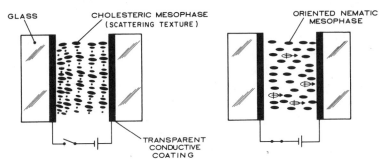

Fig. 7—Side view representation of the electric-field-induced cholesteric-to-nematic phase change.

py. The cholesteric planes are approximately perpendicular to the cell walls with zero applied field. The helical axes have random orientation and this state is strongly scattering in appearance.[42,43] As the electric field approaches the critical value, the helices begin to unwind and dilate. Above the threshold field, all the molecules, except for surface layers, are aligned parallel to the electric field. This latter condition is the perpendicular or homeotropic texture.[38] When the field is lowered below threshold, the scattering texture returns. The theoretical calculations performed by deGennes[3,40] result in the following expression for the threshold field.

$$E_{TH} = \frac{\pi^2}{P_0} \left(\frac{K_{22}}{\Delta\epsilon\epsilon_0}\right)^{1/2},$$ [4]

where P_0 is the undeformed helix pitch.

Meyer[44] performed optical measurements of the magnetically induced pitch dilation. By mixing nematic and cholesteric materials, Durand et al[45] were able to vary the pitch and to verify that the threshold magnetic field was inversely proportional to the undeformed helix pitch.

In the original experiment of Wysocki et al, the threshold field was about 10^5 V/cm.[38] Heilmeier and Goldmacher[46] reduced the threshold field to 2×10^4 V/cm by using a mixture of cholesteric and high positive dielectric anisotropy nematic materials. More recently, the threshold field has been reduced to 5×10^3 V/cm by use of the positive biphenyl compounds.[20,47]

The model presented so far is accurate, but incomplete. In actuality, the scattering texture with the helical axes parallel to the cell walls is not a stable state without an applied field. Rather, it is a metastable state that has a lifetime of from minutes to months, depending upon the surface alignment, fluid thickness, and pitch.[48] The stable state is the planar or Grandjean texture. Kahn[43] made the initial measurements of the various steps in the texture changes. Additional experiments with both electric and magnetic fields have added more detail to the model.

With no applied electric field, the liquid crystal exists in the planar texture, in which it remains until the voltage exceeds the critical field, E_H, which is proportional to $1/\sqrt{P_0 L}$.[49-51] At E_H, a square grid pattern of periodic distortions of the helical axes occurs.[42,52-54] In these perturbations, the direction of the helix axis periodically varies between being perpendicular and somewhat nonperpendicular to the cell walls. As the field increases further, the distortions grow until the helix planes become perpendicular to the walls.[42,43] With further increase in voltage, the fluid becomes perpendicularly aligned as described previously. When the voltage is shut off, the fluid returns to the metastable scattering state. The transient decay from the perpendicular texture to the scattering state is retarded by the presence of a bias voltage that is below the threshold voltage for the field-induced transition.[55,56] For no bias level, the decay to the scattering texture proceeds rapidly with the natural relaxation time. As the bias level approaches the threshold voltage, the relaxation slows considerably.

Greubel[57] has recently analyzed the cholesteric-to-nematic transition for an applied voltage. With perpendicular alignment of the liquid crystal molecules at the cell surfaces, he found that the field transition for the cholesteric-to-nematic transition was higher than the opposite nematic-to-cholesteric transition. With one mixture, the ratio of field thresholds was approximately 2.5. This work points out the importance of the proper surface orientation and cell cleanliness in achieving the best bistability.

White and Taylor[58] have described a new device that combines both the guest–host effect and the cholesteric-to-nematic transition. The pleochroic dye is added to a cholesteric material and the cell transmission changes when the cholesteric undergoes the cholesteric-to-nematic transition. Because of the rotational symmetry of the long axes of the dye molecules about the cholesteric helical axes, contrast ratios of greater than 4 to 1 were obtained without the use of a polarizer.

2.5 Dynamic Scattering

In nematic materials with negative dielectric anisotropy and electrical resistivity less than $1-2 \times 10^{10}$ ohm-cm, conduction-induced fluid flow occurs during the application of an applied voltage. The wide-angle forward-scattering phenomenon known as dynamic scattering[59,60] is the most important manifestation of the turbulence that

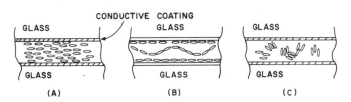

Fig. 8—Side view of the various steps in the formation of dynamic scattering for uniform parallel orientation. (A) $V = 0$, (B) $V = V_W$, and (C) $V > V_W$.

accompanies the electrohydrodynamic flow. The light scattering arises from micron-sized birefringent regions in the turbulent fluid.[61]

The various steps in the production of dynamic scattering are portrayed schematically in Fig. 8 for initial uniform parallel alignment. Below voltage V_W, no change in orientation occurs. At V_W, the fluid becomes unstable and it deforms into the periodic structure shown in Fig. 8.[62] Two similar mechanisms are responsible for the creation of the instability, one for ac voltages and a second for dc voltages. Both interactions require the presence of space charge in the fluid, but differ in the means by which the space charge is generated.

In the ac case, space-charge separation perpendicular to the applied field is caused by the anisotropy in conductivity.[62,63] The applied field produces a force upon the liquid crystal because of the space charge. This stimulus drags the fluid toward the walls. The cells walls impose boundary conditions that necessitate vortical flow of the fluid. The fluid shear torque aligns the director in the direction of the fluid flow, while the dielectric and elastic forces oppose the fluid deformation. At the threshold voltage, the fluid becomes unstable and the periodic distortion takes place.

When the cell is observed through a microscope, the periodic deformation appears as a series of alternating dark and light domains, known as Williams' domains that run perpendicular to the original homogeneous alignment. The wavelength of the periodic deformation is determined by the thickness of the cell.[64] The birefringence of the nematic fluid and the periodicity of the instability combine to form periodic cylindrical lenses in the fluid.[65] The domain lines are the result of light focused by the periodic array of lenses. This periodic lens array also acts as a transmission phase grating. Consequently, colinear diffraction spots can be observed when a laser beam propagates through the liquid crystal lens array.[61,66,67]

The presence of fluid flow in the seemingly static periodic domains has been observed by the motion of dust particles in the fluid.[65] Below the threshold voltage, the dust particles are stationary. At threshold, they move in an oscillatory pattern closely related to the domain spacing. The velocity of the particles increases with increasing voltage and with decreasing cell thickness.

Neglecting the frequency response of the electrohydrodynamic flow, Helfrich[62] calculated the threshold voltage for the domain instability. A slightly rewritten form[68] of his expression is

$$V_W^2 = \frac{V_0^2}{\zeta^2 - 1} = \frac{\dfrac{\pi^2}{\Delta\epsilon\epsilon_0}\dfrac{\epsilon_\parallel}{\epsilon_\perp}K_{33}}{\left[1 - \dfrac{\epsilon_\parallel}{\Delta\epsilon}\dfrac{1}{1 + \dfrac{\eta_0}{\gamma_1}}\right]\left[1 - \dfrac{\sigma_\perp}{\sigma_\parallel}\dfrac{\epsilon_\parallel}{\epsilon_\perp}\right] - 1},$$

[5]

where $\sigma_\perp$ and $\sigma_\parallel$ are the perpendicular and parallel components of the conductivity, and η_0 and γ are viscosity coefficients. Eq. [5] is in good agreement with the experimental data for p-azoxyanisole. The

Orsay Liquid Crystal Group[68–70] solved the electrohydrodynamic problem for a variable frequency, sinusoidal voltage source. The fluid instability occurs at the frequency-dependent threshold voltage

$$V_w^2 = \frac{V_0^2(1 + (2\pi f)^2\tau^2)}{\zeta^2 - (1 + (2\pi f)^2\tau^2)} \qquad [6]$$

where f is the frequency, $\tau = \epsilon_\parallel \epsilon_0/\sigma$ is a dielectric relaxation frequency, and V_0 and ζ are the same terms as in Eq. 5. A cutoff frequency

$$f_c = \frac{(\zeta^2 - 1)^{1/2}}{2\pi\tau}$$

results from Eq. [6]. The cutoff frequency is directly proportional to the conductivity.

The theoretical analysis predicts the existence of two regimes with different frequency dependences. For applied frequencies, $f < f_c$, the space charge in the fluid oscillates at the same frequency as the driving signal. This is the region of cellular fluid flow described in the preceding discussion. Because this region exists for frequencies less than the dielectric relaxation frequency, it is called the "conduction regime." The thickness-independent portion of the solid voltage–frequency curve in Fig. 9 is experimental verification of Eq. 6.

When the applied frequency is greater than f_c, the space charge does not oscillate. The fluid interacts with the applied field to result in the "dielectric regime." The threshold field is proportional to $f^{1/2}$ (see Fig. 9). Contrary to the low-frequency situation, the field is thickness independent. Periodic deformations, known as "chevrons," result from the field–fluid interaction. The spatial frequency of the chevron striations is a monotonically increasing function of the drive frequency.

In the calculation of the threshold voltage, Helfrich assumed that the spatial periodicity of the fluid deformation was proportional to the thickness of the cell. Recently, Penz and Ford[72,73] have solved the boundary-value problem associated with the electrohydrodynamic flow process. They have reproduced Helfrich's results and have also shown several other possible solutions that may account for the higher-order instabilities that cause turbulent fluid flow. Meyerhofer[74] has analytically solved the two-dimensional problem by making one simplifying assumption. He has been able to obtain good agreement between the experimental results and the calculated frequency dependence of the domain spacing and the threshold voltage.

Experimentally, it is observed that, when the applied voltage surpasses the threshold voltage by an increasing amount, the rotational velocity of the fluid increases.[75] The fluid gradually becomes more turbulent until the applied voltage exceeds twice the threshold voltage. The intense wide-angle forward scattering accompanying the strong turbulence is the dynamic scattering region.[59,61] Dynamic scattering only happens below the critical frequency and above the threshold voltage (see Fig. 9). As the voltage increases even further (beyond twice the threshold voltage), the scattering likewise increases and the fluid becomes even more turbulent[61] with virtually all traces of the underlying domain structure disappearing. The increasing turbulence as a function of voltage causes the scattering intensity and transient kinetics to be angle dependent.[76-78]

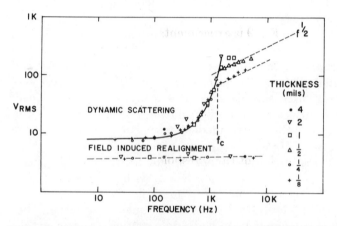

Fig. 9—Various threshold phenomena for nematic fluids with negative dielectric anisotropy and perpendicular alignment. The dashed horizontal line is the threshold voltage for induced birefringence. The curved solid line describes the frequency dependence of the threshold voltage for domains. The sloped dashed lines are the threshold plots for chevron formation. The material is MBBA at 25°C (Ref. [86]).

Though dynamic scattering usually results when the applied voltage exceeds the domain voltage, this need not be true. Dynamic scattering only seems to occur if the fluid is thick enough ($\gtrsim 6$ μm) and has sufficiently low resistivity (less than $1-2 \times 10^{10}$ ohm-cm) and for negative dielectric anisotropy. Domains have been observed without dynamic scattering when any of the preceding three conditions have been violated.[79,80] With the high resistivity and thin cells, the spatial frequency of the domains is voltage-dependent.[81] At the present time, no theory exists that defines the exact relationships between the various degrees of fluid instability.

The domain instability and dynamic scattering are also observed for both low-frequency ac and for dc applied voltage. The volume space charge, necessary for hydrodynamic motion, is not produced by the conductivity anisotropy, but by injection of charge from the electrodes.[70,82-84] Meyerhofer and Sussman[85] have measured the voltage–frequency plot for the formation of the domains from very-low-frequency ac to the cutoff frequency f_c. Below a certain frequency, which they relate to the transit time for ions, they have found that the domain threshold voltage decreases from the ac value toward the dc value. Also, the domain spacing changes below the inverse transit-time frequency. This transit-time frequency is of the order of 5 to 10 Hz. At present, the injection mechanism is unknown. It has been hypothesized that the double-layer space charge present at the electrode–liquid interface is responsible for the fluid flow.[86] Vortical fluid flow[87] and laser diffraction patterns[67] have also been observed with dc applied voltage.

Most of the discussion presented above for uniform parallel alignment is still valid with zero-field perpendicular orientation. However, there is one change. The voltage sequence for the production of dynamic scattering has an extra step. For most of the materials used at present, the voltage threshold for induced birefringence is lower than the threshold for domain formation (see Fig. 9), although there are exceptions.[11] As a function of increasing voltage, the fluid progresses from the undeformed state to the induced birefringence texture, then to the presence of domains, and finally to the occurrence of dynamic scattering.

Initial perpendicular alignment provides for greater circular symmetry in the scattering distribution. As suggested in Fig. 2, the director for the deformed fluid has only medium range order, approximately 50 μm or less. The projections of the directors in a plane parallel to the fluid are randomly oriented. Because of the circular symmetry about the axis perpendicular to the cell, the laser diffraction pattern consists of a set of circular rings instead of colinear spots.[61,66] The same symmetry is observed for the angular dependence of the dynamic scattering[76] (see Fig. 10).

2.6 Storage Mode

Optical storage effects in mixtures of nematic and cholesteric materials with negative dielectric anisotropy were first observed by Heilmeier and Goldmacher.[88] They reported the following sequence of events (see Fig. 11). Initially, with no applied voltage, the sample was in a relatively clear state. The application of a dc or low-frequency ac voltage of sufficient magnitude induced the intense scattering known as dynamic scattering. When the voltage was removed, the dynamic scattering disappeared, but a quasi-permanent, forward scattering

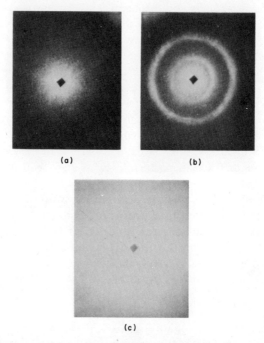

Fig. 10—Diffraction patterns for a perpendicularly oriented cell: (a) $V = 10\ V_{PP}$, very light diffuse scattering; (b) $V = 13\ V_{PP}$, diffraction rings due to domain formation; (c) $V = 16\ V_{PP}$, strong diffuse scattering characteristic of dynamic scattering. The black square in each picture is a piece of tape placed on the screen.

state remained. The reported decay time was on the order of hours at elevated temperature. The scattering texture could be returned to the clear state by the application of an audio frequency signal (greater than 500–1000 Hz). Since the original reports, other investigators have also observed the same effect.[89-91] The off-state has been identified as the Grandjean texture.[89] Due to imperfections in the Grandjean planes, the off-state is slightly scattering.

Rondelez, Gerritsma, and Arnould[54] have reported the presence of two-dimensional deformations at the threshold voltage for scattering. Electrohydrodynamic instabilities were first predicted for negative cholesteric materials by Helfrich.[50] Hurault[51] has combined Helfrich's theory with the time dependent formalism used by Dubois-Violette, deGennes, and Parodi.[68] His calculations predicted a voltage–frequency relationship similar to that observed for pure nematics of negative dielectric anisotropy (see Fig. 9). Experimental verifica-

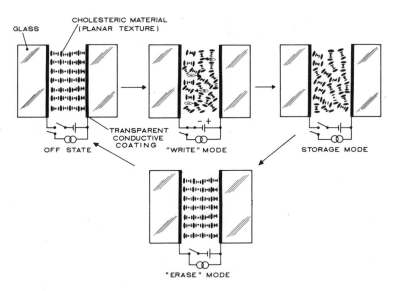

Fig. 11—Schematic illustration of the effect of low- and high-frequency signals on cholesteric fluids with negative dielectric anisotropy.

tion of both "conduction" and "dielectric" regimes has been established.[54,87] The domain periodicity is proportional to $(P_0 L)^{1/2}$ in agreement with theory, where P_0 is the zero field value of the pitch and L is the cell thickness. The threshold voltage in the conduction regime is not thickness independent, but is proportional to $(L/P_0)^{1/2}$.

The exact means by which the grid structure becomes distorted so as to form the strongly scattering state is not known. The scattering texture is approximately the same as that found in the cholesteric-nematic transition,[48,60,89] even though the mechanisms for producing the scattering states are probably quite different. As explained, in cholesteric–nematic mixtures of positive dielectric anisotropy, the dielectric forces are sufficient to tilt the fluid into the scattering state. With the materials of negative dielectric anisotropy that are used in the storage effect, the strongly scattering state must arise from the strong fluid deformations associated with dynamic scattering. The light scattered from the storage state is relatively independent of the concentration of cholesteric material, the direction of the incident light, and the cell thickness.[90]

The restoration of the planar state by the applied field is purely a dielectric interaction. It can only take place when the signal frequency is greater than cutoff frequency f_c, which is proportional to the conductivity of the liquid crystal.

2.7 Transient Response

The theoretical expressions describing the transient response of the fluid deformations are of the same general form for the different effects. A characteristic response time for the director reorientation is given approximately by[91]

$$T = \eta[\Delta\epsilon\epsilon_0 E^2 - Kq^2]^{-1}, \qquad [7]$$

where η is the proper fluid viscosity, E is the applied field, K is the appropriate elastic constant, and q is the wave-vector of the disturbance. For the phenomena occurring in pure nematics, the wave-vector is approximately π/L where L is the cell thickness. Consequently, the rise time and decay time should be of the forms

$$T_{\text{RISE}} = \eta L^2[\Delta\epsilon\epsilon_0 V^2 - K\pi^2]^{-1} \qquad [8]$$

and

$$T_{\text{DECAY}} = \frac{\eta L^2}{\pi^2 K}. \qquad [9]$$

Experimental observations for induced birefringence,[14,92] twisted nematic,[91] and dynamic scattering[60,78,84] are in reasonable agreement with the theory. For a 12-μm-thick fluid, 10 V dc, and 20°C, rise times of the order of 10 msec have been observed with the twisted nematic effect and 200 to 300 msec for dynamic scattering. Decay times are approximately 400 msec for the twisted device and 100 msec for dynamic scattering. For both dynamic scattering and the field effects, the presence of high voltages causes second-order effects to occur and the decay time becomes voltage dependent.[93,94]

The response times for both field- and conduction-induced phenomena can be changed by the presence of a second voltage source whose frequency is above some critical cutoff frequency, while the main source has a frequency less the critical frequency. For dynamic scattering, the critical frequency is f_c, which is inversely proportional to the dielectric relaxation time of the fluid. For an applied frequency $f > f_c$, the conduction torques do not affect the fluid and, through the negative dielectric anisotropy, the dielectric torque causes a return of the fluid to its nonscattering state. Consequently, the decay time can be significantly shortened as compared to the case when no high-frequency source excites the cell.[95,96]

For field-effect materials, the critical cutoff frequency occurs only in materials that have positive dielectric anisotropy at low frequencies. Above the critical frequency, the dielectric anisotropy is negative. Several materials have been developed with a critical frequency as low as a few kHz at 25°C.[97,98] As with dynamic scattering, the application of a high-frequency source can produce a decay time much shorter than the natural decay time.[98–101] The decay time is inversely proportional to the square of the amplitude of the high-frequency voltage.[98,100]

In the cholesteric–nematic phase change with $L/P_0 > 1$, the wavevector is given by π/P_0 not π/L. The experimental rise and decay times are consistent with the theory for the field-induced phase transition.[91]

The texture change from the highly scattering cholesteric state to the Grandjean texture is described by different kinetic relationships. The erasure time of the scattering state for the storage effect is proportional to V^{-m} where V is the audio frequency signal and m varies between 1 and 3 depending upon the material.[88,102] The natural decay time from the scattering to planar texture is approximately exponentially dependent on the L/P_0 ratio[65] and the inverse of the sample temperature.[102] The inverse temperature dependence suggests that the decay time of the storage state is also directly proportional to the viscosity, which is exponentially dependent on the inverse temperature.[86]

3. Display-Related Parameters

3.1 Display Life

The determination of the operating life is fraught with complications because of the difficulty of defining the conditions that describe the end of useful operation. Subjective evaluation of the steady-state cosmetic appearance and quantitative examination of the variations in response time, nematic–isotropic temperature, and power dissipation are necessary. The changes in cosmetic appearance and response times are usually manifestations of misalignment of the fluid at the liquid–solid interface. The misalignment may be caused either by the application of voltage or by chemical interaction between the fluid and the substrate surface. Time-dependent variations of the current may also arise from chemical interaction between the fluid and the cell walls. Some of the commonly used liquid-crystal materials are deleteriously affected by the presence of moisture and UV radiation. Proper cell packaging is then necessary to minimize these two unwanted agents.

Sussman[103] has examined the dc electrochemical failure mechanism in the dynamic scattering material p-methoxybenzylidene-p'-aminophenyl acetate (APAPA). Using the loss of 50% of the cell scattering area as his criterion for the end of life, he showed that the amount of charge passed through the cell determined the operating life. The failure mode was traced to the production of an insulating film at the anode. The utilization of ac drive signals, instead of dc, greatly diminishes the likelihood of failure being caused by electrochemical effects. Consequently, commercial dynamic scattering displays are driven by ac signals. AC operating life is cited as being greater than 5–10,000 hours.[76,104] Equally long operation is to be expected from field-effect displays operating with ac excitation.

3.2 Temperature Dependence

Until the resurgence in liquid-crystal research in the 1960's, most of the mesomorphic materials were solid at room temperature. Today, there are many liquid-crystal systems that exhibit the mesophase over a wide temperature range around 20°C.

The temperature variation of the fluid properties is important for display applications. The rise and decay times are directly proportional to the fluid viscosity as shown in Eqs. [8] and [9]. Since the viscosity is approximately exponentially dependent on the inverse temperature, the response times are strongly temperature dependent.[78,86] At low temperatures, even when the material is still mesomorphic, the viscosity may be so high as to preclude the operation of the liquid-crystal display because of the sluggish transient characteristics.

The threshold voltage for the field effects should be mildly temperature sensitive due to the relatively weak temperature variation of the elastic constants and the dielectric anisotropy. Experimental results on the induced birefringence phenomena confirm this statement so long as the operating temperature is less than 95% of the nematic–isotropic temperature.[92,105] Measurements show that the threshold voltage and contrast ratio in dynamic scattering devices are almost completely insensitive to temperature throughout the nematic range.[78]

The conductivity in nematics is governed by ionic equilibrium[106] and is inversely proportional to the viscosity and square root of the dissociation constant. Both the viscosity and equilibrium constant are exponentially dependent on temperature as is the conductivity.

The critical cutoff frequency for both conduction-induced and field effect phenomena is exponentially dependent on temperature. In the former case, the temperature dependence occurs because the cutoff frequency is proportional to the conductivity, whereas for field effect materials with positive dielectric anisotropy the exponential temperature dependence is a property of the molecular relaxation process which is responsible for the cutoff frequency.

The variation with temperature of the different properties of the materials must be taken into account during device design so that the driving signal can always induce the electro-optic effect over the desired temperature range with acceptable speed, wide viewing angle contrast ratio, and low power consumption.

4. Addressing Techniques

The presentation of visual information by a display requires a method or methods for exciting multiple positions in the display medium. The process of transmitting signal information throughout the display and exciting the different positions in the display medium is known as addressing. Two general approaches to addressing are useful with liquid crystals. One method involves beam steering, which includes electron-beam addressing (as performed in a cathode ray tube) and light-beam scanning. The alternative approach, which is discussed first, is that of matrix addressing or multiplexing. The two words are equivalent, but, for historical reasons, matrix addressing is used when referring to displays with a large number of elements, and multiplexing is reserved for displays with a relatively small number of elements.

4.1 Matrix Addressing

One of the strong motivating factors in liquid-crystal research is the possibility of constructing two-dimensional displays that dissipate little power. The third dimension, that of the glass–liquid-crystal–glass sandwich, is of the order of $\frac{1}{8}$ inch and is small compared to the other two dimensions. An example of a multi-element liquid-crystal display is the seven-segment, five-digit dynamic scattering display shown in Fig. 12. Each segment of each digit could be individually addressed by a driving signal whose frequency components are lower than the critical cutoff frequency, but this approach is wasteful of both driving circuitry and interconnections between the display and

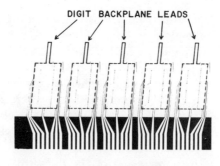

Fig. 12—Top view of a seven-segment five-digit display with electrode leads.

the circuitry. A much more economical approach is one in which the display is rearranged into an X-Y matrix as indicated in Fig. 13. One segment from the first digit is connected to its counterpart on each of the other digits. This interconnection scheme is repeated for the other six segments of the first digit and their counterparts. The seven row lines in the matrix represent the seven coordinated segment leads, while the five column lines are the same as the five digit backplane leads. Consequently, the five-digit seven-segment display with 35 leads can be electrically viewed as a five-by-seven matrix with only 12 leads. In general, a display with $M \times N$ resolution elements can be treated as a matrix array in which only $M + N$ leads are required.

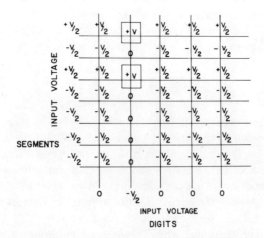

Fig. 13—Representation of the seven-segment five-digit display with leads rearranged in matrix fashion. Half-voltage selection pulses are applied.

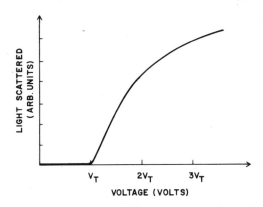

Fig. 14—Typical scattered light versus voltage curve for dynamic scattering.

For the proper operation of a matrix array, it is necessary to excite only the desired element in the matrix and no other. As an example, let us discuss the 5 × 7 array. Only the first and third segments of the second digit should be scattering light with all the other segments being nonscattering. For this situation to occur, the voltages $\pm V/2$ must be less (in an absolute magnitude sense) than the threshold voltage required to initiate light scattering. The light scattered versus voltage transfer function for a typical dynamic scattering cell is given in Fig. 14. Since the contrast ratio at $2V_T$ can be in excess of 20:1, it would appear that dynamic scattering displays can be matrix-addressed without any difficulty.

So far, the matrix array has been treated in a purely static fashion. In actuality, the segment data enters in parallel from the drive circuitry and each digit is selected sequentially in time. After all of the digits have been addressed, the cycle is repeated. The temporal dependence of an arbitrary segment is presented in Fig. 15. The seg-

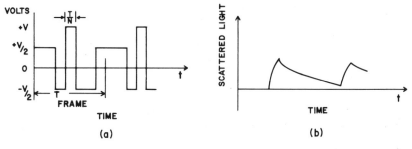

Fig. 15—(a) Applied voltage versus time for an arbitrary segment and (b) time dependence of light scattering for the same segment.

ment is only excited for a time T/N where T is the frame time of the data (typically 30 to 60 Hz) and N is the number of digits in the display. The scattered light as a function of time is modified from the voltage waveform by the finite response times of the phenomenon (see Fig. 15B).

Several devices and circuit parameters must be properly controlled to maximize the contrast ratio when operating in a scanning mode. One important consideration is that the rise time should be as short as possible. In practice, fast rise times are only achieved by using voltages that are far greater than twice the threshold voltage, and therefore elements that should be off turn on. Consequently, the contradictory requirements of short rise times and sufficient half-select capability restrict the number of digits that can be addressed in a multiplexing mode.

In order to improve the applied-voltage discrimination ratio for matrix arrays, the so-called one-third-voltage selection method shown in Fig. 16 can be used. This method clearly offers the advantage over the half-select method of applying more voltage to the "on" elements without exceeding the threshold voltage for the elements that are supposed to be nonscattering.

Second, when the decay time is longer than the frame time T, the cell integrates the successive series of input signals that occur every frame. So long as the integration property is obeyed, it has been found that both field-effect[107,108] and dynamic-scattering[109] devices respond to the driving signal in a root-mean-square fashion. As a consequence of the rms behavior, the contrast ratio for a scanned multi-

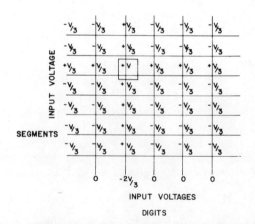

Fig. 16—Applied voltages for the one-third selection method.

plexed display can be no higher than the contrast ratio obtained with a continuous-drive waveform whose rms amplitude is the same as the rms content of the scanned waveform. Kmetz[107] has calculated the ratio of the rms value for an on element to that for an off element in a $V:V/3$ scheme to be

$$\frac{V_{ON}}{V_{OFF}} = \sqrt{\frac{8}{N} + 1}, \quad [10]$$

where N is the number of digits being scanned.

Because of the rms behavior, the contrast ratio is reduced at a fixed viewing angle. In addition the angular dependence of the contrast ratio, which is most noticeable when the applied voltage is only somewhat greater than the threshold voltage during continuous drive conditions, becomes more important. For example, in Fig. 5, the level of transmission at 4 V is very asymmetric in its angular dependence, whereas, at 10 V, it is not. In a multiplexed mode, as the number of digits increases, the rms behavior dictates that the angular dependence of the transmitted light becomes closer to that for the 4-V curve than for the 10-V curve. Consequently, the number of digits that can be multiplexed with good contrast is not only a function of the driving signal amplitudes, but also is dependent upon the viewing angle.

Another condition for the proper implementation of the present multiplexing technique is that enough of the power in the driving signal be at frequencies much lower than the cutoff frequency of the electro-optic effect being utilized. Frequency components just under the cutoff frequency are not as effective as lower-frequency components at producing the excited state, as can be seen for dynamic scattering in Fig. 9. As explained previously, signals whose frequency content is higher than the cutoff frequency cause a return to the unexcited state.

Alt and Pleshko[109] have extended Kmetz's analysis of rms-responding liquid-crystal devices. They have shown that the voltage-drive configuration that optimizes the number of scanned digits is not the $V:V/3$ scheme, but one in which the ratio of peak voltage to bias voltage is greater than 3:1. They also explicitly demonstrate that the more nonlinear the transmitted light versus voltage curve, the greater the multiplexing capability for a given set of drive-signal amplitudes.

Varying degrees of success have been reported for matrix addressing liquid-crystal displays. For displays using dynamic scattering, anywhere from 3 or 4 digits to 7 digits have been reported as the maximum addressing capability.[107,110] Twisted nematic devices possess approximately the same matrix addressing limitations as dynamic scattering.[107] Hareng, Assouline, and Leiba[111] have reported matrix addressing a 50 × 50 element array, while Schiekel and Fahrenschon[112] have successfully operated a 100 × 100 array. Both of these induced birefringence devices are two-color displays with rise times on the order of 1 second.[108,111] Due to the nature of the electro-optic process, the display appearance is a sensitive function of viewing angle, fluid thickness, voltage, and temperature. The small field of view probably limits matrix-addressed birefringence devices to projection display applications. Takata et al[110] have reported the operation of a 260 × 260 liquid-crystal display using the storage-mode effect. Because of the slow rise times at the voltage levels appropriate to matrix addressing, 10 to 20 seconds are required to adddress the whole display. The long decay times associated with the storage effect permit the maintenance of the displayed information for hours or more and the rms behavior does not occur.

The cholesteric-to-nematic phase transition effects have also been utilized in matrix-addressed displays.[55,113] Up to 28 lines have been scanned in the $V:V/3$ mode with a bias voltage of 35 V_{rms} and a contrast ratio of 15:1. The relatively large multiplexing capability is due to the long decay time produced by the bias voltage.[56]

The discussion until now has centered on the utilization of driving signals whose frequency components are much lower than the cutoff frequency. However, it is possible to implement useful multiplexing schemes with drive signals whose frequency components are both below and above the cutoff frequency. This approach will first be analyzed for dynamic-scattering devices.

Imagine a cell containing a liquid capable of dynamic scattering. Let us drive this cell with two sinusoidal voltage sources in series, one with applied frequency $f_1 \ll f_c$ and the other with frequency $f_2 \gg f_c$. The high-frequency signal retards the occurrence of dynamic scattering, and it can be shown that the low-frequency threshold voltage for the formation of dynamic scattering is related to the high-frequency signal by the following equation[114,115]

$$V_1^2 = V_0^2 + \gamma V_2^2, \qquad [11]$$

where V_o is the threshold voltage derived by Helfrich[62] for dc and very low frequency ac signals, and γ is a parameter dependent upon several material properties such as dielectric constant, shear torque, viscosity, and conductivity. For the common liquid crystal MBBA, γ = 0.5 at 32°C.[115]

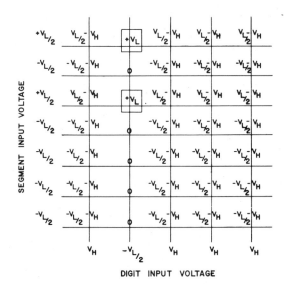

Fig. 17—Dual-frequency addressing of a matrix array.

The increase in threshold voltage for dynamic scattering induced by the high frequency can be readily utilized in a matrix display as shown in Fig. 17. $V_{L/2}$ and V_H are the zero-to-peak amplitudes of the low- and high-frequency signals. With the simultaneous application of the low- and high-frequency voltages to those elements that are not supposed to be scattering, the low-frequency signal can be increased in amplitude so that the light intensity from the "on" elements is greater than it would have been if a single low-frequency signal had been applied. With pure nematic materials, approximately 16 lines can be addressed with reasonable contrast.[116] The addition of cholesteric material to the nematic should lengthen the decay time and thereby increase the upper limit on the number of lines that can be matrix-addressed.

The two-frequency approach can also be used with field-effect materials. Bücher, Klingbiel, and Van Meter[98] have shown that the low-frequency threshold voltage for the twisted nematic effect is increased by the superposition of a signal whose drive frequency is greater than the critical frequency where the dielectric anisotropy becomes negative. They claim that

$$V_{LF}^2 = V_0^2 + \left| \frac{\Delta \epsilon_{HF}}{\Delta \epsilon_{LF}} \right| V_{HF}^2 \qquad [12]$$

where V_{LF} is the amplitude of the low-frequency signal, V_{HF} is the amplitude of the high-frequency signal, V_o is the threshold in the absence of the high-frequency signal, $\Delta\epsilon_{HF}$ is the dielectric anisotropy at the high-frequency, and $\Delta\epsilon_{LF}$ is the same quantity at the low frequency. The material they used has a crossover frequency of 2.5 kHz at 25°C, whereas other materials published in the literature have higher critical frequencies.[97,117]

Because of the combined threshold-voltage and rise-time requirements, none of the approaches described here are capable of matrix addressing a high-resolution, high-speed display. Lechner, Marlowe, Nester, and Tults[96] have investigated the application of liquid-crystal matrix displays to television and have concluded that a nonlinear threshold or isolation device, such as a diode or transistor, must be inserted in series with the liquid-crystal element at each matrix intersection to obtain the required speed and legibility for line-at-a-time addressing.

A television-rate line-at-a-time display operates in a manner similar to the small 5 × 7 matrix described previously, with the main differences being size and speed. The frame time is 30 msec and the line time is $30 \times 10^{-3}/500$ sec, or 60 μsec. A small section of a much larger matrix with diodes serving as the isolation devices is shown in Fig. 18. Assuming that the liquid crystal is exhibiting dynamic scattering, the response of a single diode–liquid-crystal combination is given in Fig. 19. Due to the inherent dielectric relaxation time, ϵ/σ, the addressing voltage across the cell is stored for a time (1 to 10 msec) sufficient to cause the excitation of the fluid deformation. The same basic description of the time response applies to all the electro-optic phenomena. Of course, ϵ/σ varies from material to material.

The design of a color-television display panel that uses the twisted nematic phenomena and polycrystalline thin-film transistors (TFT's)

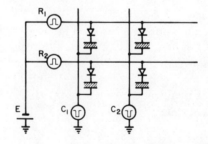

Fig. 18—A small section of a liquid-crystal–diode matrix display.

as the isolation devices has been described by Fischer, Brody, and Escott.[118] In the TFT's, CdSe serves as the semiconductor and Al_2O_3 as the gate insulator. The display has been constructed and is 6 × 6 inches in size with 14,400 TFT's in a 120 × 120 array.[119] Operation of the entire panel has recently been demonstrated, but defects were still present in the panel and the display was only black and white.[120] Good reliability is claimed for the display.

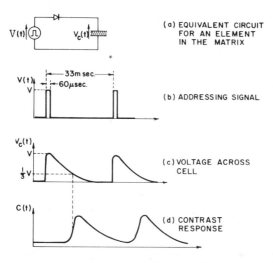

Fig. 19—Time response of a single intersection in the liquid-crystal–diode array.

Lipton and Koda[121] have also recently presented results on a CdSe TFT–liquid-crystal panel. They used relatively high-conductivity dynamic-scattering material; consequently, they added a capacitor in parallel with each TFT and display element to obtain the required electrical decay time. Brody[118,120] and co-workers were able to utilize the long dielectric relaxation time associated with the twisted nematic fluid and did not have to add the supplemental capacitance.

Liquid-crystal displays have also been constructed on a matrix of single-crystal silicon MOS FET's.[122] Pictorial-gray-scale images have been created on a 1 inch display although line defects were present. Though large arrays of TFT's should be much more economical than silicon MOS FET's, TFT's have suffered in the past from stability and reliability problems. More experimentation is necessary to prove their capabilities as the threshold devices in a liquid-crystal panel.

4.2 Beam Scanning

Images are produced on a cathode-ray tube by scanning a high-voltage electron beam across the surface of the cathodoluminescent material. Each position on the phosphor is excited sequentially as the beam is scanned by the deflection electron optics. This is an example of element-at-a-time addressing. Beam scanning in an element-at-a-time mode can be performed using either an electron beam or a light beam. Both techniques have been implemented with liquid crystals.

Van Raalte[123] was the first to describe the results of an electron-beam-scanned dynamic-scattering display. A schematic diagram of his demountable cathode-ray tube and some typical images obtained from the liquid-crystal display are presented in Fig. 20. The liquid

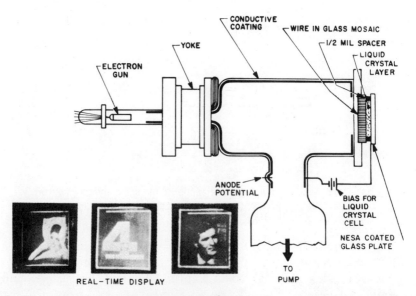

Fig. 20—Illustration of the electron-beam-scanned liquid-crystal display with some images created in the display (Ref. [123]).

crystal was sandwiched between a tin-oxide-coated glass slide and a mosaic feed-through plate constructed with fine wires inserted in glass. A segmented mirror was evaporated on the wire pin mosaic array that provided electrical contact between the electron beam and the liquid crystal. With the electron beam scanning at video rates, a maximum contrast ratio of 7.5 to 1 was obtained. Though very interesting, the display suffered from two deficiencies; (1) the liquid crystal was addressed by unipolar voltage pulses and (2) the pin mosaic was extremely difficult to manufacture at the necessary resolution. Gooch[124] and co-workers have presented a solution to the first problem. They have devised an electron-beam-scanned liquid-crystal display using bistable secondary-electron emission techniques combined with a dielectric layer to drive the liquid crystal with an ac potential. The economical construction of a hermetically sealed high-resolution wire mosaic is still very difficult, although progress has been made.[125]

The complexity of the feed-through mosaic is one of the motivating factors in the development of the light-beam-addressing approaches. In the most common variation of this second method, light illuminates a thin-film photoconductor–liquid-crystal sandwich (see Fig. 21). Margerum, Nimoy, and Wong[126] presented the operation of a

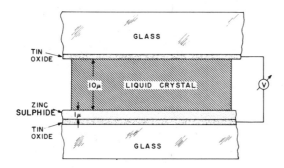

Fig. 21—Schematic diagram of a liquid-crystal–photoconductor structure.

photoconductor–liquid-crystal device. Both dynamic-scattering and storage effects were produced with light irradiating the photoconductor through the glass plate. Unilluminated, the resistance of the ZnS layer was higher than that of the liquid crystal. With UV radiation of sufficient intensity, the resistance of the photoconductor dropped below that of the liquid crystal and the voltage activated the liquid crystal. The ZnS was insensitive to the visible radiation that was used to observe the image in the liquid crystal. A sensitivity of 0.1 mJ/cm^2 was obtained.

White and Feldman[127] improved the sensitivity of the display to light by using evaporated selenium as the photoconductor. An opaque, highly reflecting light barrier was placed between the photoconductor and the liquid crystal to prevent the excitation of the photoconductor by the viewing light. Assouline, Hareng, and Leiba[128] used CdS as the photoconductor, and measured a sensitivity of 5 ×

10^{-6} J/cm². Similar results with CdS have been obtained by Jacobson et al.[29] Haas, Adams, Dir, and Mitchell[130] obtained a sensitivity of 2.5×10^{-6} J/cm² using an unspecified photoconductor and storage-effect liquid crystals.

All of the liquid-crystal–photoconductor structures discussed so far were excited by a dc voltage source. Under dc operation, various electrochemical life-degrading interactions have been observed at the liquid-crystal–photoconductor interface.[130] Beard, Bleha, and Wong[131] have reported on a photoconductor–liquid-crystal valve for projection applications that is driven by an ac voltage source (see Fig. 22). In addition, the display is operated in the reflection mode with a multilayer combination of dielectric mirror and absorbing layer to separate the CdS photoconductor from the projection light. With 200 lumens/cm² of white projection light irradiating the valve, no noticeable interaction between the projection light and the photoconductor has been observed. The light valve has been excited with the projected image from a CRT[132] as well as a static slide image. At the present time, the contrast ratio and speed are still below television standards. The above structure with the induced birefringence mode has been used as the light valve in a color projection display.[133]

The photoconductor–liquid-crystal sandwiches incorporate either a dc driving source or a somewhat complicated multilayer structure. Maydan, Melchior, and Kahn[134] have circumvented both of these aspects by utilizing the thermo-optic properties of nematic–cholesteric mixtures reported by Soref.[135] He showed that nematic–cholesteric

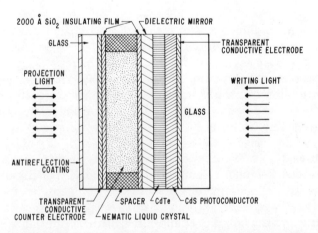

Fig. 22—Side view of an ac addressed liquid-crystal–photoconductor light valve (Ref. [131]).

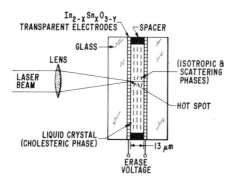

Fig. 23—A laser-beam-addressed thermo-optic liquid-crystal light valve (Ref. [134]).

mixtures can be converted from the clear planar texture to the highly scattering state by heating the material from the mesophase into the isotropic phase and then letting it cool. In their laser-beam-scanned display, Maydan and co-workers use the heat absorbed from the laser beam by the $In_{2-x}Sn_xO_{3-y}$ coatings to change the nematic-cholesteric storage material to the scattering state (see Fig. 23). The stored information is removed by exciting the liquid crystal with a high-frequency erase signal.

An improvement of this device has been described in which the nematic-cholesteric mixture is replaced by a smectic material.[136] Thermal writing induces the change of the smectic from the perpendicular to a scattering texture. Unlike the nematic–cholesteric materials, selective erasure is possible with the smectic device. The thermal writing is too slow for television-rate applications because of the thermal inertia of the glass–liquid-crystal system. With a laser-beam power of 20 mW, addressing speed is approximately 10^4 elements/sec for the smectic device. In the projection mode, the resolution is 50 lines/mm at a contrast ratio of approximately 10:1.

5. Summary

The different electro-optic phenomena have been classified into those that involve only dielectric forces and those that depend upon the interaction of conduction and dielectric torques. The field-effect phenomena possess several common properties. The resistivity of the materials may be as high as chemically practical, i.e., $\rho \gtrsim 10^{11}$ ohm-cm. For the induced birefringence, twisted nematic, and guest–host color switching effects, the threshold voltages are less than 3 or 4 volts, with an observed minimum of approximately one volt. The first two properties imply a very low power dissipation of less than 1 $\mu W/cm^2$. In ambient lighting, a diffuse reflector can be used with both the guest–host and twisted nematic effects to obtain a high-contrast and glare-free display.

Dynamic scattering and the storage effect are characterized by forward scattering, turbulent fluid motion during the presence of a field, and fluid resistivities that range between 10^8 and 10^{10} ohm-cm. Both effects can only occur with the applied frequency less than the dielectric relaxation frequency of the liquid crystal. The power dissipation is between 0.1 and 1 mW/cm^2 at an operating voltage of 15 V_{rms}. Neither a polarizer nor an analyzer are required. The static scattering can be stored in the nematic–cholesteric mixtures for anywhere from seconds to months.

Electrical addressing of liquid-crystal displays is accomplished either by X-Y matrix addressing or a scanning electron or light beam. The electron-beam approach is analogous to a cathode ray tube with the phosphor screen replaced by a liquid-crystal–pin wire mosaic feedthrough. Similarly, an amplitude-modulated scanning light beam can activate a photoconductor–liquid-crystal cell to provide the desired spatial pattern in the liquid crystal.

Small and/or slow matrix displays have been fabricated using the inherent threshold characteristics of liquid crystals. Both single-frequency and dual-frequency schemes have been described. The latter approach possesses greater multiplexing capability than the former, but at the price of higher applied voltage and larger power dissipation. However, due to speed of response and contrast ratio limitations, the construction of a 500 × 500 element liquid-crystal display operating at video rates does not appear feasible without the addition of a nonlinear device at each X-Y intersection. Thin-film transistor arrays are being fabricated to serve as the nonlinear devices. Their performance is being evaluated.

References

[1] C. W. Oseen, "The Theory of Liquid Crystals," *Trans. Faraday Soc.*, **2**, p. 833 (1933).
[2] F. C. Frank, "On the Theory of Liquid Crystals," *Disc. Faraday Soc.*, **25**, p. 19 (1958).
[3] P. Sheng, "Introduction to the Elastic Continuum Theory of Liquid Crystals, "Chapter 8.
[4] P. deGennes, *The Physics of Liquid Crystals,* Oxford University Press, London (1974).
[5] W. Helfrich, "Electric Alignment of Liquid Crystals," *Mol. Cryst. and Liq. Cryst.*, **21**, p. 187 (1973).
[6] H. Gruler, T. J. Scheffer, and G. Meier, "Elastic Constants of Nematic Liquid Crystals I. Theory of the Normal Deformation," *Z. Naturforsch. A,* **27a**, p. 966 (1972).
[7] M. F. Shiekel and K. Fahrenschon, "Deformation of Nematic Liquid Crystals with Vertical Orientation in Electrical Fields," *Appl. Phys. Lett.*, **19**, p. 393 (1971).

[8] F. J. Kahn, "Electric-Field Induced Orientational Deformation of Nematic Liquid Crystals: Tunable Birefringence," *Appl. Phys. Lett.*, **20,** p. 199 (1972).

[9] R. A. Soref and M. J. Rafuse, "Electrically Controlled Birefringence of Thin Nematic Films," *J. Appl. Phys.*, **43,** p. 2029 (1972).

[10] H. Mailer, K. L. Likins, T. R. Tayloer, and J. L. Fergason, "Effect of Ultrasound on a Nematic Liquid Crystal," *Appl. Phys. Lett.*, **18,** p. 105 (1971).

[11] H. Gruler and G. Meier, "Electric Field Induced Deformations in Oriented Liquid Crystals of the Nematic Type," *Mol. Cryst. and Liq. Cryst.*, **16,** p. 299 (1972).

[12] H. Deuling, "Deformation of Nematic Liquid Crystals in an Electric Field," *Mol. Cryst. and Liq. Cryst.* **19,** p. 123 (1972).

[13] L. T. Creagh and A. R. Kmetz, "Mechanism of Surface Alignment in Nematic Liquid Crystals," *Mol. Cryst. and Liq. Cryst.*, **24,** p. 59 (1973).

[14] J. Robert and G. Labrunie, "Transient Behavior of the Electrically Controlled Birefringence in a Nematic Liquid Crystal," *J. Appl. Phys.*, **44,** p. 4689 (1973).

[15] G. Heilmeier and J. Goldmacher, U.S. Patent No. 3,499,702 (1970).

[16] M. Schadt and W. Helfrich, "Voltage Dependent Optical Activity of a Twisted Nematic Liquid Crystal," *Appl. Phys. Lett.* **18,** p. 127 (1971).

[17] F. M. Leslie, "Distortion of Twisted Orientation Patterns in Liquid Crystals by Magnetic Fields," *Mol. Cryst. and Liq. Cryst.*, **12,** p. 57 (1970).

[18] C. J. Alder and E. P. Raynes, "Room Temperature Nematic Liquid Crystal Mixtures with Positive Dielectric Anisotropy," *J. Phys.*, **D6,** p. L33 (1973).

[19] A. Boller, H. Scherrer, and M. Schadt, "Low Electro-Optic Threshold in New Liquid Crystals," *Proc. IEEE,* **60,** p. 1002 (1972).

[20] A. Ashford, J. Constant, J. Kirton, and E. P. Raynes, "Electro-Optic Performance of a New Room Temperature Nematic Liquid Crystal," *Elec. Lett.*, **9,** p. 118 (1973).

[21] R. R. Reynolds, C. Maze and E. P. Oppenheim, "Design Considerations for Positive Dielectric Nematic Mixtures Suitable for Display Applications," Abstracts of Fifth International Liquid Crystal Conf., Stockholm, Sweden, June 1974, p. 236.

[22] C. H. Gooch and H. A. Tarry, "Optical Characteristics of Twisted Nematic Liquid Crystal Films," *Elec. Lett.*, **10,** p. 2 (1974).

[23] J. Robert and F. Gharadjedaghi, "Rotation du Plan de Polarisation de la Lumière dans une Structure Nématique en Helice," *C. R. Acad. Sc. Paris*, **278B,** p. 73 (1974).

[24] C. J. Gerritsma, W. H. DeJeu and P. VanZanten, "Distortion of a Twisted Nematic Liquid Crystal by a Magnetic Field," *Phys. Lett.*, **36A,** p. 389 (1971).

[25] C. Z. VanDoorn, "On the Magnetic Threshold for the Alignment of a Twisted Nematic Crystal," *Phys. Lett.*, **42A,** p. 537 (1973).

[26] A. I. Baise and M. M. Labes, "Effect of Dielectric Anisotropy on Twisted Nematics," *Appl. Phys. Lett.*, **24,** p. 298 (1974).

[27] D. Meyerhofer, "Electro-optic Properties of Twisted Field Effect Cells," Abstracts of Fifth International Liquid Crystal Conf., Stockholm, Sweden, June 1974, p. 220.

[28] D. W. Berreman, "Optics in Smoothly Varying Anisotropic Planar Structures: Applicated to Liquid Crystal Twist Cells," *J. Opt. Soc. Amer.*, **63,** p. 1374 (1973).

[29] S. Kobayashi and F. Takeuchi, "Multicolor Field-Effect Display Devices with Twisted Nematic Liquid Crystals," *Proc. of S.I.D.*, **14,** p. 115 (1973).

[30] C. Z. VanDoorn and J. L. A. M. Heldens, "Angular Dependent Optical Transmission of Twisted Nematic Liquid Crystal Layers," *Phys. Lett.*, **47A,** p. 135 (1974).

[31] F. Brochard, "Backflow Effects in Nematic Liquid Crystals," *Mol. Cryst. and Liq. Cryst.*, **23,** p. 51 (1973).

[32] C. J. Gerritsma, J. A. Geurst and A. M. J. Spruijt, "Magnetic-Field-Induced Motion of Disclinations in a Twisted Nematic Layer," *Phys. Lett.*, **43A,** p. 356 (1973).

[33] E. P. Raynes, "Twisted Nematic Liquid Crystal Electro-Optic Devices with Areas of Reverse Twist," *Elec. Lett.*, **9,** p. 101 (1973).

[34] E. P. Raynes, "Improved Contrast Uniformity in Twisted Nematic Liquid Crystal Electro-Optic Display Devices," *Elec. Lett.*, **10,** p. 141 (1974).

[35] P. J. Wild, "Twisted Nematic Liquid Crystal Displays with Low Threshold Voltage," *Comptes Rendus des Journées d'Electronique, EPFL*, p. 102 (1973).

[36] G. H. Heilmeier and L. A. Zanoni, "Guest-Host Interactions in Nematic Liquid Crystals—A New Electro-Optic Effect," *Appl. Phys. Lett.*, **13,** p. 91 (1968).

[37] G. H. Heilmeier, J. A. Castellano, and L. A. Zanoni, "Guest-Host Interactions in Nematic Liquid Crystals," *Mol. Cryst. and Liq. Cryst.*, **8,** p. 293 (1969).

[38] J. J. Wysocki, J. Adams, and W. Haas, "Electric-Field-Induced Phase Change in Cholesteric Liquid Crystals," *Phys. Rev. Lett.,* **20,** p. 1024 (1968).
[39] E. Sackmann, S. Meiboom and L. C. Snyder, "On the Relation of Nematic to Cholesteric Mesophases," *J. Am. Chem. Soc.,* **89,** p. 5981 (1967).
[40] P. G. deGennes, "Calcul de la Distortion D'une Structure Cholesterique Par un Champ Magnetique," *Sol. St. Commun.,* **6,** p. 163 (1968).
[41] R. B. Meyer, "Effects of Electric and Magnetic Fields on the Structures of Cholesteric Liquid Crystals," *Appl. Phys. Lett.,* **12,** p. 281 (1968).
[42] F. Rondelez and J. P. Hulin, "Distortions of a Planar Cholesteric Structure Induced by a Magnetic Field," *Sol. St. Commun.,* **10,** p. 1009 (1972).
[43] F. J. Kahn, "Electric-Field-Induced Color Changes and Pitch Dilation in Cholesteric Liquid Crystals," *Phys. Rev. Lett.,* **24,** p. 209 (1969).
[44] R. B. Meyer, "Distortion of a Cholesteric Structure by a Magnetic Field," *Appl. Phys. Lett.,* **14,** p. 208 (1969).
[45] G. Durand, L. Leger, F. Rondelez and M. Veyssie, "Magnetically Induced Cholesteric-to-Nematic Phase Transition in Liquid Crystals," *Phys. Rev. Lett.,* **22,** p. 227 (1969).
[46] G. H. Heilmeier and J. E. Goldmacher, "Electric-Field-Induced Cholesteric-Nematic Phase Change in Liquid Crystals," *J. Chem. Phys.,* **51,** p. 1258 (1969).
[47] G. W. Gray, K. J. Harrison and J. A. Nash, "New Family of Nematic Liquid Crystals for Displays," *Elec. Lett.,* **9,** p. 130 (1973).
[48] J. P. Hulin, "Parametric Study of the Optical Storage Effect in Mixed Liquid Crystal Systems," *Appl. Phys. Lett.,* **21,** p. 455 (1972).
[49] W. Helfrich, "Deformation of Cholesteric Liquid Crystals with Low Threshold Voltage," *Appl. Phys. Lett.,* **17,** p. 531 (1970).
[50] W. Helfrich, "Electrohydrodynamic and Dielectric Instabilities of Cholesteric Liquid Crystals," *J. Chem. Phys.* **55,** p. 839 (1971).
[51] J. Hurault, "Static Distortions of a Cholesteric Planar Structure Induced by Magnetic or A.C. Electric Fields," Fourth International Liquid Crystal Conf., Kent, Ohio, Aug. 1972.
[52] C. J. Gerritsma and P. VanZanten, "Periodic Perturbations in the Cholesteric Plane Texture," *Phys. Lett.,* **37A,** p. 47 (1971).
[53] T. J. Scheffer, "Electric and Magnetic Field Investigations of the Periodic Gridlike Deformation of a Cholesteric Liquid Crystal," *Phys. Rev. Lett.* **28,** p. 598 (1972).
[54] F. Rondelez, H. Arnould and C. J. Gerritsma, "Electrohydrodynamic Effects in Cholesteric Liquid Crystals Under AC Electric Fields," *Phys. Rev. Lett.,* **28,** p. 735 (1972).
[55] J. J. Wysocki et al., "Cholesteric-Nematic Phase Transition Displays," *Proc. SID,* **13,** p. 115 (1972).
[56] T. Ohtsuka and M. Tsukamoto, "AC Electric-Field-Induced Cholesteric-Nematic Phase Transition in Mixed Liquid Crystal Films," *Jap. J. Appl. Phys.,* **12,** p. 22 (1973).
[57] W. F. Greubel, "Bistability Behavior of Texture in Cholesteric Liquid Crystals in an Electric Field," *Appl. Phys. Lett.,* **25,** p. 5 (1974).
[58] D. L. White and G. N. Taylor, "A New Absorptive Mode Reflective Liquid Crystal Display Device," *J. Appl. Phys.,* **45,** p. 4718 (1974).
[59] G. H. Heilmeier, L. A. Zanoni and L. A. Barton, "Dynamic Scattering in Nematic Liquid Crystals," *Appl. Phys. Lett.,* **13,** p. 46 (1968).
[60] G. H. Heilmeier, L. A. Zanoni and L. A. Barton, "Dynamic Scattering: A New Electro-Optic Effect in Certain Classes of Nematic Liquid Crystals," *Proc. IEEE,* **56,** p. 1162 (1968).
[61] C. Deutsch and P. N. Keating, "Scattering of Coherent Light from Nematic Liquid Crystals in the Dynamic Scattering Mode," *J. Appl. Phys.,* **40,** p. 4049 (1969).
[62] W. Helfrich, "Conduction-Induced Alignment of Nematic Liquid Crystals: Basic Model and Stability Considerations," *J. Chem. Phys.,* **51,** p. 4092 (1969).
[63] E. F. Carr, "Ordering in Liquid Crystals Owing to Electric and Magnetic Fields," *Advan. Chem. Ser.,* **63,** p. 76 (1967).
[64] R. Williams, "Domains in Liquid Crystals," *J. Chem. Phys.,* **39,** p. 384 (1963).
[65] P. A. Penz, "Voltage-Induced Vorticity and Optical Focusing in Liquid Crystals," *Phys. Rev. Lett.,* **24,** p. 1405 (1970).
[66] T. O. Carroll, "Liquid Crystal Diffraction Grating," *J. Appl. Phys.,* **43,** p. 767 (1972).
[67] G. Assouline, A. Dmitrieff, M. Hareng, and E. Leiba, "Diffraction d'un Faisceau Laser par un Cristal Liquide Nématique Soumis à un Champ Électrique," *C. R. Acad. Sci. Paris,* **271B,** p. 857 (1970).
[68] E. Dubois-Vilette, P. G. deGennes, and O. Parodi, "Hydrodynamic Instabilities of Nematic Liquid Crystals Under AC Electric Fields," *J. Physique,* **32,** p. 305 (1971).

[69] Orsay Liquid Crystal Group, "Hydrodynamic Instabilities in Nematic Liquids Under AC Electric Fields," *Phys. Rev. Lett.*, **25**, p. 1642 (1970).

[70] P. G. deGennes, "Electrohydrodynamic Effects in Nematics," *Comments Sol. St. Phys.*, **3**, p. 148 (1971).

[71] R. A. Kashnow and H. S. Cole, "Electrohydrodynamic Instabilities in a High-Purity Nematic Liquid Crystal," *J. Appl. Phys.*, **42**, p. 2134 (1971).

[72] P. A. Penz and G. W. Ford, "Electrohydrodynamic Solutions for Nematic Liquid Crystals," *Appl. Phys. Lett.*, **20**, p. 415 (1972).

[73] P. A. Penz and G. W. Ford, "Electromagnetic Hydrodynamics of Liquid Crystals," *Phys. Rev.*, **6A**, p. 414 (1972).

[74] D. Meyerhofer, "Electro Hydrodynamic Instabilities in Nematic Liquid Crystals," Chapter 9.

[75] T. O. Carroll, "Dependence of Conduction-Induced Alignment of Nematic Liquid Crystals Upon Voltage Above Threshold," *J. Appl. Phys.*, **43**, p. 1342 (1972).

[76] L. Goodman, "Light Scattering in Electric-Field Driven Nematic Liquid Crystals," *Proc. SID*, **13**, p. 121 (1972).

[77] L. Cosentino, "On the Transient Scattering of Light by Pulsed Liquid Crystal Cells," *IEEE Trans. Electron Devices*, **ED-1**, p. 1192 (1971).

[78] L. Creagh, A. Kmetz and R. Reynolds, "Performance Characteristics of Nematic Liquid Crystal Display Devices," *IEEE Trans. Electron Devices*, **ED-18**, p. 672 (1971).

[79] W. F. Greubel and U. W. Wolff, "Electrically Controllable Domains in Nematic Liquid Crystals," *Appl. Phys. Lett.*, **19**, p. 213 (1971).

[80] W. H. DeJeu, C. J. Gerritsma, and A. M. VanBoxtel, "Electrohydrodynamic Instabilities in Nematic Liquid Crystals," *Phys. Lett.*, **34A**, p. 203 (1971).

[81] L. K. Vistin, "Electrostructural Effect and Optical Properties of a Certain Class of Liquid Crystals and Their Binary Mixtures," *Sov. Phys. Crystallogr.*, **15**, p. 514 (1970).

[82] N. Felici, "Phénomènes Hydro et Aerodynamiques dans la Conduction des Dielectrique Fluide," *Rev. Gen. Elec.*, **78**, p. 717 (1969).

[83] Orsay Liquid Crystal Group, "AC and DC Regimes of the Electrohydrodynamic Instabilities in Nematic Liquid Crystals," *Mol. Cryst. and Liq. Cryst.*, **12**, p. 251 (1971).

[84] H. Koelmans and A. M. VanBoxtel, "Electrohydrodynamic Flow in Nematic Liquid Crystals," *Mol. Cryst. and Liq. Cryst.*, **12**, p. 185 (1971).

[85] D. Meyerhofer and A. Sussman, "The Electrohydrodynamic Threshold in Nematic Liquid Crystals in Low Frequency Fields," *Appl. Phys. Lett.*, **20**, p. 337 (1972).

[86] A. Sussman, "Electro-Optic Liquid Crystal Devices: Principles and Applications," *IEEE Trans. Parts, Hybrids and Packaging*, **PHP-8**, p. 28 (1972).

[87] G. Durand, M. Veyssie, F. Rondelez and L. Leger, "Effet Électrohydrodynamique dans un Cristal Liquide Nématique," *C. R. Acad. Sc. Paris*, **270B**, p. 97 (1970).

[88] G. H. Heilmeier and J. E. Goldmacher, "A New Electric Field Controlled Reflective Optical Storage Effect in Mixed Liquid Crystal Systems," *Proc. IEEE* **57**, p. 34 (1969).

[89] G. Dir et al., "Cholesteric Liquid Crystal Texture Change Displays," *Proc. SID*, **13**, p. 105 (1972).

[90] D. Meyerhofer and E. F. Pasierb, "Light Scattering Characteristics in Liquid Crystal Storage Materials," *Mol. Cryst. and Liq. Cryst.*, **20**, p. 279 (1973).

[91] E. Jakeman and E. P. Raynes, "Electro-Optic Response Times in Liquid Crystals," *Phys. Lett.*, **39A**, p. 69 (1972).

[92] J. Robert, G. Labrunie and J. Borel, "Static and Transient Electric Field Effect on Homeotropic Thin Layers," *Mol. Cryst. and Liq. Cryst.*, **23**, p. 197 (1973).

[93] A. Sussman, "Secondary Hydrodynamic Structure in Dynamic Scattering," *Appl. Phys. Lett.*, **21**, p. 269 (1972).

[94] C. J. Gerritsma, C. Z. VanDoorn and P. VanZanten, "Transient Effects in the Electrically Controlled Light Transmission of a Twisted Nematic Layer," *Phys. Lett.*, **48A**, p. 263 (1974).

[95] C. H. Gooch and H. A. Tarry, "Dynamic Scattering in the Homeotropic and Homogeneous Textures of a Nematic Liquid Crystal," *J. of Phys. D. Appl. Phys.*, **5**, p. L25 (1972).

[96] B. J. Lechner, F. Marlowe, E. Nester, and J. Tults, "Liquid Crystal Displays," *Proc. IEEE*, **59**, p. 1566 (1971).

[97] W. H. DeJeu, C. J. Gerristma, P. VanZanten, and W. J. A. Gossens, "Relaxation of the Dielectric Constant Electrohydrodynamic Instabilities in a Liquid Crystal," *Phys. Lett.*, **39A**, p. 355 (1972).

[98] H. K. Bücher, R. T. Klingbiel, and J. P. VanMeter, "Frequency-Addressed Liquid Crystal Field Effect," *Appl. Phys. Lett.*, **25**, p. 186 (1974).

[99] E. P. Raynes and I. A. Shanks, "Fast Switching Twisted Nematic Electro-Optical Shutter and Color-Filter," *Elec. Lett.*, **10**, p. 114 (1974).

[100] T. S. Chang and E. E. Loebner, "Crossover Frequencies and Turn-Off Time Reduction Scheme for Twisted Nematic Liquid Crystal Displays," *Appl. Phys. Lett.*, **25**, p. 1 (1974).

[101] G. Baur, A. Stieb, and G. Meier, "Controlled Decay of Electrically Induced Deformations in Nematic Liquid Crystals," *Appl. Phys.*, **2**, p. 349 (1973).

[102] B. Kellenevich and A. Coche, "Relaxation of Light Scattering in Nematic-Cholesteric Mixtures," *Mol. Cryst. and Liq. Cryst.*, **24**, p. 113 (1973).

[103] A. Sussman, "Dynamic Scattering Life in the Nematic Compound p-Methoxybenzylidene-p-Amino Phenyl Acetate as Influenced by Current Density," *Appl. Phys. Lett.*, **21**, p. 126 (1972).

[104] L. Pohl, R. Steinsträsser, and B. Hampel, "Performance of Nematic Phase V and VA in Liquid Crystal Displays," Fourth Internat. Liq. Cryst. Conf., Kent, Ohio; Aug. 1972, Paper No. 144.

[105] I. Haller, "Elastic Constants of the Nematic Liquid Crystalline Phase of p-Methoxybenzylidene-p-n-Butylaniline (MBBA)," *J. Chem. Phys.*, **57**, p. 1400 (1972).

[106] A. Sussman, "Electrochemistry in Nematic Liquid-Crystal Solvents," Chapter 17.

[107] A. R. Kmetz, "Liquid Crystal Displays Prospects in Perspective," *IEEE Trans. Elec. Dev.*, **ED-20**, p. 954 (1973).

[108] M. Hareng, G. Assouline and E. Leiba, "La Biréfringence Électriquement Contrôlée dans les Cristaux Liquides Nématiques," *Appl. Opt.*, **11** p. 2920 (1972).

[109] P. M. Alt and P. Pleshko, "Scanning Limitations of Liquid Crystal Displays," *IEEE Trans. Elec. Dev.*, **ED-21**, p. 146 (1974).

[110] H. Takata, O. Kogure, and K. Murase, "Matrix-Addressed Liquid Crystal Display," *IEEE Trans. Elec. Dev.*, **ED-20**, p. 990 (1973).

[111] M. Hareng, G. Assouline, and E. Leiba, "Liquid Crystal Matrix Display by Electrically Controlled Birefringence," *Proc. IEEE*, **60**, p. 913 (1972).

[112] M. F. Schiekel and K. Fahrenschon, "Multicolor Matrix Displays Based on Deformation of Vertically Aligned Nematic Liquid Crystal Phases," Digest 1972 Soc. for Information Display International Symp., San Francisco, Calif., p. 98.

[113] T. Ohtsuka, M. Tsukamoto, and M. Tsuchiya, "Liquid Crystal Matrix Display," *Jap. J. Appl. Phys.*, **12**, p. 371 (1973).

[114] C. R. Stein and R. A. Kashnow, "A Two Frequency Coincidence Addressing Scheme for Nematic Liquid Crystal Displays," *Appl. Phys. Lett.*, **19**, p. 343 (1971).

[115] P. J. Wild and J. Nehring, "An Improved Matrix Addressed Liquid Crystal Display," *Appl. Phys. Lett.*, **19**, p. 335 (1971).

[116] C. R. Stein and R. A. Kashnow, "Recent Advances in Frequency Coincidence Matrix Addressing of Liquid Crystal Displays," Digest 1972 Soc. for Information Display International Symp., San Francisco, Calif. p. 64.

[117] M. Schadt, "Dielectric Properties of Some Nematic Liquid Crystals with Strong Positive Dielectric Anisotropy," *J. Chem. Phys.*, **56**, p. 1494 (1972).

[118] A. G. Fischer, T. P. Brody, and W. S. Escott, "Design of a Liquid Crystal Color TV Panel," IEEE Conf. Record 1972 Conf. on Display Devices, New York, NY, p. 64.

[119] T. P. Brody, J. Asars and G. D. Dixon, "A 6 × 6 Inch 20 Lines per Inch Liquid Crystal Display Panel," *IEEE Trans. Elec. Dev.*, **ED-20**, p. 995 (1973).

[120] T. F. Brody, F. C. Luo, D. H. Vavies, and E. W. Greeneich, "Operational Characteristics of a 6 × 6 Inch, TFT Matrix Array, Liquid Crystal Display," Digest 1974 Soc. for Information Display International Symp., San Diego, Calif., p. 166.

[121] L. Lipton and N. Koda, "Liquid Crystal Matrix Display for Video Applications," *Proc. SID*, **14**, p. 127 (1973).

[122] M. Ernstoff, A. M. Leupp, M. J. Little and H. T. Peterson, "Liquid Crystal Pictorial Display," Technical Digest 1973 International Electron Devices Meeting, Washington, D.C., p. 548.

[123] J. A. van Raalte, "Reflective Liquid Crystal Television Display," *Proc. IEEE*, **56**, p. 2146 (1968).

[124] C. H. Gooch et al., "A Storage Cathode-Ray Tube with Liquid Crystal Display," *J. Phys.*, **D6**, p. 1664 (1974).

[125] C. Burrowes, "Electrical Fiber Plates—A New Tool For Storage and Display," *IEEE Conf. Record 1970 IEEE Conf. on Display Devices*, New York, NY, p. 126.

[126] J. D. Margerum, J. Nimoy and S.-Y. Wong, "Reversible Ultraviolet Imaging with Liquid Crystals," *Appl. Phys. Lett.*, **17**, p. 51 (1970).

[127] D. L. White and M. Feldman, "Liquid Crystal Light Valves," *Elec. Lett.*, **6**, p. 837 (1970).

[128] G. Assouline, M. Hareng, and E. Leiba, "Liquid Crystal and Photoconductor Image Converter," *Proc. IEEE*, **59**, p. 1355 (1971).

[129] A. Jacobson et al., "Photoactivated Liquid Crystal Light Valve," Digest 1972 SID International Symp., San Francisco, Calif., p. 70.

[130] W. Haas, J. Adams, G. Dir and C. Mitchell, "Liquid Crystal Memory Panels," *Proc. SID,* **14,** p. 121 (1973).

[131] T. D. Beard, W. P. Bleha and S.-Y. Wong, "Alternating Current Liquid Crystal Light Valve," *Appl. Phys. Lett.,* **22,** p. 90 (1973).

[132] W. P. Bleha, J. Grinberg, and A. D. Jacobson, "AC Driven Photoactivated Liquid Crystal Light Valve," Digest 1973 SID International Symp., New York, NY, p. 42.

[133] J. Grinberg et al., "Photoactivated Liquid Crystal Light Valve for Color Symbology Display," *Conf. Record 1974 IEEE-SID Conf. on Display Devices and Systems,* New York, p. 47.

[134] D. Maydan, H. Melchior and F. Kahn, "Thermally Addressed Electrically Erased High-Resolution Liquid Crystal," *Appl. Phys. Lett.,* **21,** p. 392 (1972).

[135] R. A. Soref, "Thermo-Optic Effects in Nematic-Cholesteric Mixtures," *J. Appl. Phys.,* **41,** p. 3022 (1970).

[136] F. J. Kahn, "IR-Laser-Addressed Thermo-Optic Smectic Liquid Crystal Storage Displays," *Appl. Phys. Lett.,* **22,** p. 111 (1973).

15

Liquid-Crystal Optical Waveguides

D. J. Channin

RCA Laboratories, Princeton, N. J. 08540

1. Introduction

In recent years a new optical technology has developed in the field of light guiding in thin films and fibers. Liquid-crystal materials have yet to contribute to this technology to the extent that they have to display devices and optical beam processing; nevertheless, examples of liquid-crystal waveguide modulators,[1,2] switches,[3] and deflectors[4] have been demonstrated. Furthermore, optical waveguiding holds potential as a tool for investigating physical and chemical processes in thin films,[5] including liquid crystals.

This chapter reviews the basic theory of optical waveguiding in planar structures, with emphasis on the general concepts of mode spectra, energy distribution within the guide, phase matching, and scattering. Experimental techniques appropriate to liquid crystals are discussed, and some experimental results on attenuation measurements and electro-optic effects are described. More detailed discussions of optical waveguide theory and technology (often called integrated optics) are available in recent reviews and books.[6-8]

2. Guided Optical Waves

The simplest optical waveguide structure is an infinite planar slab of perfectly transparent, isotropic, dielectric material with refractive index n_g bounded on both surfaces by similar material of index $n_o < n_g$ (see Fig. 1). According to Snell's law, light rays in the slab will be totally internally reflected if they have an incidence angle θ with a surface satisfying

$$\sin\theta > \frac{n_0}{n_g}. \qquad [1]$$

Since both surfaces of the slab are parallel, the light will undergo an identical reflection at the opposite boundary and be trapped within

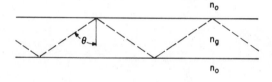

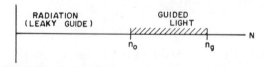

Fig. 1—Total internal reflection of light in slab waveguide. $N = c/v$ is the effective refractive index for phase velocity v.

the slab. The phase velocity v of the light as it travels within the slab is given by

$$v = \frac{c}{n_g \sin\theta}, \qquad [2]$$

where c is the speed of light in free space. From Eqs. [1] and [2] it is apparent that the guided light may have phase velocities within the range

$$\frac{c}{n_0} < v < \frac{c}{n_g}. \qquad [3]$$

Propagation under such conditions is called waveguiding.

If θ is too small to satisfy Eq. [1], some light is transmitted out of the slab at each encounter with the surface. Nevertheless, sufficient reflection may occur that the light in the slab can be characterized as a "leaky" wave, subject to loss by radiation away from the slab. It is convenient to deal with the effective refractive index $N \equiv c/v$ to characterize both leaky and truly guided light.

It has been tacitly assumed that the thickness of the slab is much greater than an optical wavelength, so that geometric optics is applicable. Should the thickness become comparable to the wavelength, destructive interference between the multiple reflections prevents guided propagation except at specific incidence angles for which the interference is constructive. Waveguiding then occurs only for a discrete set of guided modes, with effective indices N_j determined by the guide thickness d as well as the two refractive indices n_g and n_o. This situation, though retaining many features of the geometric optics limit, is properly described by wave optics.

Suppose now that the slab and surroundings are replaced by a medium that is translationally invariant in the y-z plane, but has refractive index $n(x)$ that varies with position in the x direction. The conditions

$$1 \le n_0 \le n(x) \le n_g \qquad [4a]$$

$$n(\pm\infty) = n_0 \qquad [4b]$$

are imposed on $n(x)$.

Consider light propagating parallel to the z axis, translationally invariant along the y axis, and having an electric field ϵ_y. Such light satisfies the wave equation

$$\frac{\partial^2 \epsilon_y}{\partial x^2} + \frac{\partial^2 \epsilon_y}{\partial z^2} + K_0^2 n^2(x) \epsilon_y = 0. \qquad [5]$$

Separating the variables by coordinates yields

$$\epsilon_y = \epsilon(x) \exp\{iKz\} \qquad [6]$$

$$\frac{\partial^2 \epsilon}{\partial x^2} + K_0^2 \left[n^2(x) - \frac{K^2}{K_0^2} \right] \epsilon = 0. \qquad [7]$$

In these equations $K_o = 2\pi/\lambda_o$, where λ_o is the free-space optical wavelength. It is convenient to make again the identification of $K/K_o = c/v = N$, the effective refractive index of the guided modes.

Equations formally identical to Eq. [7] arise in many physical problems, such as the Schrödinger equation for a particle in a potential well. N is the eigenvalue for this equation and, in accordance with Eqs. [4a] and [4b], will take on a discrete and finite spectrum N_o, $N_1, \ldots N_j, \ldots N_{\max}$ within the range $n_o < N_j < n_g$, and a continuous spectrum in the range $1 < N < n_o$. The former characterizes the guided modes and the latter the so-called radiation modes. The two kinds of modes together are associated with a complete orthogonal set of eigenfunctions of Eq. [7]. The leaky modes are solutions of Eq. [7] but are not part of the complete set of eigenfunctions. They have a discrete spectrum within the range of N spanned by the continuous radiation mode spectrum. The ranges spanned by the different modes are shown in Fig. 2.

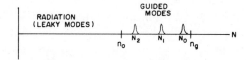

Fig. 2—Ranges of effective indices for guided and radiation modes.

We return now to the particular case of the slab waveguide defined by

$$n(x) = \begin{cases} n_g, |x| < d/2 \\ n_0, |x| > d/2 \end{cases} \quad [8]$$

The guided modes have effective indices N_j determined by the solutions of the dispersion relation

$$\tan\left(\frac{K_0 d}{2}\sqrt{n_g^2 - N_j^2}\right) = \begin{cases} -\sqrt{\dfrac{n_g^2 - N_j^2}{N_j^2 - n_0^2}}, & j = 1,3,\ldots \\ \sqrt{\dfrac{N_j^2 - n_0^2}{n_g^2 - N_j^2}}, & j = 0,2,\ldots \end{cases} \quad [9]$$

The number of modes increases with the guide thickness d. The lowest-order mode N_o is closest in effective index to n_g, while the highest order mode has effective index $N_{\max}$ closest to n_o.

The spatial distributions of the electric fields for these modes are given by

$$\epsilon_y(x,z) = A_j \exp(iK_0 N_j z) \begin{cases} \sin(xK_0\sqrt{n_g^2 - N_j^2}), \\ j = 1,3,\ldots |x| < d/2 \\ \cos(xK_0\sqrt{n_g^2 - N_j^2}), \\ j = 0,2,\ldots \end{cases}$$

$$\epsilon_y(x,z) = B_j \exp(iK_0 N_j z) \begin{cases} \exp(-xK_0\sqrt{N_j^2 - n_0^2}), \\ x > d/2 \\ \exp(+xK_0\sqrt{N_j^2 - n_0^2}) \\ x < d/2 \end{cases}$$

[10]

The exponentially decaying field outside the slab is called the evanescent field. Since the electric field ϵ_y lies in the waveguide plane the modes are called transverse electric or TE. A corresponding set of transverse magnetic or TM modes also exist. These satisfy a wave equation for H_y similar to Eq. [7]. The dispersion relationship for TM modes is

$$\tan\left(\frac{K_0 d}{2}\sqrt{n_g^2 - N_j^2}\right) = \begin{cases} -\left(\frac{n_0}{n_g}\right)^2 \sqrt{\frac{n_g^2 - N_j^2}{N_j^2 - n_0^2}}, \ j = 1,3,\ldots \\ \left(\frac{n_g}{n_0}\right)^2 \sqrt{\frac{N_j^2 - n_0^2}{n_g^2 - N_j^2}}, \ j = 0,2,\ldots \end{cases}$$

[11]

Eq. [11] differs slightly from Eq. [9], indicating differing phase velocities for TE and TM waves despite the assumption of isotropic materials. The velocity differences come from the differing phase shifts on reflection at the interfaces for light polarized parallel to (TE) or perpendicular to (TM) the reflecting surface. Should the waveguide be composed of anisotropic material there will be an additional velocity difference due to the variation of guide index n_g with optical polarization.

In practice, many optical waveguides are asymmetric, comprising thin films with substrate on one side and air on the other. Others have graded continuous index distributions produced by diffusion of atoms into or out of a bulk material. For such waveguides the range of effective indices spanned by the guided modes is

$$(n_0)_{\max} < N_j < n_g,$$

[12]

where $(n_o)_{\max}$ is the *greatest* of the indices surrounding the guide. The spacing of the modes and the light distribution within the guide is determined by the specific refractive index profile, though some features are common to all guides.

In general, the optical energy distribution is concentrated in the highest-index material for low-order modes, and spreads out into the lower-index material as the mode order increases. Thus, the optical field of a multimode waveguide will sample different regions of the guide as different modes are excited. The techniques of optical waveguiding are therefore potentially useful in studying the structure of thin films and layers.

3. Phase Matching and Coupling

The recent surge of activity in planar optical waveguides was initiated by the development of practical and efficient ways to couple light between guided modes and beams in free space.[9,10] These techniques are based on the introduction of an additional element to the wave-

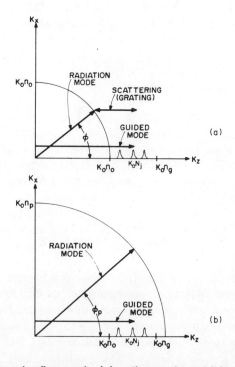

Fig. 3—Wave vector diagrams for (a) grating coupler and (b) prism coupler.

guide to match the phase velocities of the guided modes to those of the radiation modes.

Fig. 3 represents the extension of Figs. 1 and 2 to include the plane perpendicular to the waveguide. The axes now represent wave vectors, the magnitudes of which are related to the effective indices by multiplication by K_o. Fig. 3a shows the basis for a grating coupler. A small periodic modulation of the guide thickness or refractive index phase matches the guided mode with wave vector $K_o N_j$ to radiation in the form of a ray at angle ϕ to the guide surface. The periodic modulation has wavelength λ_s such that

$$K_s = \frac{2\pi}{\lambda_s} = K_0(N_j - n_0 \cos\phi). \qquad [13]$$

The basis for a prism coupler is shown in Fig. 3b. A small part of the waveguide is bounded on one side by material of index $n_p > n_g$. The condition

$$n_p \cos\phi_p = N_j \qquad [14]$$

establishes coupling between the guided mode N_j and a ray with incidence angle ϕ_p inside the high-index material. In practice the high-index material takes the form of a prism that refracts the internal ray into free space at a convenient angle.

Both kinds of couplers allow selective excitation of particular waveguide modes. TE or TM modes are determined by setting the polarization of the incident beam parallel to or perpendicular to the waveguide plane. The particular effective index N_j is determined by varying the coupling angle after having chosen a prism index n_p or periodic modulation spacing λ_s. In output coupling, a waveguide with one or more modes excited will cause beams of light to be coupled out at angles corresponding to the effective indices of the excited modes.

4. Scattering

Inhomogeneities in the waveguide material will scatter the guided light and thereby attenuate the waveguide modes. Such scattering may put light into other guided modes or into radiation modes. For waveguides with sharp boundaries such as the uniform slab, a distinction is made between scattering centers located at the material interfaces (surface scattering) and scattering centers distributed throughout the waveguide material (bulk scattering). In both cases the attenuation is a function of N_j because the different spatial distributions of light in the guide have differing overlap of the regions containing scattering centers.[8,11]

For TE modes in a symmetric slab guide with lossless surroundings, the bulk loss α_b is given by

$$\alpha_b = \alpha_{bo} \frac{n_g\sqrt{N_j^2 - n_0^2}}{N_j} \frac{\sqrt{N_j^2 - n_0^2} + K_0 d(n_g^2 - n_0^2)}{(n_g^2 - n_0^2)(1 + K_0 d\sqrt{N_j^2 - n_0^2})},$$

[15]

where α_{bo} is the bulk scattering coefficient of the waveguide material. If the scattering itself is wavelength dependent, α_{bo} will also vary with N_j. For example, if the attenuation is due to Rayleigh scattering, α_{bo} is proportional to $(N_j)^{-4}$. Eq. [15] shows that when α_{bo} is constant, α_b is maximum at the lowest order modes ($j = 0$), and decreases rapidly at the highest modes, where most of the light is in the evanescent field and out of the scattering material.

The surface loss α_s for TE modes of a symmetric slab waveguide is given by

$$\alpha_s = \alpha_{so} \frac{(n_g^2 - N_j^2)\sqrt{N_j^2 - n_0^2}}{N_j(n_g^2 - n_0^2)(1 + Kd\sqrt{N_j^2 - n_0^2})}.$$

[16]

If the surface-scattering coefficient α_{so} is constant, α_s is seen to approach zero for the lowest-order mode as $N_{j=0}$ approaches n_g. The increase in loss for higher mode numbers comes from the increasing optical field at the scattering interface. For the highest-order modes the scattering goes again to zero, since the light is distributed over a large volume in the evanescent field.

The dependence of bulk and surface scattering on N in multimode waveguides is sketched in Fig. 4. The total scattering is the sum of the two contributions, and this of course is what is determined when the attenuation of the various modes is measured. In many cases it is possible to resolve the total scattering into the bulk and surface components. This is based on the observation that (1) the surface scattering is zero and the factor multiplying α_{bo} is unity for $N = n_g$ and (2) the bulk scattering is nearly constant except near $N = n_o$, so that the rate of increase in scattering for $N < n_g$ is proportional to the surface scattering. The second condition fails if either α_{bo} or α_{so} depends on N. A more detailed discussion of waveguide scattering has been published elsewhere.[12]

5. Liquid-Crystal Waveguides

Optical waveguides have been made of nematic liquid-crystal layers. There is no apparent reason why light could not be guided in other mesophases as well, but to date this has not been reported, and the subsequent discussion is restricted to nematics. Since the liquid-crystal layer must be confined on both surfaces and since the bounding material can easily be identical on both sides, the symmetric slab model is appropriate for such waveguides. The high birefringence of nematics makes it necessary that the materials be well aligned and free from domains of differing orientation. The resulting uniaxial layer is of course different from the isotropic materials of the previ-

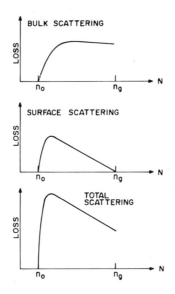

Fig. 4—Dependence of scattering loss on attenuation for multimode slab waveguides.

ous discussion. However, if propagation and polarization directions are restricted to the principle axes, the theory of the isotropic case is applicable if the appropriate value of n_g is chosen for the particular mode polarization (TE or TM) involved.

The technique used for coupling light into the liquid-crystal waveguide must be compatible with the need to confine the layer on both surfaces. We have found it possible to combine the optical coupling prism and one bounding surface into a unitized structure. Alternatively, light may be coupled into a solid-film waveguide that terminates at the liquid crystal, but this technique makes it difficult to selectively excite specific modes in the liquid-crystal layer.

Fig. 5 shows some coupling arrangements we have used. In Fig. 5a the prism with index $n_p = 1.95$ is bonded to low-index glass or fused quartz of index $n_o = 1.48$–1.51. The base of the structure is optically polished. The laser beam is focused just to the left of the prism corner. Leaky modes excited in the liquid-crystal film under the high-index prism become true guided modes after passing into the film bounded by low-index material. A second prism bonded on the other end of the low-index glass may be used to couple the guided light out again.

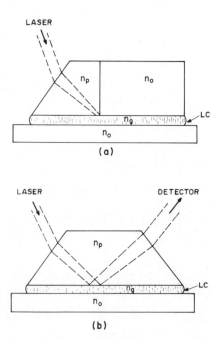

Fig. 5—Prism coupler for exciting (a) guided or (b) leaky modes in liquid-crystal (LC) layers.

A leaky wave coupler is shown in Fig. 5b. Here light coupled into guided modes of the liquid-crystal layer is directly coupled out again. This structure is useful for measuring the mode structure of the liquid-crystal film by measuring the discrete angles for which light rays are coupled into and out of the particular waveguide modes.

The attenuation by scattering of light guided in the arrangement of Fig. 5a is determined by observing the intensity of light scattered from the guided modes into radiation. The streak of scattered light is imaged on a slit mounted in front of a photomultiplier, and this assembly is translated to record the decrement of scattered light intensity as a function of distance from the input coupling point.

Fig. 6 shows attenuation data obtained at room temperature for

TE and TM modes in guides of 6- to 12-μm thickness. The liquid-crystal layers were MBBA aligned with lecithin to have optic axis perpendicular to the bounding surface. For TM modes, $n_g = n_{ext} = 1.75$, while for TE modes, $n_g = n_{ord} = 1.54$. Each data point represents a separate guided mode.

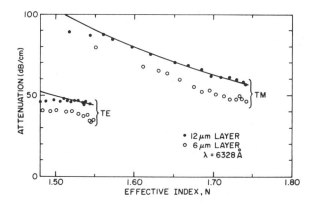

Fig. 6—Attenuation measured for nematic waveguides consisting of MBBA layers with hemeotropic alignment.

That the measured attenuation is due to scattering rather than absorption was established from the absence of significant loss when the liquid crystal was heated above the nematic–isotropic phase transition. The waveguide layers were well aligned, homogeneous, and free from disclinations. The scattering results from thermal fluctuations in molecular order[13] over volumes comparable to optical wavelengths.

As described previously, the scattering of the lowest order modes is equal to the bulk scattering coefficient of the waveguide material. This was in the range 35–45 dB/cm for TE modes and 45–60 dB/cm for TM modes. According to Eq. [15], the bulk scattering strength α_b is independent of waveguide thickness d in the limit $N_j \rightarrow n_g$. The differences in attenuation must be attributed to differences in α_{bo} due, perhaps, to different sample purities or to experimental inaccuracy. The measured attenuation in all samples measured, including samples with molecular alignment parallel to the waveguide plane, was much greater than that reported by Sheridan et al.[2] The very low attenuations they report (less than 1 dB/cm) could not be reproduced in any of our experiments.

The TM mode attenuation increases with increasing mode number. The rate of increase with decreasing N_j is seen to be independent of sample thickness. Surface scattering strength increases with mode number, but is inversely proportional to the waveguide thickness (see Eq. [16]). Since the thickness dependence is not observed, the mode dependence of the TM scattering is attributed to wavelength dependence in the bulk scattering. The solid curve was fitted to the TM_0 attenuation measured for the 12-μm guide, and then scaled by the factor $(N_j/N_o)^{-4}$ to represent the wavelength dependence of Rayleigh scattering. The fit to the data is obviously very good. Further work is needed, however to understand in detail the scattering mechanisms. It may be particularly rewarding to try and resolve any surface scattering effects resulting from various techniques of molecular alignment.

Electro-optic effects were produced in nematic waveguides by applying pulsed electric fields. In one series of experiments, interdigital electrodes were deposited on one of the bounding surfaces and driven with voltage pulses. The light coupled out of the guide downstream from the electrodes was monitored with a photomultiplier and the response viewed on an oscilloscope.

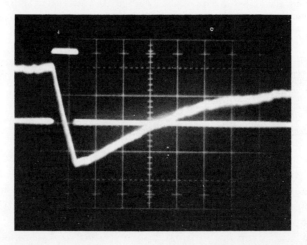

Fig. 7—Transmission of light in nematic waveguide is diminished in response to a pulsed electric field (rectangular pulse) and recovers when field is off (drive voltage of 1-msec duration and 50-volt amplitude).

Fig. 7 shows a voltage pulse (1 msec duration and 50 V amplitude) and its effect on waveguide propagation. Transmission drops during the pulse duration and recovers exponentially afterwards. The light transmission may be completely turned off by this process. Response times T_{off} and recovery times T_{on} are shown in Fig. 8. It is presumed that the modulation mechanism is refraction of light out of the waveguide by the index modulation caused by molecular rotation in response to the applied electric field.

As in the case of the attenuation, a more detailed investigation of these phenomena is warranted. For example, the variation of the re-

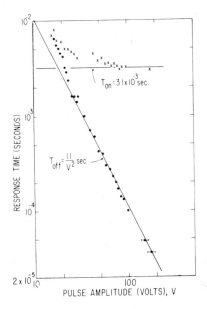

Fig. 8—Response times to application of field T_{off} and recovery times after removal of field T_{on} for waveguide transmission shown in Fig. 7.

covery time T_{on} with pulse voltage is quite unexpected, and suggests that complex patterns of molecular rotation may be excited by the interdigital electrodes. Such effects are relevant to display-device technology, and optical-waveguide techniques appear to be an advantageous way of studying them.

Dynamic scattering has also been used to modulate guided light. In this experiment a liquid-crystal layer was established on the surface of a passive thin-film waveguide. Electrically excited turbulance scattered light in the evanescent field of the thin-film waveguide, which extended into the liquid crystal. The extinction of waveguide light as a function of ac and dc dynamic-scattering excitation voltages is seen in Fig. 9. Very high voltages are necessary, since only a small part of the optical field interacts with the liquid crystal.

6. Conclusions

Optical-waveguide propagation in liquid-crystal layers has been achieved, and basic electro-optic device phenomena have been demonstrated. The usefulness of such devices has yet to be established, particularly in view of the considerable progress being made with other active waveguides, such as single-crystal electro-optic films.[14] Liquid-crystal materials offer the advantages of high electro-optic index changes and a variety of birefringence and scattering operation

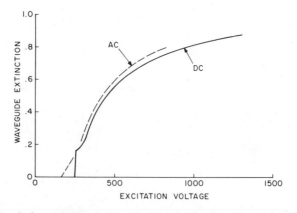

Fig. 9—Transmission reduction in thin-film waveguide when dynamic scattering is excited in nematic overlay.

modes. They are handicapped at present by slow response and recovery speeds, high scattering loss, and the need for confinement in sealed cells. Possibly the most fruitful area of application may be for side-illuminated display panels, should useful materials with low scattering loss be developed.

Optical waveguiding may be most useful as a technique for studies of liquid-crystal phenomena. Accurate measurements of the mode spectrum can be used as a probe of transient changes in the refractive index profile of the liquid-crystal layer. Light scattering into radiation modes and between guided modes determines the spatial frequency spectrum of the alignment inhomogeneities and fluctuations. Scattering between closely spaced modes is a sensitive measure of very-small-angle forward scattering. These same techniques may be useful in studying physical phenomena in other liquid layers as well as liquid crystals.

References

[1] D. J. Channin, "Optical Waveguide Modulation Using Nematic Liquid Crystal," *Appl. Phys. Lett.* **22**, p. 365 (1973).

[2] J. P. Sheridan, J. M. Schnur, and T. C. Giallorenzi, "Electro-Optic Switching in Low-Loss Liquid Crystal Waveguides," *Appl. Phys. Lett.*, **22**, p. 561 (1973).

[3] J. P. Sheridan, "Liquid Crystals in Integrated Optics," OSA Topical Meeting on Integrated Optics, New Orleans, Jan., 1974.

[4] Chenming Hu, John R. Winnery, and Nabil M. Amer, "Optical Deflection in Thin-Film Nematic-Liquid-Crystal Waveguides," *IEEE J. Quan. Elect.*, **QE-10**, p. 218 (1974).

[5] H. A. Weakliem, D. J. Channin, and A. Bloom, "Determination of Refractive Index Changes in Photosensitive Polymer Films by an Optical Technique," *Applied Optics*, **14**, 560 (1975).

[6] P. K. Tien, "Light Waves in Thin Films and Integrated Optics," *Appl. Opt.*, **10**, p. 2395 (1971).

[7] Dietrich Marcuse, ed., *Integrated Optics*, IEEE Press, New York (1973).

[8] Dietrich Marcuse, *Light Transmission Optics*, Van Nostrand Reinhold Co., New York (1972).

[9] P. K. Tien, R. Ulrich, and R. J. Martin, "Modes of Propagating Light Waves in Thin Deposited Semiconductor Films," *Appl. Phys. Lett.*, **14**, p. 291 (1969).

[10] M. L. Dakss, L. Kuhn, P. F. Heidrich, and B. A. Scott, "Grating Coupler for Efficient Excitation of Optical Guided Waves in Thin Films," *Appl. Phys. Lett.*, **16**, p. 523 (1970).

[11] J. Kane and H. Osterberg, "Optical Characteristics of Planar Guided Modes," *J. Opt. Soc. Am.*, **54**, p. 347 (1964).

[12] D. J. Channin, J. M. Hammer, and M. T. Duffy, "Scattering in ZnO-Sapphire Optical Waveguides," *Applied Optics*, **14**, 923 (1975).

[13] P. G. deGennes, *The Physics of Liquid Crystals*, Oxford University Press, London (1974).

[14] J. M. Hammer and W. Phillips, "Low-Loss Single-Mode Optical Waveguides and Efficient High-Speed Modulators of $LiNb_xTa_{1-x}O_3$ on $LiTaO_3$," *Applied Phys. Lett.*, **24**, p. 545 (1974).

16

The Electro-Optic Transfer Function in Nematic Liquids*

Alan Sussman
RCA Solid State Division, Somerville, N.J. 08876

1. Introduction

When an electric field is applied across transparent plane-parallel electrodes containing mesomorphic liquids, many complex phenomena occur that depend on the optical, dielectric, and elastic properties of the liquid, the geometry of the test situation, and the nature of the electrical signal.[1] The electro-optic transfer function is a way of specifying such optical changes. An ever increasing interest in liquid-

---

* Presented in part as an invited paper at the Optical Society of America annual meeting, Rochester, N.Y. Oct. 1973.

crystal electro-optic phenomena particularly, but not entirely, in the field of display devices has caused a corresponding growth of the literature[2]; hence, this chapter is limited to steady-state properties of nematic liquids.

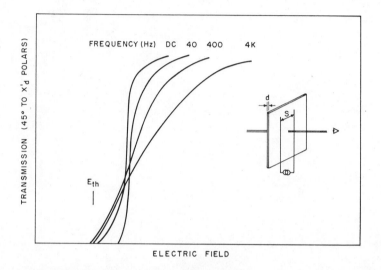

Fig. 1—Field effect (lateral electrode geometry). The change in optical orientation properties with frequency is a result of changing field penetration. The dielectric relaxation frequency is 80 Hz. This material of positive dielectric anisotropy does not show any turbulent flow (R.A. Soref[9]).

It is convenient to divide nematic liquids into two classes depending on the dielectric anisotropy—positive for those in which the molecular orientation described by a vector, called the director, tends to align parallel to an electric field, and negative when that alignment is perpendicular to the field. Those phenomena in which reorientation is the major result are considered "field effects". This designation is not arbitrary, since it also separates these dielectric effects from those whose mechanism has additionally a hydrodynamic flow, i.e., realignment or disruption of the liquid by material transport. Sometimes, material parameters completely specify electro-optical performance, while in other cases, the two regimes are selectable by choosing the frequency of excitation.

In materials of positive dielectric anisotropy, most electro-optic phenomena are frequency independent field effects; hydrodynamic effects, occurring with certain boundary conditions result in stable (laminar) flow,[3] and no turbulent-flow reorientation is observed.[4]

In the storage effect,[12] on the other hand, a write mode is obtained below the dielectric relaxation frequency. This is a hydrodynamically produced turbulent flow that disturbs the (cholesteric) order. An erase mode is produced at high frequencies, where the ordered structure is restored by dielectric reorientation forces.

Depending on the geometry, dielectric reorientations are sometimes accompanied by hydrodynamic transients, i.e., a backflow.[13] (The reorientation caused by hydrodynamic flow is well known.[14]) Such backflow transients, under certain conditions, can give rise to turbulent effects similar to dynamic scattering; the latter term should be reserved specifically for the sequence: domains, domain instability, and scattering, produced by conductance hydrodynamics.

Another important frequency is the inverse of the time it takes for an ion to traverse the whole of the way across the electrode distance $t_{tr} = V^2/\mu d$, where V is the voltage, μ the mobility, and d the electrode spacing.[15] Unlike the dielectric relaxation frequency, f(transit^{-1}) depends on voltage and cell thickness. It effectively separates ac from dc effects whose hydrodynamic mechanisms are related to electrode effects.[16] Note that for high enough voltages and/or small enough spacing, f(transit^{-1}) can exceed the electrical operating frequency. This condition is particularly favorable for loss of conducting ions by electrosorption; such effects have been noted, especially in nematics with low (initial) conductivity.[17]

Differences in threshold behavior when other than sinusoidal excitation is used in dynamic scattering[18] have been observed.[19]

Modifications to the threshold-voltage characteristic may also be obtained by adding a high-frequency signal to the low-frequency drive.[20] This adds a dielectric orienting field that tends to suppress hydrodynamic flow, thereby increasing the low-frequency voltage required to cause a given scattering as compared to that required without the high frequency.

2. Geometric Considerations in Optical Measurements

Careful consideration must be given to using an appropriate geometry when taking measurements, particularly if the results are to be used to evaluate the visual performance of a display system.

In dielectric reorientation effects (such as voltage-controlled birefringence,[21] twisted nematic structure effects,[22] and guest-host effects[23]), the optical changes are basically variations in either optical density or wavelength of absorption. As such, the optical measurements are quite straightforward. The plane-parallel-geometry structures may be considered neutral density filters, with angle-dependent transmission or wavelength characteristics. It is necessary to apply corrections for refraction because of index mismatch,[24] which is particularly important for measurements on optical parameters where a liquid crystal is used as a switch to couple into a waveguide,[25] to modulate and switch in a waveguide,[26] or to control total reflectance at a prism/liquid-crystal interface.[27]

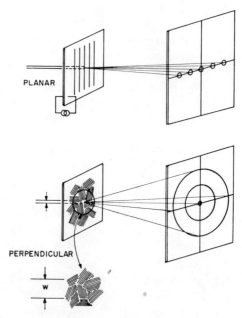

Fig. 2—Diffraction by domains. A nematic orientation that is originally planar usually results in a well-ordered domain structure that behaves like a phase grating, and a diffraction pattern consisting of a series of spots is obtained (top). When the orientation is originally perpendicular, clusters of domains may give rise to diffraction spots or rings depending on the relative diameter of the probe (w) and the dimensions of the domain clusters.

When scattering or diffraction phenomena are studied, however, the geometry becomes particularly important, both because of the rich nature of the optical processes and because of the more varied ways in which such processes are applied to display devices. In the case of diffraction by domains,[28] different results may be obtained depending on the original orientation of the fluid and the relation between the beam probe size and the domain cluster size.

When domains are observed in configurations where the surfaces induce planar orientation,[29] they exhibit the properties of a phase grating, and unpolarized laser illumination at normal incidence produces a spot diffraction pattern[30] (Fig. 2, upper part).

If, however, the orientation is originally perpendicular, the first electro-optic effect is the conversion to a planar orientation in the bulk, followed at higher voltage by the domains, whose cylindrical axes may not be correlated over large distances, and "clusters" result. The diffraction pattern will be a ring if the probe diameter (w) is large compared to the cluster size (Fig. 2, lower part).

To measure the angle dependence of scattering in a transmission-type display, three general schemes may be employed. In Fig. 3 they are illustrated schematically, and qualitative electro-optic transfer curves are given for each.

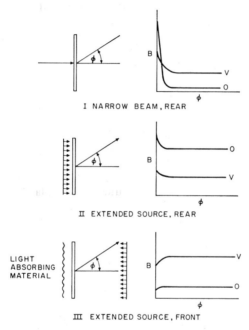

Fig. 3—Influence of geometrical parameters on scattering electro-optic transfer function measurements. Three cases are illustrated, with their corresponding electro-optic transfer functions (highly schematic).

Type I—Narrow Beam, Rear

In type I, a laser beam is used as a coherent probe; such a configuration has given a good deal of information on the ordering of the nematic liquid,[31] on domains,[32] and on the structural modifications that occur during dynamic scattering.[33] Both the angle of incidence and the detector angle can be varied; such data can be reported in compact graphical form.[34] This configuration is also applicable to optical data processing[35] and holographic information-storage techniques.[36]

With illumination normal to the plane of the structure (Fig. 3, top) and no voltage applied (0 curve), only a small inherent scattering may be measured off-axis.[31] On-axis, of course, the unscattered beam is virtually undiminished in intensity. With the application of voltage V, there is finite scattering into off-axis angles (increase in signal), but an attenuation of the on-axis brightness signal; a contrast ratio as high as 10^4 has been measured for spatial filtering applications.[37]

Type II—Extended Source, Rear

Type II has an extended-area source behind the cell, with the detector in front. The liquid-crystal plane when excited reduces the intensity, and may be thought of as a variable density screen, similar to type I before the crossover, with the liquid crystal plane integrating the light, which arrives at all angles. This is a rather uninteresting configuration, although some knowledge of the angle dependence of the scattering is needed when the liquid-crystal plane is used for a viewing screen,[38] as a variable optical stop, or as an image plane for projection.

Type III—Extended Source, Front

In type III the illumination is from the viewing side, with a light-absorbing screen behind. This configuration is not used frequently in digital displays, because the scattering to the rear is not particularly efficient.[17] By illuminating the screen from the back, full use is made of the scattering in the propagation direction, i.e., toward the viewer. A unique configuration for display devices is to illuminate from the rear through a screen consisting of blackened vanes adjusted so as to give the maximum viewing angle with minimum total thickness of the system.[39]

A similar configuration, but with a mirror surface as the rear electrode, has a wide application in watch and clock displays. Best results are obtained when the viewer sees reflected in the mirror a light-absorbing surface. The on-state is particularly bright because both transmitted light and scattered light are reflected back toward the viewer. A type III electro-optic transfer curve is thereby obtained.

If, however, the mirror reflects a bright surface, it is possible that contrast reversal will occur, i.e., the type II configuration. For some angles and brightnesses, a no-contrast result is possible. In the case of field effects, particularly the twisted structure, type III effects are especially subtle when the angle of incidence is varied.

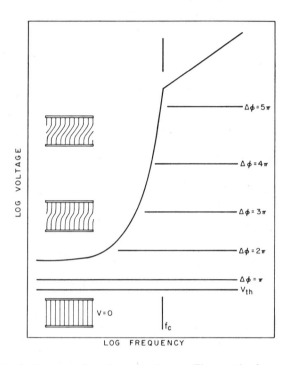

Fig. 4—Retardation as a function of voltage. The continuing reorientation of the structure (left side) with increasing voltage results in an ever increasing retardation, i.e., birefringence. The retardation $\Delta\phi$ continues to increase unless another electro-optic regime ensues; viz, below the dielectric relaxation frequency, domains, and above, chevron distorions (M. Hareng, et al[21]).

3. Field Effects—Negative Dielectric Anisotropy

With an orientation that is originally perpendicular to the electrodes, and material of negative dielectric anisotropy, an electric field interacts with the fluid to cause a tilt in the optical axis, beginning at the center of the cell, when a threshold voltage is reached.[21] This tilt results in birefringence that continues to increase with increasing voltage, unless, of course, the threshold voltage for another phenomenon is exceeded, e.g., domains at low frequency or "chevrons" at high frequency[40] (Fig. 4). As the field increases, the tilt angle continues to increase as does the retardation, measured as a phase difference $\Delta\phi$.[41] To get uniform retardation over large areas, the surface may be prepared to give a slight bias, i.e., a uniform but small tilt off-normal.[42]

The electro-optic transfer function at a fixed angle between crossed polarizers is shown in Fig. 5. By swinging between two voltages, it is possible to modulate the color equivalent to a phase retardation at one wavelength (first voltage) to produce a color equivalent to a different wavelength (second voltage). This is the basis for a matrix-

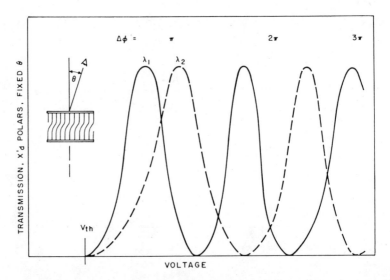

Fig. 5—Electro-optic transfer as a function of voltage. At a fixed viewing angle, the transmission of monochromatic light between crossed polarizers undergoes a series of maxima, which occur with increasing voltage and represent phase retardations of π, 2π, ... etc (M. Hareng, et al[21]).

addressed color crossed-array display[43] where the angle dependence of the color produced by the phase retardation (or transmission of light between crossed polarizers when monochromatic light is used, as shown in Fig. 6) has been avoided by using a projection system. Interdigitated electrodes may be used to create small areas of different retardation, the overall area being angle independent.[44]

The angle dependence can be used to make accurate measurements of the electrode spacing in liquid crystal cells.[45]

Advantage may be taken of the property of certain pleichroic dyes in which the wavelength of absorption in the crystal is different along different crystalline axes. If such a dye is dissolved in and follows the alignment of the liquid crystal of the previous example[46] with the analyzer removed, the absorption in the voltage-off state would depend on the extinction coefficient perpendicular to the dye crystal axis; when the structure is changed by application of voltage, the absorption becomes converted to that related to the extinction coefficient parallel to the crystal axis. If one of these absorptions is not in the visible, no polarizer is required to see the color change, and the angle dependence results in a change in optical density at the visible wavelength.

Discussion of the chevron regime, although also a field effect, is described in the section on diffraction.

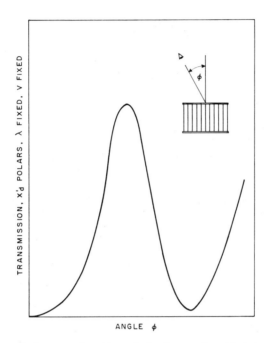

Fig. 6—Transmission as a function of viewing angle. At a constant voltage, and for a fixed wavelength, the transmission between crossed polarizers shows an angular dependence. At angles other than normal, a correction for refraction by the glass is necessary (zero voltage is illustrated).[44]

4. Field Effects—Positive Dielectric Anisotropy

Since materials of positive anisotropy turn their orientation parallel to the direction of a field, a necessary requirement is an off-state of uniform planar alignment. Then, a change analogous to that just discussed may be obtained by the application of a field; above a threshold voltage, the parallel orientation (beginning at the center) eventually becomes indistinguishable optically from one that is perpendicularly aligned. Between crossed polarizers, the result is a lessening of the retardation with increasing voltage, with an electro-optic transfer function similar to that of Fig. 5. The change is maximum when the polarizer axes are normal to the planar axes. At any voltage, the transmission is angle dependent,[47] with a symmetry related to whether or not the angle is varied in a plane containing the undisturbed liquid orientation.

Incorporation of a pleiochroic dye and elimination of the analyzer results in voltage-controlled optical absorption, the guest-host effect.[48]

The "twisted" nematic structure has unique optical properties: the plane of linearly polarized light, of wavelength shorter than the distance over which the twist makes one revolution, is rotated following the twist.[49] In static patterns a 90° rotation is maximum, greater angles being unstable. Placed between crossed polarizers, such a cell would cause maximum transmission. The converse would be true if the polarizers were initially parallel.

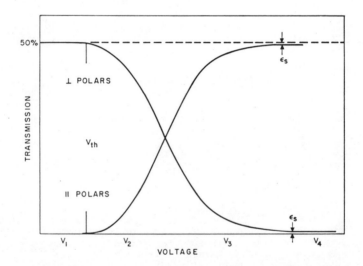

Fig. 7—Electro-optic transfer function of twisted structure. Above threshold, increasing the voltage casuses the structure to lose the ability to rotate the plane of polarized light; eventually, the optical behavior of the structure approaches that of a perpendicular homeotropic nematic. The difference (ϵ_s), however, is still finite at three times threshold.[52]

If the material has a positive dielectric anisotropy, then an electric field can convert the twisted structure to one that eventually has the optical properties of a perpendicularly oriented one.[22] A threshold voltage, as shown in Fig. 7, is present; only at sufficiently high voltages is the difference between the still-twisted and the non-twisted structure undetectable. As in the voltage-controlled birefringence, the tilt of the orientation continues with increasing voltage, being greatest at the midpoint between the electrodes (Fig. 8, top). This causes birefringent color changes, but an overshadowing optical effect occurs because of the modifications to the twist. Initially, the twist may be considered uniform throughout the cell thickness (Fig. 8, bottom). With increas-

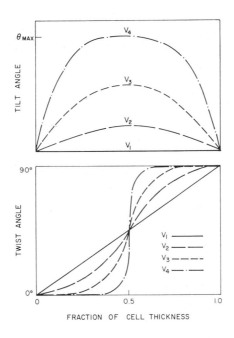

Fig. 8—Variation of tilt and twist angles throughout twisted structure as a function of voltage. Top figure shows symmetric tilt occuring as the voltage increases (as in Fig. 7). Bottom figure shows the symmetric twist that, while uniform at voltages below threshold, becomes distorted with increasing voltage. When most of the twist occurs over a distance small compared to the wavelength of light, the structure can no longer rotate the plane of polarization.[52]

ing voltage, the twist becomes distorted. Above that value of voltage corresponding to the optical threshold, the total distributed twist is converted to distortions near the wall, so that the rotation conditions cannot be met[50] and the optical density of the structure changes. Eventually, increasing voltages result in almost total loss of rotation; for a cell between crossed polarizers initially, the optical transmission would increase from that of the parallel polarizers to one that approaches that of the crossed polarizers. The coherence of light so rotated is not affected greatly.[51]

Fig. 9 shows calculated curves[52] for angle-dependent transmission

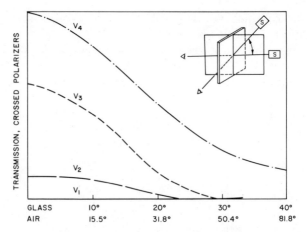

Fig. 9—Electro-optic transfer coefficient along surface orientation for twisted structure. With the source (S) in a plane perpendicular to the electrodes and parallel to the optic axis of the nematic orientation at that electrode, the optical transmission is a monotonic function of angle and voltage (calculated curves).[52]

in which the light being measured emerges perpendicular to the electrode face, while the source (S) is rotated in a plane that is perpendicular to the electrode and contains the orientation of one of the electrodes. Note that the results are relatively monotonic. If however, the source is in a plane rotated so as to be 45° to the orientation direction in the electrode face, the results are complex (Fig. 10).* Type I, II, and III measurements may be expected to yield such complex results.

Type II is at present the basis for most digital applications. If a specular back is used (behind the polarizer), the reflected brightness is maximized, but at the expense of curtailed viewing angle; in Type III configuration, with a scattering back plane acting as a (partially depolarized) source of illumination, the optical results would be expected to be similar to those with Type II.

5. Hydrodynamic Effects—Diffraction by Domains

Liquid-crystal hydrodynamic phenomena are extremely complex, and the mechanisms for their production are far from being completely

---

* The reason is again that the light in the probe beam now is resolved into birefringent components. In these calculations, the index of the glass was assumed to be equal to the higher index of the nematic liquid and the necessary refraction for other than normal incidence corrected for. Note that this results in a considerable difference between the geometrical beam/glass angle and the actual beam/liquid angle.

elucidated. The differences between positive and negative dielectric anisotropy material are more distinct than in the field effects. The negative materials are more likely to produce striking opto-hydrodynamic changes. There are, however, indications that positive materials may participate in similar although more subdued effects.[53]

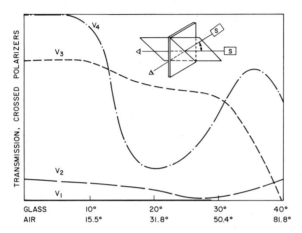

Fig. 10—Electro-optic transfer coefficient at 45° to surface orientation for twisted structure. When the source (S) is in a plane perpendicular to the electrodes but not parallel to the optic axis of the nematic in the electrode (director), the variation of transmission between crossed polarizers as a function of angle and voltage is no longer simple, because of anisotropic propagation within the liquid (calculated curves).[52]

The following is a typical series of events when a material of negative dielectric anisotropy, originally in the planar orientation, is subject to an electric field of a frequency between the dielectric relaxation frequency and the inverse transit-time frequency. Dielectrically, the field should cause no change; above a threshold voltage, however, a periodic pattern of light and dark lines appears. These domains,[54] associated with vortical liquid flow,[55] have a spacing proportional to the electrode separation. The refractive properties are not unlike an array of cylindrical lenses, with alternate convergent and divergent bands, which are the domain boundaries, visible in ordinary light. This focal length decreases with increasing voltage and represents the change in curvature of each lens due to increased velocity of the fluid. (The number density of domains need not change.)

The periodic pattern also exhibits the properties of a phase grating, and a diffraction pattern may be observed (Fig. 2). When the voltage is raised, the steady-state result is a higher domain periodicity, i.e., new domains have made their appearance. There being no smooth way for a new domain to decrease the periodicity, a transient diffraction occurs, which relaxes into a new equilibrium steady-state pattern whose intensity distribution may be correlated with the spatial frequency of the domain.[30] Such new domains, by the nature of the symmetry of the electrode configuration, might be expected to also have a symmetrical flow pattern. With further increase in voltage, the new domains apparently cannot enter into an orderly array, and the turbulent properties of dynamic scattering begin to appear. Optically speaking, the structure behaves like a deep phase screen, i.e., at each position there is a phase retardation resulting in a path difference which may be several wavelengths.[33]

At frequencies below the inverse transit time, the electrode effects may no longer be geometrically symmetric;[17] the domain periodicity is nonuniform because a second diffraction pattern, harmonically unrelated to the original set, has made its appearance.[15]

The stability of the hydrodynamic modes of cylindrical domains has been investigated using a model with the appropriate boundary conditions.[56] The results indicate a stability curve with two branches—one on which the domain spacing remains constant with voltage, the other on which the spacing varies inversely with the voltage (as just described). These appear in Figure 11. Under certain conditions, which may be depend on material parameters as well as electrode spacing, dc excitation does not result in dynamic scattering, but only in a continuing increase in domain periodicity.[57] This result is shown in Fig. 12. Here, the simplified experimental data show the intensity distribution of a diffraction ring. The original orientation was perpendicular to the electrodes, so that the domain clusters would be expected. In this case, an unusual domain structure was observed.[58] A tree-like pattern, as shown in the upper inset of Fig. 11, allowed an almost continuous variation in domain spacing as the voltage changed. The new domains appeared as new branches and the growth was quite smooth. When the voltage was increased, the diffraction pattern would become diffuse for an instant, then relax into the sharp diffraction rings, as shown in Fig. 12. The intensity profile of the highly reflected domains was a sharp as that for less diffracted ones, indicating an improvement in diffraction efficiency with increasing voltage.

At frequencies above the dielectric relaxation frequency, another diffraction with a threshold makes its appearance (see Fig. 4). This is the "chevron" regime,[59] which is not hydrodynamic in origin but depends on interaction between the anisotropic dielectric properties of the fluid and the field. It represents a sinusoidal modulation of the

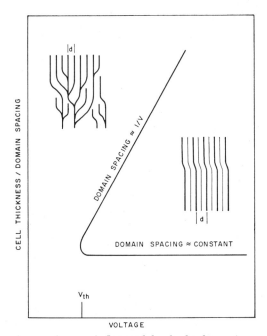

Fig. 11—Domain spacing as influenced by hydrodynamic regime. The periodicity of domains shows a voltage variation that depends on hydrodynamic modes, represented by two branches of the stability curves.[56] Insets: right, domains are cylindrical; left, a tree structure allows a continuous variation of the spacing d.[58]

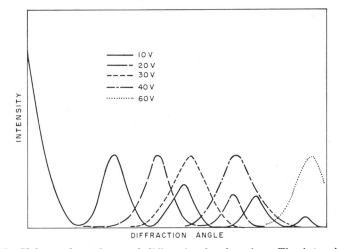

Fig. 12—Voltage dependence of diffraction by domains. The intensity distribution of the diffraction rings (for a single wavelength) with increasing dc voltage demonstrates a constant intensity profile even at 60 volts where the diffraction angle is nearly 50°. No dynamic scattering occurs.[57]

director, with a threshold field proportional to the square root of the frequency, and results in a phase-grating condition.[40] The spatial frequency of these domains is much greater than that of the low-frequency domains; the diffraction also gives rise to angle-dependent chromatic effects. The model based on a symmetric periodic slit grating, which is satisfactory for low-frequency domains, must be corrected for refraction within the liquid.[60] An added stabilizing field, at frequencies above threshold for chevrons, causes an increase in period.[61]

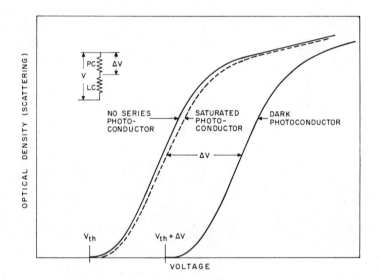

Fig. 13—Photoconductor control of dynamic scattering. Electro-optic transfer function is shown in the dark and under saturated photoconductor conditions. For the liquid crystal to be below threshold voltage in the dark, the value of the applied voltage must not exceed the original threshold voltage (V_{th}) plus the voltage drop across the dark photoconductor (ΔV). With suitable choice of photoconductor parameters, the light-saturated photoconductor allows the liquid-crystal response to behave almost as if no photoconductor were present.

6. Dynamic Scattering

As domains become increasingly unstable with increasing voltage, the phenomenon known as dynamic scattering[18] occurs. Fig. 13 shows a a typical electro-optic transfer function. Note the lack of saturation of scattering with voltage in the type I case. At higher voltages, changes in flow patterns occur at definite thresholds, and result in changed optical, electrical, and kinetic properties.[62] Another phenomenon, caused by certain patterns of flow, particularly near the dielectric relaxation frequency, converts an originally perpendicularly oriented off-state to one that is (temporarily) planar.[63]

The angular dependence of the scattering has been observed as a function of voltage,[64] and the results in Fig. 14 are typical. (The

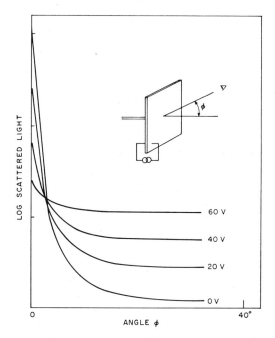

Fig. 14—Angular dependence of dynamic scattering as a function of voltage. Under narrow beam condition (Type I), the scattering of the original beam near the optic axis results in a reduction in transmission; off-axis, however, the scattering results in an increase in light detected. (The crossovers do not necessarily occur at the same angle.)

crossover, however, is not always found at the same angle for every voltage.) The effect of increasing cell thickness is an increase in the voltage at which equivalent scattering occurs; the threshold is unchanged. The angular dependence is also relatively independent of thickness.[34] The scattering at a particular value of thickness, angle, and voltage has a direct dependence on the optical anisotropy of the nematic compound under study.[65] The scattering for any compound is independent of the current density[66] as long as the measurement is made sufficiently far from the dielectric relaxation frequency.

When polarized light is used, the angular scattering dependence is a function of the angle between the polarizer and the direction of the (original) planar orientation.[31]

7. Photoconductor Control

The incorporation of a photoconductor layer in series[67] with the liquid crystal layer allows direct conversion from optical input to optical output; such devices hold promise of usefulness in the field of image and data processing,[43] optical conversion,[68] and storage.[69] As of this review, photoconductor control of dynamic scattering and storage effects only have been reported. Optimum performance of such devices is achieved when the resistance of the photoconductor is matched to that of the liquid crystal, so that the voltage drop across the photoconductor, ΔV, leaves the voltage across the liquid crystal below the threshold value when the photoconductor is unilluminated; correspondingly, when the photoconductor is illuminated, the voltage drop should be as low as possible, allowing full scattering operation of the liquid crystal. Fig. 13 shows the electro-optic transfer functions for the dynamic scattering discussed above for the series photoconductor saturated and in the dark. The optimum applied voltage, as will be seen later, is $V_{th} + \Delta V$.

The relationship between the photoconductor and liquid resistance in the dark and under illumination has been treated for the case of pure resistance. There are reported cases in which the electro-optic response depends on whether the illumination falling on the photoconductor reaches it from the liquid-crystal side or from the transparent-conductor side. This rectification is not observed in all liquid-crystal photonconductor systems.[70] However, the generality of the above discussion is not affected.

To achieve sufficient control over the photoconductor, with either a bulk or barrier photoconductor, requires a photon flux of $10^{11}/cm^2$.[71] This value is comparable to the flux required for other processes, such as Xerographic or electroluminescent outputs. When the photoconductor is nonohmic, higher minimum exposures may be required.

Operation in the transmission mode allows direct insertion of the image plane into an optical crystal. This, of course, requires the photoconductor to operate outside the spectral range of the projection source. In reflectance modes, the photoconductor may be isolated from the secondary illumination by means of an insulating and/or opaque dielectric layer.[72] Such a configuration requires ac operation, which happily circumvents some of the electrochemical problems associated with dc.[73]

Using the somewhat artificial assumption that equal increments of illumination result in equal changes in ΔV, Fig. 15 was developed. The liquid-crystal scattering is plotted against the illumination up to the value at which the photoconductor "saturates", i.e., when increasing light no longer significantly changes ΔV. If the values are properly

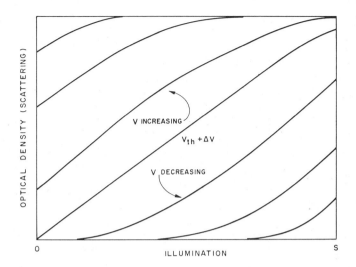

Fig. 15—Dynamic scattering range as a function of light intensity and applied voltage. The operating voltage $V_{th} + \Delta V$ may be optimized. At lower than optimum voltages, the maximum scattering (S) with a properly selected photoconductor, is reduced. At even lower voltage, the scattering reaches a saturation value, but is not zero even at low light levels (calculated curves).

chosen, the behavior of the liquid crystal should approach that observed with no series photoconductor. At voltages above and below the optimum, $V_{th} + \Delta V$, the scattering curve was estimated from Fig. 13. Note that for voltages lower than optmium, maximum scattering cannot be obtained; at the lowest voltages, the scattering is not zero even when the photoconductor is unilluminated. The shape of the original scattering–voltage curve has a significant effect on the dynamic range. For example, a steep electro-optic function, which is desirable for matrix addressing or digital display, will reduce the halftone range in the photoconductor combination, as might be expected.

References

[1] Several review articles, emphasizing electro-optic properties of liquid crystals are available: R. A. Soref, "Liquid Crystal Light Control Experiments," in **The Physics of Opto-Electronic Materials,** ed. W. A. Albers, Jr., Plenum Press, New York (1971); A. Sussman, "Electro-Optic Liquid Crystal Devices: Principles and Applications," **IEEE Trans. Parts, Hybrids, and Packaging,** Vol. PHP8, p. 24 (1972); and A. Sussman, "Liquid Crystals in Display Systems," in **Liquid Crystalline Systems,** ed. G. W. Gray and P. A. Winsor, Ellis Horwood, London (in press).

[2] For example in 1969, there were 70 papers listed in Physics Abstracts. In 1970, 250; in 1971, 315; and in 1972, 390.

[3] P. DeGennes, "Electrohydrodynamic Effects in Nematic Liquid Crystals I. DC Effects," in **Comments on Solid State Physics,** Vol. 3, p. 35 (1970); and P. A. Penz, Electrohydrodynamic Solutions for Nematic Liquid Crystals with Positive Dielectric Anisotropy," **Mol. Cryst.,** Vol. 23, p. 1 (1973).

[4] W. H. DeJeu, C. J. Gerritsma, and Th. W. Lathouwers, "Instabilities in Electric Fields of Nematic Liquid Crystals with Positive Dielectric Anisotropy: Domains, Loop Domains, and Reorientation," **Chem. Phys. Lett.**, Vol. 14, p. 503 (1972).

[5] G. H. Heilmeier and W. Helfrich, "Orientational Oscillations in Nematic Liquid Crystals," **Appl. Phys. Lett.**, Vol. 16, p. 155 (1970).

[6] H. S. Harned and B. B. Owen, **The Physical Chemistry of Electrolytic Solutions**, 2nd ed., Reinhold Publishing Co., New York, 1950, Chap. 4.

[7] A. Sussman, "Ionic Equilibrium and Ionic Conductance in the System Tetra-iso-pentyl Ammonium Nitrate p-Azoxyanisole," **Mol. Cryst. and Liq. Cryst.**, Vol. 14, p. 182 (1971).

[8] D. Meyerhofer, A. Sussman, and R. Williams, "Electro-Optic and Hydrodynamic Properties of Nematic Liquid Films with Free Surfaces," **J. Appl. Phys.**, Vol. 43, p. 3685 (1972).

[9] R. A. Soref, "Transverse Field Effects in Nematic Liquid Crystals," **Appl. Phys. Lett.**, Vol. 22, p. 165 (1973); and N. V. Madhusudana, P. P. Karat, and S. Chandrasekhar, "Some Electrohydrodynamic Distortion Patterns in Nematic Liquid Crystals," **Current Science**, Vol. 42, p. 147 (1973).

[10] W. Haas, J. Adams, and J. B. Flannery, "New Electro-Optic Effect in a Room-Temperature Nematic Liquid Crystal," **Phys. Rev. Lett.**, Vol. 25, p. 326 (1970).

[11] W. Helfrich, "A Simple Method to Observe the Piezoelectricity of Liquid Crystals," **Phys. Lett.**, Vol. 35A, p. 393 (1971).

[12] G. H. Heilmeier and J. Goldmacher, "A New Electric Field Controlled Reflective Optical Storage Effect in Mixed Liquid Crystal Systems," **Proc. IEEE**, Vol. 57, p. 34 (1969).

[13] F. Brochard, "Backflow Effects in Nematic Liquid Crystals," **Mol. Cryst. and Liq. Cryst.**, Vol. 23, p. 51 (1973).

[14] W. Helfrich, "Molecular Theory of Flow Alignment of Nematic Liquid Crystals," **J. Chem. Phys.**, Vol. 50, p. 100 (1969); and "Conduction-Induced Alignment of Nematic Liquid Crystals: Basic Model and Stability Considerations," **J. Chem. Phys.**, Vol. 51, p. 4092 (1969).

[15] D. Meyerhofer and A. Sussman, "The Electrohydrodynamic Instabilities in Nematic Liquid Crystals in Low-Frequency Fields," **Appl. Phys. Lett.**, Vol. 20, p. 337 (1972).

[16] N. Felici, "Phenomenes Hydro et Aerodynamiques dans la Conduction des Dielectric Fluides," **Rev. Gen. Elect.**, Vol. 78, p. 717 (1969); A. Sussman, "Contribution of the Ionic Double Layer to the DC Hydrodynamic Instabilities in Nematic Liquids," Paper presented at Fourth International Liquid Crystal Conf., Kent, Ohio, Aug. 1972; and R. J. Turnbull, "Theory of Electrohydrodynamic Behaviour of Nematic Liquids in a Constant Field," **J. Phys. D: Appl. Phys.**, Vol. 6, p. 1745 (1973).

[17] A. Derzhanski and A. G. Petrov, "Inverse Currents and Contact Behaviour of Some Nematic Liquid Crystals," **Phys. Lett.**, Vol. 36A, p. 307 (1971).

[18] G. H. Heilmeier, L. A. Zanoni, and L. A. Barton, "Dynamic Scattering: A New Electro-Optic Effect in Certain Classes of Nematic Liquid Crystals," **Proc. IEEE**, Vol. 56, p. 1162 (1968).

[19] W. H. DeJeu, "Instabilities of Nematic Liquid Crystals in Pulsating Electric Fields," **Phys. Lett.**, Vol. 37A, p. 365 (1971); and Orsay Liquid Crystal Group, "Transition Between Conduction and Dielectric Regimes of the Electrohydrodynamic Instabilities in a Nematic Liquid Crystal," **Phys. Lett.**, Vol. 39A, p. 181 (1972).

[20] P. Wild and J. Nehring, "Turn-on Time Reduction and Contrast Enhancement in Matrix-addressed Liquid Crystal Valves," **Appl. Phys. Lett.**, Vol. 19, p. 335 (1971); and C. Stein and R. Kashnow, "A Two-Frequency Coincidence Addressing Scheme for Nematic-Liquid-Crystal Display," **Appl. Phys. Lett.**, Vol. 19, p. 343 (1971).

[21] M. Schiekel and K. Fahrenschon, "Deformation of Nematic Liquid Crystals with Vertical Orientation in Electrical Fields," **Appl. Phys. Lett.**, Vol. 19, p. 391 (1971); R. A. Soref and M. J. Rafuse, "Electrically Controlled Birefringence of Thin Nematic Films," **J. Appl. Phys.**, Vol. 43, p. 2029 (1972); F. J. Kahn, "Electric-field-induced Orientational Deformation of Nematic Liquid Crystals: Tunable Birefringence," **Appl. Phys. Lett.**, Vol. 20, p. 199, (1972); and M. Hareng, E. Leiba, and G. Assouline, "Effet du Champ Electrique sur la Biréfringence de Cristaux Liquides Nématiques," **Mol. Cryst. and Liq. Cryst.**, Vol. 17, p. 361 (1972).

[22] M. Schadt and W. Helfrich, "Voltage-Dependent Optical Activity of a Twisted Nematic Liquid Crystal," **Appl. Phys. Lett.**, Vol. 28, p. 127 (1971).

[23] G. H. Heilmeier and L. A. Zanoni, "Guest-Host Interactions in Nematic Liquid Crystals. A New Electro-Optic Effect," **Appl. Phys. Lett.**, Vol. 13, p. 91 (1968).

[24] I. Haller, H. A. Huggins, and M. J. Freiser, "On the Measurement of Indices of Refraction of Nematic Liquids," **Mol. Cryst. and Liq. Cryst.**, Vol. 16, p. 53 (1972).

[25] D. J. Channin, "Optical Waveguide Modulation Using Nematic Liquid Crystals," **Appl. Phys. Lett.**, Vol. 22, p. 365 (1973).

[26] J. P. Sheridan, J. M. Schnur, and T. G. Giallorenzi, "Electro-Optic Switching in Low-Loss Liquid Crystal Waveguides," **Appl. Phys. Lett.**, Vol. 22, p. 560 (1973).

[27] R. A. Kashnow and C. R. Stein, "Total-Reflection Liquid-Crystal Electro-Optic Device," **Appl. Optics**, Vol. 12, p. 2309 (1973).

[28] G. Assouline, A. Dmitrieff, M. Hareng, and E. Leiba, "Diffraction d'un Faisceau Laser par un Cristal Liquide Nématique Souvris à un champ Electrique," **C. R. Acad. Sci.**, Vol. B271, p. 857 (1970).

[29] W. Helfrich, "Orientation Pattern of Domains in Nematic p-azoxyanisole," **J. Chem. Phys.**, Vol. 51, p. 2755 (1969).

[30] T. O. Carroll, "Liquid-Crystal Diffraction Grating," **J. Appl. Phys.**, Vol. 43, p. 767 (1972).

[31] C. Deutsch and P. N. Keating, "Scattering of Coherent Light from Nematic Liquid Crystals in the Dynamic Scattering Mode," **J. Appl. Phys.**, Vol. 40, p. 4049 (1969); and Orsay Liquid Crystal Group, "Viscosity Measurements by Quasi-Elastic Light Scattering in p-azoxyanisole," **Mol. Cryst. and Liq. Cryst.**, Vol. 13, p. 187 (1971).

[32] P. A. Penz, "Order Parameter Distribution for the Electrohydrodynamic Mode of a Nematic Liquid Crystal," **Mol. Cryst. and Liq. Cryst.**, Vol. 15, p. 151 (1971).

[33] E. Jakeman and P. N. Pusey, "Light Scattering from Electrohydrodynamic Turbulence in Liquid Crystals," **Phys. Lett.**, Vol. 44A, p. 456 (1973); and F. Scudieri, M. Bertolotti, and R. Bartolino, "Light Scattered by a Liquid Crystal: A New Quasi-Themal Source," **Applied Optics**, Vol. 13, p. 181 (1974).

[34] D. Meyerhofer and E. F. Pasierb, "Light Scattering Characteristics in Liquid Crystal Storage Materials," **Mol. Cryst. and Liq. Cryst.**, Vol. 20, p. 279 (1973).

[35] R. B. MacAnally, "Liquid Crystal Displays for Matched Filtering," **Appl. Phys. Lett.**, Vol. 18, p. 54 (1971).

[36] G. W. Taylor and W. F. Kosonocky, "Ferroelectric Light Valves for Optical Memories," **Ferroelectrics**, Vol. 3, p. 81 (1972).

[37] H. J. Caulfield and R. A. Soref, "Optical Contrast Enhancement in Liquid Crystal Devices by Spatial Filtering," **Appl. Phys. Lett.**, Vol. 18, p. 5 (1971).

[38] E. Tomkins, "Liquid Crystal Viewing Screen," Opt. Soc. Am. Meeting, Tucson, Ariz. (1971).

[39] A. Sussman, "Illumination Scheme for Liquid Crystal Displays," U.S. Patent pending.

[40] R. A. Kashnow and H. S. Cole, "Electrohydrodynamic Instabilities in a High-Purity Nematic Liquid Crystal," **J. Appl. Phys.**, Vol. 42, p. 2134 (1971).

[41] M. Hareng, G. Assouline, and E. Leiba, "La Biréfringence Electriquement Contrôlée dans les Cristaux Liquides Nématiques," **Appl. Optics**, Vol. 11, p. 2920 (1972).

[42] L. T. Creagh and A. R. Kmetz, "Performance Advantages of Liquid Crystal Displays with Surfactant-produced Homogeneous Alignment," Soc. for Information Display, 1972 International Symp. Dig. Tech. Papers (Lewis Winner, New York), p. 90; and F. J. Kahn, "Orientation of Liquid Crystals by Surface Coupling Agents," **Appl. Phys. Lett.**, Vol. 22, p. 386 (1973).

[43] G. Assouline, M. Hareng, and E. Leiba, "Liquid Crystal and Photoconductor Image Converter," **Proc. IEEE**, Vol. 59, p. 1355 (1971); and M. Hareng, G. Assouline, and E. Leiba, "Affichage Bicolore à Cristal Liquide (Two Color Liquid-Crystal Display)," **Electron. Lett.**, Vol. 7, p. 699 (1971).

[44] T. Shimojo, K. Matsuda, and K. Kasano, "Singular Electro-Optical Characteristics of Liquid Crystal Display with Interdigital Electrodes," **S.I.D. International Symp. Digest**, 1973 (p. 36).

[45] R. A. Kashnow, "Thickness Measurements of Nematic Liquid Layers," **Rev. Sci. Inst.**, Vol. 43, p. 1837 (1972).

[46] J. A. Castellano and M. T. McCaffrey, "Liquid Crystals IV. Electro-Optic Effects in p-alkoxybenzylidene-p'-aminoalkyphenones and Related Compounds," in **Liquid Crystals and Ordered Fluids**, ed. J. F. Johnson and R. S. Porter, Plenum Press, New York, p. 293 (1970).

[47] U. Bonne and D. P. Cummings, "Properties and Limitations of Liquid Crystals for Aircraft Displays," Contract #N00014-71-C-0262, ONR Task No. NR 215-173, Honeywell, Inc., Oct. 1972, Chap. VII.

[48] G. H. Heilmeier, J. A. Castellano, and L. A. Zanoni, "Guest-Hose Interactions in Nematic Liquid Crystals," **Mol. Cryst. and Liq. Cryst.**, Vol. 8, p. 293 (1969).

[49] J. Dryer, "Liquid Crystal Optical Devices," Reported at Second International Liq. Cryst. Conf., Kent, Ohio, 1968.

[50] H. DeVries, "Rotary Power and Other Properties of Certain Liquid Crystals," **Acta. Cryst.**, Vol. 4, p. 219 (1951).

[51] C. B. Burckhardt, M. Schadt, and W. Helfrich, "Holographic Recording with an Electro-Optic Liquid Crystal Cell," **Appl. Optics**, Vol. 10, p. 2196 (1971).

[52] D. W. Berreman, "Optics in Smoothly Varying Anisotropic Planar Structures: Application to Liquid-Crystal Twist Cells," **J. Opt. Soc. Am.**, Vol. 63, p. 1374 (1973).

[53] G. H. Heilmeier, "Some Cooperative Effects in Butyl p-Anisylidene-p-Amino Cinnamate," in **Ordered Fluids and Liquid Crystals**, Advances in Chemistry Series #63, p. 68, American Chemical Society, Washington, D.C. (1967); A. Takase, S. Sakagami, and M. Nakamizo, "Light Diffraction in a Nematic Liquid Crystal with Positive Dielectric Anisotropy," **Japan J. Appl. Phys.**, Vol. 12, p. 1255 (1973); W. H. DeJeu and C. J. Gerritsma, "Electrohydrodynamic Instabilities in Some Nematic Azoxy Compounds with Dielectric Anisotropies of Different Sign," **J. Chem. Phys.**, Vol. 56, p. 4752 (1972); and Ref. 4.

[54] R. Williams, "Domains in Liquid Crystals," **J. Chem. Phys.**, Vol. 39, p. 384 (1963).

[55] P. A. Penz, "Voltage-Induced Vorticity and Optical Focusing in Liquid Crystals," **Phys. Rev. Lett.**, Vol. 24, p. 1405 (1970).

[56] P. A. Penz and G. W. Ford, "Electromagnetic Hydrodynamics of Liquid Crystals," **Phys. Rev.**, Vol. 6A, p. 414 (1972).

[57] H. Greubel and U. Wolff, "Electrically Controllable Domains in Nematic Liquid Crystals," **Appl. Phys. Lett.**, Vol. 19, p. 213 (1971); and L. K. Vistin, "New Electrostructural Phenomenon in Liquid Crystals of Nematic Type," **Sov. Phys. Crys.**, Vol. 15, p. 514 (1970).

[58] A. Sussman, unpublished results.

[59] Orsay Liquid Crystal Group, "Hydrodynamic Instabilities in Nematic Liquids Under ac Electric Fields," **Phys. Rev. Lett.**, Vol. 25, p. 1642 (1970).

[60] R. A. Kashnow and J. E. Bigelow, "Diffraction from a Liquid Crystal Phase Grating," **Appl. Optics**, Vol. 12, p. 2302 (1973).

[61] Y. Galerne, G. Durand, M. Veyssie, and V. Pontikis, "Electrohydrodynamic Instability in a Nematic Liquid Crystal: Effect of an Additional Stabilizing ac Electric Field on the Spatial Period of 'Chevrons'," **Phys. Lett.**, Vol. 38A, p. 449 (1972).

[62] A. Sussman, "Secondary Hydrodynamic Structure in Dynamic Scattering," **Appl. Phys. Lett.**, Vol. 21, p. 269 (1972).

[63] J. Nehring and M. S. Petty, "The Formation of Threads in the Dynamic Scattering Mode of Nematic Liquid Crystals," **Phys. Lett.**, Vol. 40A, p. 307 (1972).

[64] L. Goodman, "Light Scattering in Electric-Field Driven Nematic Liquid Crystals," Soc. for Information Display 1971 International Symp. Dig. Tech. Papers, (Lewis Winner, New York), p. 124. See also References [31] and [47].

[65] L. T. Creagh, "Nematic Liquid Crystal Materials for Displays," **Proc. IEEE**, Vol. 61, 814 (1973); and Eastman Liquid Crystal Products, Bulletin JJ-14 (1973).

[66] G. Assouline and E. Leiba, "Cristaux Liquides," **Rev. Tech. CSF**, Vol. 1, p. 483 (1969).

[67] J. D. Margerum, J. Nimov, and S.-Y. Wong, "Reversible Ultraviolet Imaging with Liquid Crystals," **Appl. Phys. Lett.**, Vol. 17, p. 51 (1970).

[68] D. H. White and M. Feldman, "Liquid Crystal Light Valves," **Electron Letters**, Vol. 6, p. 837 (1970).

[69] J. D. Margerum, T. D. Beard, W. P. Bleha, Jr., and S.-Y. Wong, "Transparent Phase Images in Photoactivated Liquid Crystals," **Appl. Phys. Lett.**, Vol. 19, p. 216 (1971).

[70] A. D. Jacobson, "Photo-Activated Liquid Crystal Valve," Soc. for Information Display, 1972 International Symp., Dig. Tech. Papers (Lewis Winner, New York), p. 70.

[71] A. Rose, "The Role of Space-Charge-Limited Currents in Photoconductivity-controlled Devices," **IEEE Trans. on Electron Dev.**, Vol. ED19, p. 430 (1972).

[72] T. O. Beard, W. P. Bleha, and S.-Y. Wong, "AC Liquid Crystal Light Valve," **Appl. Phys. Lett.**, Vol. 22, p. 90 (1973).

[73] A. Sussman, "Dynamic Scattering Life in the Nematic Compound p-Methoxybenzylidene-p-amino phenyl acetate as Influenced by the Current Density," **Appl. Phys. Lett.**, Vol. 21, p. 126 (1972).

17

Electrochemistry in Nematic Liquid-Crystal Solvents

Alan Sussman

RCA Solid State Division, Somerville, N. J. 08876

1. Introduction

This chapter discusses the electrolytic-solution properties of low-dielectric-constant nematic solvents. Dissolved substances, if electrolytes, can contribute only a fraction of their ions to the conductance because of equilibrium between the free ions and ion pairs. If the solute forms ions through intermediate charge-transfer reactions, additional equilibria must be considered. For nematics, the solvent fluidity is anisotropic, and the conductance depends on the direction of current flow with respect to the orientation of the fluid. The variation of the conductance with temperature is directly related to the variation with temperature of both the ionic equilibrium and the fluidity.

A considerable background of both theoretical and experimental work is available.

The properties of the interface between conductors and ordinary electrolytic solutions are exceedingly complex; for low-dielectric-constant solvents, details of the double layer are lacking, but dimensions may be estimated from simple theory. In cells of the dimensions of the usual liquid-crystal devices, many of the properties of the interface can assume an increased significance as the usual dimensional differences between the bulk and the interface become less distinct.

Many problems of charge transport are incompletely solved, but through the use of carefully purified solvents, specially prepared electrodes, and well-defined experimental conditions, it is possible to separate the contributions of bulk processes, electrode processes, and diffusion. Some kinetic studies of transport phenomena, operating life, and a few electrochemical reactions are discussed.

The relationship between electrolytic and hydrodynamic solution properties is still under intensive study and is not treated in this paper. Many instances of specific electrolyte–low-dielectric solvent interaction need to be investigated fully. The equivalent problems in anisotropic solvents are not completely understood. This review is presented with that thought in mind.

2. Equilibrium Properties of Bulk Solutions

The equivalent conductance of a solution is a convenient chemical quantity. It is defined as the hypothetical conductance of one chemical equivalent of a dissolved substance; $\Lambda = \alpha N° \mu e$, where α is the fraction of the dissolved substance (solute) in the ionic form, and $N°$, μ, and e are, respectively, Avogadro's number, the ionic mobility, and the elementary charge. The equivalent conductance is related to the conductance $\sigma = \mu e$ by the relation $\Lambda = 1000 \, \sigma/c$, where c is the solute concentration in moles per liter. With solvents of dielectric constant greater than 30, solutions of simple electrolytes generally may be expectd to be fully ionized at all concentrations, i.e., $\alpha = 1$. Upon dilution of concentrated solutions in which the mobility of the ions is reduced by interionic forces, the variation of Λ with concentration follows the general limiting law:[1]

$$\Lambda = \alpha(\Lambda_0 - S\sqrt{\alpha c}) \qquad [1]$$

where S is the limiting slope and $\alpha \to 1$. Λ_0 is obtained by extrapolation of Eq. [1] to "infinite" dilution; Λ_0 is about 50% greater than Λ at 0.01 m/l in a typical simple electrolyte.

In solvents of low dielectric constant, on the other hand, the principal variation of Λ on dilution results from an increase in free ion concentration due to the dissociation of ion pairs:[2]

$$A^+B^- \rightleftharpoons A^+ + B^- \qquad [2]$$

where A^+, B^- and A^+B^- are the positive, negative, and paired ions, respectively. The pairs are the direct result of interionic attraction, since the electrostatic force between ions is shielded less in the low dielectric constant solvent than in the more usual solvents. There is no charge transfer in the formation of ion pairs. The value of α is less than unity, and the variation of Λ is almost entirely controlled by the dependence of the free-ion concentration on the solution concentration.

Writing the equilibrium constant for the reaction of Eq. [2], using brackets to denote concentrations,

$$K_1 = [A^+B^-]/[A^+][B^-] = \frac{c(1-\alpha)}{\alpha^2 c^2} \approx \frac{1}{\alpha^2 c}, \; \alpha \ll 1 \qquad [3]$$

$$\alpha = \sqrt{\frac{1}{K_1 c}}$$

we see that the fraction in ionic form depends inversely on the square root of the solute concentration. The simplifying assumption of $\alpha \ll 1$ breaks down at concentrations near K^{-1}; this usually occurs at dilutions outside experimental range.

A second set of reactions involving the ion pairs and free ions gives conducting triplets and still further clustering at higher concentrations, resulting in a nonlinear increase in the equivalent conductance; the presence of these reactions results in a minimum in the conductance curve (see Ref. [2], Chap. 8). Taking the formation of ion triplets into account, the equivalent conductance may be written

$$\Lambda = \Lambda_0 \sqrt{1/K_1 c} + \lambda_0 \frac{\sqrt{c/K_1}}{K_2}, \qquad [4]$$

where λ_0 is the limiting equivalent conductance of the ion triplets, and K_2 is the equilibrium constant for the reactions $AB + A^+ = A_2B^+$ and $AB + B^- = AB_2^-$, assumed for simplicity to have identical equilibrium properties. At concentrations below the minimum, and for $c \geq 1/K_1$, the functional variation of Λ is

$$\Lambda = \Lambda_0 \sqrt{1/K_1 c}. \tag{6}$$

Substituting Eq. [6] into the relation between σ and Λ, we find the current; it is proportional to the square root of the solute concentration,

$$J = \sqrt{\frac{c}{K}} \frac{\Lambda_0}{1000} \frac{VA}{d}, \tag{7}$$

where V is the voltage, A the area, and d the cell thickness. Fig. 1 shows the equivalent conductance as a function of concentration for

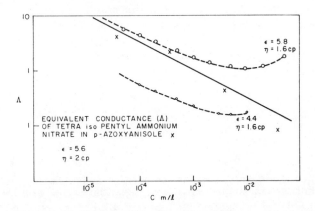

Fig. 1—Equivalent conductance of tetra-*iso*-pentyl ammonium nitrate in isotropic *p*-azoxyanisole at 152°C, (solid line). The equilibrium constant for ion-pair formation is 2×10^{-6} m/l. The data is bracketed between calculated values of Eq. [4] for two values of the dielectric constant. The variation of the equivalent conductance with dielectric constant is found in Eq. [8]. (Ref. [3])

the solute tetra-*iso*-pentyl ammonium nitrate in the nematic *p*-azoxyanisole.[6] The problem of obtaining values of Λ_0 in this case by extrapolation to infinite dilution are experimentally complicated because of the very low concentrations needed to insure that $\alpha = 1$. Use can therefore be made of the semi-empirical Walden's rule:

The product of the limiting equivalent conductance and the viscosity is a constant for each solute, almost independent of temperature.

It is particularly accurate for large ions, but must be corrected slightly for dielectric constant.[4] In an anisotropic solvent, the constant may be calculated by using the appropriate viscosity. Fig. 2 shows[5], for *p*-azoxyanisole, the reciprocal viscosity (fluidity) parallel to the orientation, perpendicular to the orientation, and for a nonoriented sample plotted against the reciprocal temperature.

In the nematic range, the large variations in fluidity on the nematic side of the nematic–isotropic transition may be related to pretransitional phenomena; when the parallel orientation becomes more disordered, the flow becomes more difficult, while for the perpendicular

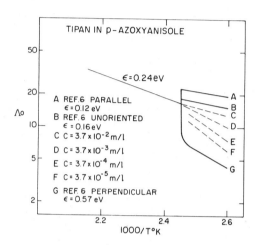

Fig. 2—Fluidity of *p*-azoxyanisole versus the reciprocal temperature for flow parallel to the orientation, perpendicular to the orientation, and for a nonoriented sample. The dashed line shows the current data of Fig. 3 normalized to the isotropic range. It is suggestive to consider that the variation of the activation energy of conduction with concentration depends on the solute influence on the orientation. (Ref. [5])

orientation, disordering makes the flow easier. This change in orientation can be noticed in the current–reciprocal-temperature graph of Fig. 3, for the tetra-*iso*-pentyl ammonium nitrate/*p*-azoxyanisole system. Some effects on the fluidity may be related to the concentration of solute. During the measurements the sample birefringence indicated that the solvent was ordered perpendicular to the electric field, requiring that the appropriate mobility be for flow perpendicular to the orientation. With increasing solute concentration, the fluidity properties begin to resemble these for flow parallel to the orientation. This may be either a direct interaction between the ions and the solvent or may be due to the influence of the ions on the way the surface affects the bulk orientation.

Ionic equilibrium was also observed in 1-cm-thick cells in the system tetra-*iso*-pentyl ammonium tetra-phenyl borate/methoxy-benzylidene p-n-butylaniline (MBBA).[6] Walden's rule gave good agreement in the isotropic region, i.e., the product of the conductance and the viscosity was constant. The samples were in the nonaligned state. In the nematic range, the experimental values of the mobility near the transition temperature were five times too low, increasing to ten times too low at the lowest temperature, compared to the calculated values. Although no attempt to order the sample was made, ordering

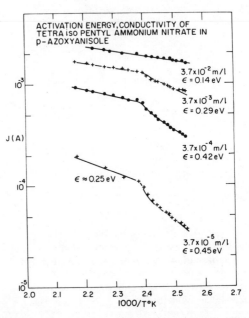

Fig. 3—Current as a function of the reciprocal temperature in the system tetra-*iso*-pentyl ammonium nitrate/*p*-azoxyanisole. Note how the activation energy in the nematic range decreases with increasing solute concentration. Compare this variation with the influence of the orientation (Fig. 2). (Ref. [3])

of such samples has been observed under similar experimental conditions.[7] Nonapplicability of Walden's rule has been noted under conditions of unipolar charge injection, where measured values were up to ten times lower than the calculated ones.[8]

The conductance of a solution and, therefore, the current will depend on the square root of the solute concentration over the range for which ion-pair equilibrium operates. In those devices for which a finite conductance is required, such as dynamic scattering and the storage-effect devices,[10] departures from Ohm's law may be observed because of the reorientation of the fluid by field and hydrodynamic effects. If the voltage is slowly raised in a dynamic scattering cell that was originally in the perpendicular homeotropic orientation, the current will be controlled by the fluidity parallel to the field, $\mu_\parallel$, until the threshold voltage for the reorientation to a birefringent condition is exceeded. The current than becomes dependent on the value of the fluidity perpendicular to the field, $\mu_\perp$. Between the threshold for reorientation (a dielectric effect) and the onset of domain formation (a hydrodynamic effect) at another threshold voltage, the application of a voltage pulse will result in a current transient, as shown by the arrows in Fig. 4. When the voltage is raised still further, the domains

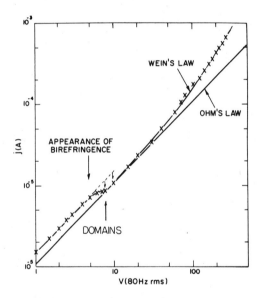

Fig. 4—Current-voltage characteristic for a dynamic scattering device, original orientation perpendicular homeotropic. The Ohm's law line is drawn for mobility based on fluidity perpendicular to the orientation, $\mu_\perp$.

become unstable and the turbulent dynamic scattering regime is entered. An average mobility,

$$\mu_{ave} = \frac{1}{3}(2\mu_\perp + \mu_\parallel)$$

now controls the current. At sufficiently high fields, the ion-pair equilibrium is disturbed in favor of dissociation. This, the second Wein effect, is most easily observed in solvents of low dielectric constant (see Ref. [1], Chap. 4, Sec. 7).

Measurements of conductance are usually made using alternating voltage of a frequency that falls between the dielectric relaxation frequency (see Ref. 1, Chap. 4, Sec. 1), above which ions cannot contribute to the current, and the inverse transit time,[11] below which, as will be seen later, there are complications due to polarization of the electrodes.

Variation of the conductance with temperature, both in the tetra-*iso*-pentyl ammonium nitrate/*p*-azoxyanisole and tetra-*iso*-pentyl ammonium phenyl borate/MBBA systems, depends almost entirely on the fluidity,[3] since ionic equilibrium contributes only a small factor over the ranges of the studies. Because this may not always be the case, the variation of the equilibrium constant with temperature will be considered.

For the association–dissociation reaction Eq. [1], the equilibrium constant is given approximately as[4]

$$K_1^{-1} \approx \text{const. } \exp\left(\frac{e^2}{a\epsilon kT}\right) \qquad [8]$$

The exponential term is the ratio of the electrostatic to thermal energy, with a the distance at which the ions can be considered paired, k is Boltzmann's constant, and ϵ is the average dielectric constant of the solvent ϵ_{ave}. The ion size parameter a is a constant with temperature showing little variation with solvent. The variation of K_1^{-1} does depend on the variation of the dielectric constant(s) with temperature. Once the values for a, $\epsilon(T)$, $\mu(T)$, and μ (orientation) are known for a given solute/solvent system, the current as a function of temperature and concentration may be considered determined. Experimental and calculated results for the system tetra-*iso*-pentyl ammonium bromide in a mixed solvent containing compounds of alkoxy benzylidine-*p*-amine phenyl esters (APAPA family)[12] are shown in Fig. 5.

There is another equilibrium system that can lead to ionic conduction. When an electron-accepting compound is introduced into a solvent that is an electron donor, the reaction to form a donor–acceptor pair ensues: $D + A \rightleftharpoons (DA)$. This donor–acceptor pair may subsequently dissociate to give a pair of ions: $(DA) \rightleftharpoons D^+ + A^-$. This reaction is distinct from ion-pair reactions, since charge transfer does occur formally, but the role of the low-dielectric solvent is similar in affecting the equilibrium. A typical electron acceptor, chloranil (tetrachloro-1,4-benzoquinone), in reacting with MBBA in the dual role of solvent and electron donor, was able to alter the conductance.[13] In another case, equal parts of the donor hydroquinone and the acceptor p-benzoquinone were dissolved in MBBA;[14] this resulted in the formation of ionizable charge transfer complex: HQ + p-BQ $\rightleftharpoons$ HQ $\cdot$ p-BQ $\rightleftharpoons$ HQ$^+$ + p-BQ$^-$. The conductance of the solution became sufficient to produce dynamic scattering. In a 1:1 mixture of MBBA and p-methoxy p-n-hexylaniline, carefully purified, the current remained constant at 0.15 μA/cm^2 for 4000 hours in hermetically sealed cells of thickness 25 μm, probably because the conducting species can undergo oxidation and reduction reactions by electron transfer alone at the appropriate electrodes, thereby revers-

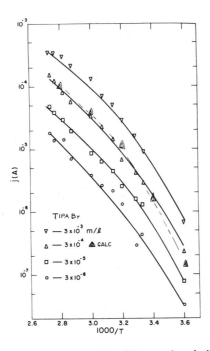

Fig. 5—Current as a function of the reciprocal temperature in the system tetra-*iso*-pentyl ammonium bromide/APAPA RT mixture, with concentration as a parameter. The calculated current shows good agreement with the experimental data.

ing the formation reactions. The accumulation of neutral products at the respective electrodes results in a concentration gradient, and eventually a steady state is set up, allowing reformation of the charge transfer complex in the bulk. No irreversible reactions can be expected except as side reactions. With unpurified MBBA, increased currents were obtained with continued operation.

3. Electrochemical Reactions

Kinetic studies of some electrochemical reactions in nematic solvents have been made, as well as studies of those solvents dissolved in more polar solvents. It was noticed early that under dc excitation of dynamic-scattering cells,[15] the currents remained relatively constant and the cells continued to operate for orders of magnitude longer than would be expected for a simple faradic process, i.e., where an electrochemical reaction was controlled by the net charge passed through the cell.

In p-methoxybenzylidene-p-aminophenyl acetate, the dynamic scattering life was limited by a faradic reaction of low efficiency which produced polymeric anode films, probably by a free radical mechanism.[16] The probable source of the ionic conduction was ionization of intrinsic impurities. Electron injection by a Schottky mechanism was proposed,[9] but could not be correlated with the activation energy of the conductance. A possible paradox exists, since in order to have a high enough field at the electrodes to cause charge injection, a concentrated electrolytic solution is required, and any injection would be masked by ionic currents. A dilute solution, with no overshadowing ionic currents, would naturally have a lower field across the double layer, so in order to achieve high enough fields for injection, large voltages must be applied to the entire cell.

The thickness[17] of the double layer, $K^{-1} = [4\pi \alpha c /(\epsilon kT)]^{-1/2}$, for a solvent of dielectric constant 5.5 appears in Fig. 6 as a function of concentration, assuming an equilibrium constant for ion-pair dissociation of 10^{-7}, which is appropriate for MBBA[6] or p-azoxyanisole.[3] Estimates of the double-layer field can be made assuming that the double-layer potential saturates at a few volts. Note that the double-layer thickness, which depends on $(\alpha c)^{-1/2} = c^{-1/4}$ is almost constant and that its thickness is not insignificant when compared to typical liquid-crystal-device cell thicknesses (10^{-3} cm).

An alternative theory to Schottky emission is the creation of negative and positive ions from the solvent by oxidation and reduction reactions at the appropriate electrodes, followed by recombination after diffusion into the bulk. This mechanism was suggested to explain the long life of dynamic scattering in compounds of the APAPA series.[18] The role of added ionic compounds was unstated.

In undoped purified samples of MBBA, low-field conductance was attributed to thermal dissociation of trace impurities,[19] but at fields greater than 1500 V/cm, electrode processes begin to interfere. Through the use of ion exchange membranes as an electrode coating, injection effects were supressed. Then one observes at low fields an ohmic current due to the natural impurities, if the dissociation rate of the impurities is fast enough to overcome the rate at which the ions are deposited on the electrodes. Eventually, the current saturates as the deposition rate (related to the inverse transit time) exceeds the generation rate, and at the high fields, the current again begins to increase because of protons injected from the membrane, giving an imine which is believed to be indentical to that occuring in the first step of MBBA hydrolysis. That species may be one of the conducting impurities in undried MBBA

The effects of direct current on MBBA containing 300 parts per million of water and, therefore, traces of the hydrolysis products p-n-butylaniline and p-anisaldehyde were that, first, the p-n butylaniline disappeared by anodic oxidation, producing blackening and

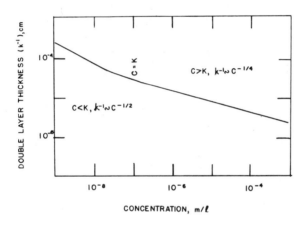

Fig. 6—Double layer thickness K^{-1} as a function of concentration for dielectric constant 5.5 at room temperature. The effect of ion pairing has been included. The thickness of the double layer over the range of interest varies only as the fourth root of the solute concentration.

then a rapid rise in current.[20] The reaction rate depended on the current density. Oxidation–reduction reactions of MBBA itself and of the hydrolysis products were studied in the solvent acetonitrile. It was found that the oxidation of MBBA and p-n-butylaniline is irreversible, while the anisaldehyde is not oxidized; the reduction of MBBA and the aldehyde produce anion radicals by acceptance of single electrons in a reversible reaction, while the p-butylaniline is not reduced. In the solvent dimethyl formamide,[24] reduction of the MBBA resulted in a radical ion with a measured half-life of 4 seconds.

Observations of phenomena related to space-charge accumulation at the electrodes during dc operation of dynamic scattering was reported and attributed to nonspecific electrode processes.[22] The electrode charging and discharging transients, i.e., the ionic charge accumulated in the double layers, was found to be approximately 10^{-5} coulomb per cm^2. The kinetics of the discharge were shown to depend on diffusion from the double layers into the bulk; the charge did not leave the cell via the electrodes. Fig. 7 shows the charging transients

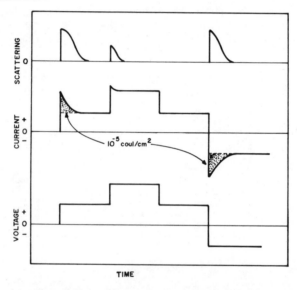

Fig. 7—Current and dynamic scattering transients upon application of voltage steps and voltage reversal (typical results for a low conductivity sample). The excess current above steady state corresponds to electrode charge of about 10^{-5} coulomb/cm^2.

most easily observed when the conductance of the fluid is less than 10^{-10} (ohm-cm)$^{-1}$. Then the dynamic scattering disappears, also as the charge accumulates in the double layers. If the voltage is increased or if the polarity is reversed, another transient is produced, the latter causing a reversal of the potential of the double layer and releasing the charge, which is transported to the other double layer. At high enough voltage, dynamic scattering resumes, either because of injection from the electrodes or field-assisted dissociation in the bulk.

References

[1] H. S. Harned and B. B. Owen, *The Physical Chemistry of Electrolytic Solutions*, 3rd edition, Reinhold Publishing Co., New York (1958). (A general reference for bulk properties of electrolytic solutions.)

[2] R. M. Fuoss and F. Assascina, *Electrolytic Conductance*, Chap. XVI And XVII, Interscience Pub., Inc., New York (1959).

[3] A. Sussman, "Ionic Equilibrium and Ionic Conduction in the System Tetra-*iso*-pentyl Ammonium Nitrate *p*-Azoxyanisole," *Mol. Cryst. and Liq. Cryst.*, **14**, p. 182 (1971).

[4] J. Barthel, "Conductance of Electrolyte Solutions," *Agew. Chem. Internat. Edit.*, **7**, p. 260 (1968) (special reference to nonaqueous solvents).

[5] R. S. Porter and J. F. Johnson, "Orientation of Nematic Mesophases," *J. Phys. Chem.*, **66**, p. 1826 (1962).

[6] A. Denat, B. Gosse and J. P. Gosse, "Étude du Cristal Liquide *p*-Méthoxybenzilidène *p*-Butylaniline," *J. Chim. Phys;* **2**, p. 319 (1973).

[7] F. Gaspard, R. Herino and F. Mondon, "Electrohydrodynamic Instabilities in DC fields of a Nematic Liquid Crystal with Negative Dielectric Anisotropy," *Chem. Phys. Lett.*, **25**, p. 449 (1974).

[8] J. C. Lacroix and R. Teoazeon, "Sur la Mesure de Mobilitiés Ioniques dans un Cristal Liquide Nematique," *Comptes Rendus*, **278**, p. 623 (1974).

[9] G. H. Heilmeier, L. A. Zanoni, and L. A. Barton, "Dynamic Scattering—A New Electro-optic Effect in Certain Classes of Nematic Liquid Crystals," *Proc. IEEE*, **56**, p. 1162 (1968).

[10] G. H. Heilmeier and J. Goldmacher, "A New Electric Field Controlled Reflective Optical Storage Effect in Mixed Liquid Crystal Systems," *Proc. IEEE*, **57**, p. 34 (1969).

[11] A. Sussman, "The Electro-optic Transfer Function in Nematic Liquids," Chapter 16.

[12] A. Sussman, unpublished results.

[13] A. I. Baise, I. Teucher, and M. M. Labes, "Effect of Charge-Transfer Acceptors on Dynamic Scattering in a Nematic Liquid Crystal," *Appl. Phys. Lett.*, **21**, p. 142 (1972). See however: F. Gaspard and R. Herino, "Comments on 'Effect of Charge-transfer Acceptors on Dynamic Scattering in a Nematic Liquid Crystal'," *Appl. Phys. Lett.*, **24**, p. 252 (1974).

[14] Y. Ohnishi and M. Ozutsumi, "Properties of Nematic Liquid Crystals Doped with Hydroquinine and *p*-Benzoquinone: Long-term Dynamic Scattering Under DC Excitation," *Appl. Phys. Lett.*, **24**, p. 213 (1974).

[15] G. H. Heilmeier, L. A. Zanoni, and L. A. Barton, "Further Studies of the Dynamic Scattering Mode in Liquid Crystals and Related Topics," *IEEE Trans. Elec. Dev.* **ED 17**, p. 22 (1970).

[16] A. Sussman, "Dynamic Scattering Life in the Nematic Compound *p*-Methoxy-benzylidene-*p*-Amino Phenyl Acetate as Influenced by the Current Density," *Appl. Phys. Lett.*, **21**, p. 126 (1972).

[17] P. Delahay, *Double Layer and Electrode Kinetics*, Chap. 3, Interscience–John Wiley & Sons, Inc. New York (1965).

[18] M. Voinov and J. S. Dunnett, "Electrochemistry of Nematic Liquid Crystals," *J. Electrochem. Soc.*, **120**, p. 922 (1973).

[19] G. Brière, R. Herino, and F. Mondon, "Correlation Between Chemical and Electrochemical Reactivity in the Isotropic Phase of a Liquid Crystalline *p*-Methoxybenzilidene *p-n*-Butylaniline," *Mol. Cryst.*, **19**, p. 157 (1972).

[20] A. Denat, B. Gosse, and J. Gossé, "Chemical and Electrochemical Stability of *p*-Methoxybenzilidene-*p-n*-Butylaniline," *Chem. Phys. Lett.*, **18**, p. 235 (1973).

[21] A. Lomax, R. Hirasawa, and A. J. Bard, "The Electrochemistry of the Liquid Crystal N-(*p*-Methoxybenzilidene)-*p-n*-Butylanaline (MBBA)," *J. Electrochem. Soc.*, **119**, p. 1679 (1972).

[22] A. Derzhanski and A. G. Petrov, "Inverse Currents and Contact Behavior of Some Nematic Liquid Crystals," *Phys. Lett.*, **36A**, p. 307 (1971).

18

Lyotropic Liquid Crystals and Biological Membranes: The Crucial Role of Water

Peter J. Wojtowicz

RCA Laboratories, Princeton, N. J. 08540

1. Introduction

No collection of papers on liquid crystals would be complete without some discussion of lyotropic liquid crystals and biological membranes. At the present time these two subjects constitute areas of intensive research effort providing a literature of rapidly increasing size. The high current interest mandates a discussion of these topics, but at the same time makes it very difficult to select the material to be presented in the limited space available. The discussion in this chapter will therefore be confined to two main topics, (1) the composition and structure of lyotropic liquid crystals and biological membranes and (2) the nature of the principal interactions that give rise to their existence and stability.

Lyotropic liquid crystals and biological membranes are similar to the thermotropic liquid crystals described in previous chapters in that they are fluid phases that possess considerable molecular order. They are quite different, however, in that they are necessarily systems of two or more components being composed of large organic molecules dissolved in a highly polar solvent, most often water. They are also different because it is not the intermolecular interaction between the molecules partaking in the order that is responsible for the formation of the basic structures of lyotropic liquid crystals and biological membranes. Rather it is the interaction of the organic molecules with the aqueous solvent that is most crucial in providing the stability of these ordered phases. This interaction with the water, the so-called *hydrophobic* interaction, was first recognized in the study of the solubility of simple hydrocarbons in water.[1] The principles involved were then successfully applied to the elucidation of the native conformation of the complex proteins.[2] Very recently these principles have also aided in the understanding of the structure and function of biological membranes.[3] The importance of the hydrophobic interaction, however, has not yet been fully appreciated in the case of lyotropic liquid crystals.[4] One of the intentions of this chapter, therefore, is to help call attention to the decisive role of the hydrophobic interaction in providing the stability of lyotropic liquid crystals.

The following sections of this chapter will briefly review the composition and structure of lyotropic liquid crystals and biological membranes. We will then consider the different interactions that are important in determining the structure of these phases. The hydrophobic interaction will be examined in some detail; the properties of the water itself will be shown to be the predominant driving force leading to the existence of lyotropic liquid crystals and biological membranes. Finally, we will discuss an unusual characteristic of these ordered phases. Unlike most of the other ordered phases encountered in physics and chemistry, the lyotropics and the membranes derive their stability not from a competition between the energy and entropy, but from a competition between two different kinds of entropy.

2. Lyotropic Liquid Crystals

The literature on the composition and structure of lyotropic liquid crystals has been extensively reviewed by Winsor.[5] A somewhat shorter review has been given by Brown, Doane and Neff.[6] Our discussions in this section will be based principally on these two papers; both are recommended to the reader seeking additional information or further detail.

2.1 Constituents of Lyotropics

Lyotropic liquid crystals are chemical systems composed of two or more components. Specifically, they are mixtures of amphiphilic compounds and a polar solvent, most frequently water. *Amphiphilic* compounds are characterized by having in the same molecule two groups that differ greatly in their solubility properties. One part of the molecule will be *hydrophilic,* highly soluble in water or other polar solvents, while the other portion will be *lipophilic,* highly soluble in hydrocarbon or nonpolar solvents. In the context of our following discussions of the interactions responsible for the stability of lyotropics it is perhaps more proper to call these latter groups *hydrophobic,* emphasizing their insolubility in water rather than their hydrocarbon solubility. Typical hydrophilic groups are -OH, -CO_2H, -CO_2Na, -SO_3K, -O(CH_2-CH_2-O)$_n$H, -N(CH_3)$_3$Br, and -PO_4-CH_2CH_2-NH_2. Typical lipophilic or hydrophobic groups are -$C_n H_{2n+1}$, -C_6H_4-$C_n H_{2n+1}$, and any other radicals containing long hydrocarbon chains, with or without aromatic rings included. An example of an amphiphilic compound that has been studied extensively is sodium laurate, whose molecular structure is displayed in Fig. 1a. The hydrophilic portion is the carboxylic acid group (shown ionized as it is in aqueous solution), while the hydrophobic part is the long straight chain hydrocarbon group. In describing the structures of aggregates of such molecules it is convenient (as shown in Fig. 1a) to represent the hydrophilic "head" by a black dot and the hydrophobic "tail" by the zig-zag line.

Depending on the relative strengths of the hydrophilic and lipophilic tendencies of the two different parts of the molecule, amphiphilic compounds can vary widely in their solubility behavior. They can be predominantly hydrophilic, water soluble and hydrocarbon insoluble (such as $C_n H_{2n+1}$-CO_2K with $n = 1,2,3$) or predominantly lipophilic, hydrocarbon soluble and water insoluble (such as $C_n H_{2n+1}$-OH with $n > 12$). The most striking amphiphilic properties (solubilization, formation of micelles, formation of liquid crystals) occurs when the hydrophilic and hydrophobic tendencies are both strong but evenly balanced (such as in $C_n H_{2n+1}$-CO_2Na with $n = 8$ to 20).

Because of their dual characteristics, amphiphilic compounds are capable of displaying remarkable solubility properties. They are soluble in both water and hydrocarbons, and show strong co-solvent or solubilization effects. Moderately concentrated soap solutions, for example, can dissolve many different kinds or organic compounds that will not dissolve in water alone. Similarly, some amphiphilic compounds (such as Aerosol OT) dissolved in hydrocarbon can solubilize water or other polar compounds that do not ordinarily dissolve in hydrocarbons. The proper way to describe both situations is to say that the amphiphilic compound acts as co-solvent for both the water and the hydrocarbon.

2.2 Micelles

At extreme dilution, amphiphilic molecules are distributed randomly and uniformly throughout the solution. As the concentration of the

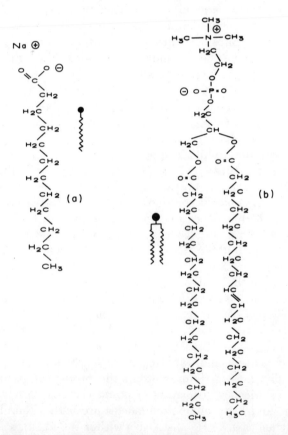

Fig. 1—(a) Molecular structure of a typical amphiphilic molecule, sodium laurate. (b) Molecular structure of a typical lipid found in membranes, phosphatidylcholine (part b after Ref. [10]).

amphiphilic molecules is increased, however, aggregates of molecules begin to form. Groupings of molecules called micelles arise in which like is associated with like. Fig. 2 displays the structure of spherical and cylindrical micelles. In both forms, the hydrophilic heads are associated with each other on the outer periphery of the aggregates, while the hydrophobic tails are grouped together in the fluid-like interior of the micelles. Of even greater significance, however, is the observation that the hydrophilic heads are placed in close association with the aqueous solvent, while the hydrophobic tails are sequestered in the interior of the micelles completely out of contact with the water.

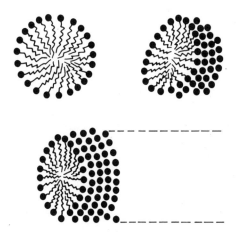

Fig. 2—Schematic representation of the structures of spherical (upper drawings) and cylindrical (lower drawing) micelles.

Micelles are not entities composed of fixed numbers of molecules having a fixed geometrical shape. They must be regarded as statistical in nature, in equilibrium with the surrounding amphiphilic molecules, and fluctuating constantly in size and shape in response to temperature. On dilution of the mixture, micelles dissociate rapidly, while on concentrating the solution, more extended micellar structures appear, eventually forming the many different lyotropic liquid-crystal phases.

The observed structure of micelles permits the rationalization of the solubilization property of aqueous solutions of amphiphilic compounds. The interior of the micelles can be thought of as small pockets of essentially pure liquid hydrocarbon. The interiors of the micelles, the pockets of hydrocarbon, are then capable of dissolving other lipophilic or hydrophobic molecules added to the solution. The action of soap in cleansing materials of oily or greasy dirt or soil is precisely of this nature.

2.3. Structures of Lyotropics

As the proportion of amphiphilic compound to water increases, an impressive variety of different lyotropic liquid crystal phases are observed. A systematic classification of the different types and their structure is presented in References [5] and [6] to which the reader is referred for the extensive details. In this chapter we shall only attempt to give the flavor of the situation.

One type of lyotropic structure is the so-called "isotropic" phase. In this phase spherical micelles form the basic unit of the liquid crystal structure. These are then deployed in either a face-centered cubic or body-centered cubic arrangement within the fluid aqueous medium. Although the micelles are essentially arranged on a lattice, they do not touch and the structure is not rigid as in a true crystal. The presence of the intervening aqueous solvent provides sufficient fluidity so that this structure (as well as the others to be described below), though highly ordered, is still properly classified as a liquid. Another broad class of structures is the so-called "middle" phase. In this phase the basic structural unit is the rod-like cylindrical micelle of essentially infinite length. These are then disposed in a hexagonal arrangement within the aqueous medium. Related to this category are the phases composed of rod-like micelles of rectangular cross section arranged in square or rectangular packings. In all of these structures the common feature is, of course, the sequestering of the hydrophobic tails away from contact with the water while allowing the hydrophilic heads to reside within the aqueous solvent.

When the proportion of water to amphiphilic compound becomes low, structures of the "reversed" type occur. The basic units here are spherical or cylindrical reversed micelles in which the hydrophobic tails are on the outside, while the hydrophilic heads line the interior which contains the water. Liquid crystalline arrangements of the reversed micelles then occur in the several ways outlined above. In all cases the hydrophobic tails on the exterior of the reversed micelles are in contact only with each other; the water is sequestered in the polar interiors.

Among the most interesting of the lyotropic phases are the "lamellar" structures. The most important of these is the so-called "neat" phase, whose structure is depicted in Fig. 3. In this structure we find the amphiphilic molecules arranged in double layers of essentially infinite extent in two dimensions. The internal structure of the double layer is such that the hydrophobic tails occupy the interior out of contact with water, while the hydrophilic heads line the exterior in contact with the fluid aqueous medium (regions denoted by A in Fig. 3). The double layers then stack periodically along the third dimension alternating with layers of the aqueous solvent. The thickness of the double layer is somewhat less than twice the length of the amphiphilic molecules (layer thickness is thus about 30 to 40 Å). The thickness of the intervening aqueous layers is about 20 Å. The hydrocarbon region within the double layers is essentially fluid. Some experiments, however, show a gradual transition from a fluid-like property of the hydrocarbon chains to a more rigid type of behavior at lower temperatures.[7] The overall structure of the neat phase is rather analogous to that of the smectic liquid crystals.

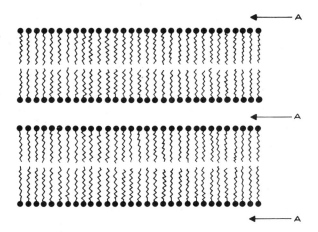

Fig. 3—Schematic representation of the structure of a portion of the neat phase. The regions containing the aqueous solvent are denoted by the A's.

Transitions from one kind of phase to another occur with changes in both the temperature and the concentration of the amphiphilic compound in water. A typical phase diagram is that for sodium laurate shown in Fig. 4. The T_c line gives the temperatures of transition from solid to liquid or liquid crystalline phases. It can be thought of as representing the depression of the melting point of sodium laurate by the water. The T_i line gives the temperatures of transition from liquid crystalline to normal liquid phases. The regions of stability of the "neat" and "middle" phases are shown. The cross-hatched areas are the regions of stability of the "isotropic" structures. The intermediate regions labelled I are most probably conjugate mixtures of the adjacent stable phases.

3. Biological Membranes

Studies of the structure and functions of biological membranes currently constitute an area of very active and intensive research. The literature on this subject is extensive and continually growing. The material selected for this section is primarily taken from articles by Rothfield[8] and by Singer.[3] The reader interested in further information is encouraged to read these articles as well as others appearing in the same volume. Also recommended are the more popular articles by Singer[9] and by Capaldi,[10] plus the recent review by Singer.[11]

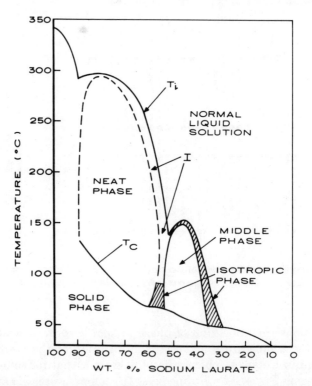

Fig. 4—Temperature-composition phase diagram of the sodium-laurate–water system (after Refs. [5] and [6]).

In spite of the extensive knowledge that has been obtained, it is still difficult to define a biological membrane. Rothfield,[8] however, has provided a particularly succinct description. Reproduced verbatim it reads: "Biological membranes are continuous structures separating two aqueous phases. They are relatively impermeable to water-soluble compounds, show a characteristic trilaminar appearance when fixed sections are examined by electron microscopy, and contain significant amounts of lipids and proteins." We shall enlarge on this description in the following two sections.

3.1 Constituents of Membranes

The principal constituents of all biological membranes are lipids, proteins, and oligosaccharides. The aqueous environment surrounding the membrane should properly also be considered one of the major components of membranes. The constituents that give the membranes their primary structure (and account for approximately half their mass) are the lipids. A bewildering diversity of lipids is found in membranes, and any one membrane will contain several different lipids. The lipids observed to be present belong to a variety of classes including phosphatidylcholine, phosphatidylethanolamine, sphingomyelin, glycolipids, cholesterol, etc.

The molecular structure of one example of lipid, phosphatidylcholine, is shown in Fig. 1b. The molecule is obviously recognized to be amphiphilic. At the top of the drawing we find the hydrophilic "head." In the aqueous environment of biological systems, the head is ionized as shown; the zwitterion (hybrid ion charged both positively and negatively) in this case is composed of phosphate and trimethylamine groups. In some lipids the hydrophilic head may, however, be an un-ionized or neutral group. Below the head is the glycerol group (sometimes called the "backbone"). Attached to the backbone are the twin hydrophobic "tails." The tails consist of long hydrocarbon chains. In any one class of lipids the chains will appear in many different lengths and several degrees of saturation (number of carbon-carbon double bonds). Their means of attachment can consist of a variety of covalent linkages to the glyceryl phosphate moiety. In discussing the structure of biological membranes it is convenient (as shown in Fig. 1b) to represent the hydrophilic head as a large dark dot and the twin hydrophobic tails as a pair of zig-zag lines.

The second major constituent of biological membranes (accounting for approximately half their mass) is the proteins. As with the lipids, a perplexing variety of different proteins are found. Membranes from different sources are observed to contain large numbers of proteins of different molecular weights, ranging from less than 15,000 to over 100,000. No single type of protein dominates. While the complete role of the proteins in the total function of the membranes is still not completely understood, it does seem clear that they are of lesser consequence in determining the primary structure to be described below.

We will, therefore, not give the proteins any further consideration here; the same holds true for the oligosaccharides.

3.2 Structure of Membranes

In spite of the significant heterogeneity in the molecular structure of lipids occurring in biological membranes, one important property is common: all membrane lipids are strongly amphiphilic. Because of this, one should expect that in the aqueous environment of biological systems these molecules will aggregate into structures similar to the lyotropic liquid crystals described above. This is indeed the case. Aqueous solutions of phospholipids and synthetic mixtures of naturally occurring lipids dissolved in water display various forms of liquid-crystal behavior.

The most common structure observed is the so-called "bimolecular leaflet" or phospholipid bilayer. The structure is analogous to the neat phase of the lyotropic liquid crystals and is shown schematically in Fig. 5. The most prominent feature of this structure is the arrange-

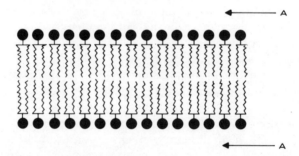

Fig. 5—Schematic representation of the structure of a portion of the lipid bilayer component of biological membranes. The regions containing the aqueous environment are denoted by the A's.

ment of the lipids into two contiguous layers of essentially infinite extent, such that the hydrophobic tails occupy the interior of the bilayers out of contact with the aqueous environment (regions denoted by A in Fig. 5), while the hydrophilic heads line the exterior in contact with the water. The thickness of the bilayer is approximately twice the length of the individual lipid molecules (40–50 Å).

The bimolecular leaflet is not a rigid structural entity. In keeping with its liquid-crystalline nature, the hydrocarbon interior of the bilayer is quite fluid. The fluidity or mobility can be viewed simply as random movements of the individual hydrocarbon chains of the molecules in the bilayer. In general, mobility is favored at higher temperatures, by greater degrees of unsaturation of the hydrocarbon chains, and by shorter chain lengths. Gradual transitions from more-fluid states to less-fluid states are seen in several experiments when the temperature is lowered. The implications of this fluidity (and its changes) to the biological function of membranes are many but cannot be examined here.

The structure of actual biological membranes cannot be as simple as the phospholipid bilayer. Membranes do, after all, contain large numbers of proteins of varying weights, shapes, and sizes. One of the major problems of membrane biology is the question of the detailed incorporation of both the lipids and the proteins into the overall membrane structure. The experimental situation is very difficult and no clearly observed membrane structure is available. Many models, however, have been proposed. The basic philosophy is that a general pattern of organization exists and that the heterogeneity and distinctiveness of different membranes can be understood as variations on a common structural theme. A particularly appealing model of this kind has been presented by Singer and others.[3,9,10]

The model is variously called the Lipid-Globular Protein Mosaic model or the Fluid Mosaic model. The basic structure of the membrane is assumed to be the bimolecular leaflet of lipid molecules. The leaflet is not considered to be continuous, however. The globular integral proteins and patches of the lipid bilayer are assumed to be arranged in an alternating mosaic pattern throughout the membrane. The hydrophobic portion of the lipids and a large fraction of the nonpolar amino acid residues of the proteins are sequestered from contact with water, mainly in the interior of the membrane. The hydrophilic groups of the lipids and the ionic residues of the proteins are in direct contact with the aqueous environment on the exterior of the membrane. Some of the proteins lie predominantly near the surface of the membrane, others penetrate the interior. Some of the latter may penetrate a short distance into the inside; others may extend clear through the entire thickness of the membrane. The saccharide components, being hydrophilic, presumably reside on the surface in direct contact with the aqueous environment.

The Lipid-Globular Protein Mosaic model is based on experimental data on the conformation of proteins in intact membranes and on general thermodynamic considerations (maximization of hydrophilic interactions and minimization of hydrophobic interactions of the lipids and proteins with the water). At the present time, there do not seem to be any data or experimental results that are clearly inconsistent with this model. While there is not yet any direct experimental evidence for this structural pattern, its consistency and thermodynamic feasibility recommend this model as a working hypothesis for further investigation.

4. Interaction of Amphiphilic Compounds with Water

Lyotropic liquid crystals and biological membranes are ordered assemblies of amphiphilic molecules situated in an aqueous environment. By analogy with the other ordered or condensed phases encountered in physics and chemistry, one might suspect that the stability of lyotropics and membranes derive from favorable attractive interactions between constituent amphiphilic molecules, and that the water only serves to provide the medium in which the ordered aggregates can reside. This notion is, however, incorrect. The participation of the water is far from passive. The role of the water is, in fact, crucial to the formation and stability of lyotropics and membranes. The chief mechanism by which the water acts to promote the various ordered structures is the *hydrophobic effect*.

Simplistically stated, the hydrophobic effect may be defined as the tendency of water to reject any contact with substances of a nonpolar or hydrocarbon nature. The existence of this effect was first recognized in the study of the extremely low solubility of hydrocarbons in water.[1] The principles involved were later successfully applied to the elucidation of the native conformation of protein molecules by Kauzmann.[2] The application of these ideas to the study of membrane structures has been advanced by Singer.[3] Recently, Tanford[4] published an entire book on the hydrophobic effect, including the influence of this interaction on the formation of micelles, lipid bilayers, membranes and other ordered structures. Aside from Singer's[3] and Tanford's[4] statements on the decisive role of the hydrophobic effect on lyotropics, the lyotropic liquid-crystal literature seems peculiarly unaware of this phenomenon. Winsor's[5] extensive review with its systematic analysis (R-theory) of the many lyotropic phases does not take the hydrophobic effect into account. More recent reviews[6,7] of lyotropic liquid crystals do not mention the phenomenon. We hope that the present discussion will help to advance the realization of the importance of the hydrophobic effect to lyotropics. The material of the following sections is taken chiefly from Ref. [3] with some assistance from Refs. [2] and [4].

4.1 Solubility of Hydrocarbons in Water

Most of the thermodynamic data concerning the hydrophobic interaction comes from studies of the solubility of hydrocarbons in water. An examination of these data provides an understanding of the nature and magnitude of the hydrophobic effect. The knowledge gained here can then be qualitatively applied to the more complex systems of interest.

The thermodynamics of solute–solvent interactions is most conveniently described in terms of *unitary* quantities.[12] The unitary free energy and unitary entropy changes accompanying some process (such as the transfer of hydrocarbon from nonpolar solvent to water or the transfer of hydrocarbon from pure hydrocarbon to water) are the standard free-energy and entropy changes corrected for any translational entropy terms (the *cratic* entropy) that are not intrinsic to the interaction under consideration. The cratic entropy is simply the entropy of mixing the solute and solvent into an ideal solution. With the cratic contribution removed, the unitary free energy and entropy contain only contributions to the thermodynamics of the process that come from the interaction of the individual solute molecules with the solvent.

Table 1—Thermodynamic Changes in the Transfer of Hydrocarbons from Nonpolar Solvents to Water at 25°C (After Ref. [2])

Process	ΔF_u (cal/mole)	ΔH (cal/mole)	ΔS_u (cal/mole °K)
CH_4 in benzene → CH_4 in H_2O	+2600	−2800	−18
CH_4 in ether → CH_4 in H_2O	+3300	−2400	−19
CH_4 in CCl_4 → CH_4 in H_2O	+2900	−2500	−18
C_2H_6 in benzene → C_2H_6 in H_2O	+3800	−2200	−20
C_2H_6 in CCl_4 → C_2H_6 in H_2O	+3700	−1700	−18
C_2H_4 in benzene → C_2H_4 in H_2O	+2920	−1610	−15
C_2H_2 in benzene → C_2H_2 in H_2O	+1870	−190	−7
Liq. propane → C_3H_8 in H_2O	+5050	−1800	−23
Liq. n-butane → C_4H_{10} in H_2O	+5850	−1000	−23

Table 1 contains a compilation of thermodynamic data for the transfer of simple hydrocarbons from nonpolar solvents to water. The unitary free energy, enthalpy, and unitary entropy changes are denoted by ΔF_u, ΔH and ΔS_u, respectively; the temperature is 25°C in all cases. ΔF_u is unfavorably positive in all examples. This is in keeping with the empirical observation that hydrocarbons do not dissolve in water to any appreciable extent. The ΔH are, however, exothermic, demonstrating that the intermolecular interaction energies favor the solution of hydrocarbon molecules in water. The unfavorable positive ΔF_u can therefore only arise because of a strongly unfavorable unitary entropy ($\Delta F_u = \Delta H - T\Delta S_u$). Table 1 shows that this is indeed the case. For all examples, ΔS_u is large and negative.

The low solubility of hydrocarbons in water is therefore a consequence of the large decrease in unitary entropy accompanying the introduction of such molecules into an aqueous environment. In their classic investigation, Frank and Evans[1] concluded that this large entropy decrease must be due to some kind of ordering of the water molecules around the hydrocarbon molecules dissolved in the water. The water molecules at the surface of the cavity created by the introduction of the hydrocarbon molecule must be capable of rearranging themselves in order to regenerate broken hydrogen bonds. In doing so, however, they create a higher degree of order than existed in the undisturbed water; the result of this ordering is the decrease in the unitary entropy. It is important to realize that it is the property of the water alone (its highly structured nature resulting from the considerable degree of hydrogen bonding) that is responsible for the hydrophobic interaction. The hydrophobic effect is relatively insensitive to the precise nature of the nonpolar solutes involved, and, furthermore, is not nearly so pronounced in the case of other polar solvents. Expressing the description of the hydrophobic effect in simple terms, water rejects contact with hydrocarbon and other nonpolar groups because not to do so would require the water to increase the local order of its structure, thereby reducing the entropy of the system.

4.2 Solubility of Ionized Species in Water

The interaction of the hydrophilic portions of amphiphilic molecules with the aqueous solvent is another important factor in the stabilization of the structures of lyotropics and membranes. Information on the hydrophilic interaction may be obtained from thermodynamic studies of the solubility of simple ionized molecules in water and other polar solvents.

Simple electrostatic arguments suggest that the free energy of ionized species is inversely proportional to the dielectric constant of the medium in which the ions reside. This should be so because ions interact more strongly with the molecules of polar solvents than with those of nonpolar solvents. This trend is indeed observed in thermodynamic data. Table 2 contains a compilation of the solubilities of a typical charged molecule, the zwitterion of glycine, $^+H_3N\text{-}CH_2\text{-}CO_2^-$, in various solvents of decreasing polar character. Also included is the unitary free energy of transfer of this species from water to the other solvents. ΔF_u is seen to be strongly positive for all nonaqueous solvents. It is clear that this ionized molecule is at a much lower free energy in contact with water than with any other solvents.

Table 2—Solubility and Free Energy of Transfer of Glycine in Various Solvents at 25°C (After Ref. [3])

Solvent	Solubility (mole/liter)	ΔF_u (cal/mole)
Water	2.886	—
Formamide	0.0838	1680
Methanol	0.00426	3430
Ethanol	0.00039	4630
Butanol	0.0000959	5190
Acetone	0.0000305	6000

The data in Table 2 reveal another of the unique properties of the solvent water. The interaction of the water with the ions is not simply attributable to the high dielectric constant; formamide has almost the same dielectric constant as water, yet it requires 1680 cal/mole of free energy to transfer glycine from water to formamide. This suggests that, as with the nonpolar solutes, ionic species induce significant changes in the local order of the water structure.

Another estimate of the importance of the hydrophilic interaction may be obtained from a consideration of the free-energy difference between the ionized species in water versus the uncharged species in nonpolar solvent. The thermodynamic data show that it is much more favorable for the molecules to exist as ionized species in water. For example, in the case of the carboxyl group, $\text{-}CO_2^-$, it takes a unitary free energy of 3300 cal/mole at 25°C just to protonate it (change it into the neutral group, $\text{-}CO_2H$) in water at pH 7. These considerations demonstrate that it is a thermodynamic necessity for the hydrophilic groups of amphiphilic molecules to be ionized and in direct contact with the aqueous environment.

4.3 Aggregation of Amphiphilic Compounds

The two fundamental thermodynamic principles described in the previous sections may now be applied to the question of the stability of lyotropic liquid crystals and biological membranes.

A very large cratic entropy is associated with the uniform dispersal of amphiphilic molecules throughout the aqueous medium; at extreme dilution this contribution to the free energy will stabilize isolated molecules. But, as the concentration of amphiphilic species is increased, large amounts of negative unitary entropy are produced by the hydrophobic effect. The free energy of the system can then be more effectively minimized by having the amphiphilic molecules aggregate. The cratic entropy is lost, but far larger amounts of unitary entropy are gained by forming micelles, bilayers, or other lyotropic structures in which the hydrophobic groups are completely sequestered from contact with the water. We noted in previous sections that all the structures observed did indeed sequester the hydrophobic groups. These structures, moreover, also arranged all the hydrophilic groups onto the exterior of the aggregates in direct contact with the water and away from association with the nonpolar hydrophobic parts. Thus, the observed structures simultaneously satisfy both of the important thermodynamic requirements: (1) minimization of the hydrophobic interaction and (2) maximization of the hydrophilic interaction with the aqueous solvent.

This interpretation is consistent with experiments involving the addition of nonaqueous solvents to aqueous solutions of amphiphilic compounds. The addition of 20 to 30 mole % of ethanol to solutions of amphiphilic compounds significantly reduces the stability of micelles; isolated molecules become much more favored. The primary effect of the ethanol is to reduce the hydrophobic effect, thereby diminishing the stability of aggregates. In addition, conductance measurements show that the net charge on the amphiphilic molecules is decreased in 30% ethanol because of formation of ion pairs (with the small cations) in the medium of lower dielectric constant. Thus, the hydrophilic interaction is also reduced on addition of the less polar solvent, and this also contributes to the destabilization of micelles.

Several other interactions play an important but secondary role in determining the stability of aggregates of amphiphilic molecules. These include van der Waals attraction among the hydrocarbon tails, electrostatic repulsion between similarly charged hydrophilic heads, electrostatic attraction between zwitterionic heads, and possible hydrogen bonding among the polar portions of the molecules. At any given concentration and temperature, the interplay of these forces plus the major hydrophobic and hydrophilic interactions determines the precise form of aggregation of the amphiphilic molecules. A quantitative theory of the composition and temperature dependence of the various structures of the lyotropic phases that includes all of the above considerations is not yet available. A reconstruction of the R-theory[5] to explicitly include the hydrophobic effect along the lines given by Kauzmann,[2] Singer,[3] and Tanford[4] would be a highly desirable beginning.

5. Conclusion

The composition and structure of lyotropic liquid crystals and biological membranes have been examined. An understanding of the varied structures was obtained in terms of the hydrophobic and hydrophilic interactions. That is, aggregates of amphiphilic molecules in aqueous solution always form in such a manner as to minimize the hydrophobic interaction between the hydrocarbon tails and the water, while simultaneously maximizing the hydrophilic interaction of the polar heads with the aqueous solvent. In this way, we saw that the water was not merely the medium in which these phenomena take place. Instead, it became clear that it is the unique properties of the water itself that give rise to the crucial interactions that stabilize lyotropics and membranes.

The ordered structures discussed here are also unusual from another point of view. While most of the ordered phases encountered in physics and chemistry owe their existence to a successful competition between the energy and the entropy, the ordered structures discussed here derive their stability from a competition between two different kinds of entropy. In magnetism, superconductivity, ferroelectricity, thermotropic liquid crystals, solid-liquid-vapor equilibrium, multicomponent phase separation, etc., the ordered phase is always one of low energy and low entropy, while the disordered phase has high energy and high entropy. At low temperatures the free energy (given by $F = E - TS$) will be minimized by the low energy ordered state. At higher temperatures the TS term gains in importance and eventually the free energy will be minimized by the high entropy disordered state.

In the case of micelles, lyotropics and membranes, however, the ordered state is characterized by having a low cratic entropy and a high unitary entropy. The disordered state with the molecules dispersed throughout the solvent, on the other hand, is a state of high cratic entropy and low unitary entropy (because of the hydrophobic effect). Since the cratic and unitary entropies have different dependencies on composition and temperature, the minimization of the free energy will sometimes be accomplished by maximizing the cratic entropy, and sometimes by maximizing the unitary entropy, depending on the exact conditions.

The outcome of the competition between cratic and unitary entropies, as the temperature and concentration are varied, thus gives rise to the broad features of the phase diagram shown in Fig. 4. This situation is somewhat analogous to existence of an ordered nematic state in the hard rod model.[13] In this case, the two kinds of entropy in competition are the cratic and orientational entropies.

References

[1] H. S. Frank and M. W. Evans, "Free Volume and Entropy in Condensed Systems III," *J. Chem. Phys.*, **13**, p. 507 (1945).

[2] W. Kauzmann, "Some Factors in the Interpretation of Protein Denaturation," *Adv. Protein Chem.*, **14**, p. 1 (1959).

[3] S. J. Singer, "The Molecular Organization of Biological Membranes," in *Structure and Function of Biological Membranes*, ed. by L. I. Rothfield, Academic Press, N. Y. (1971).

[4] C. Tanford, *The Hydrophobic Effect*, John Wiley and Sons, N. Y. (1973).

[5] P. A. Winsor, "Binary and Multicomponent Solutions of Amphiphilic Compounds," *Chem. Reviews*, **68**, p. 1 (1968).

[6] G. H. Brown, J. W. Doane and V. D. Neff, *A Review of the Structure and Physical Properties of Liquid Crystals*, CRC Press, Cleveland, Ohio (1971).

[7] A. Saupe, "Liquid Crystals," in *Annual Reviews of Phys. Chem.*, ed. by H. Eyring, Annual Reviews, Inc., Palo Alto, Vol. 24 (1973).

[8] L. I. Rothfield, "Biological Membranes: An Overview at the Molecular Level," in *Structure and Function of Biological Membranes*, ed. by L. I. Rothfield, Academic Press, N. Y. (1971).

[9] S. J. Singer, "The Fluid Mosaic Model of the Structure of Cell Membranes," *Science*, **175**, p. 720 (1972).

[10] R. A. Capaldi, "A Dyanmic Model of Cell Membranes," *Scientific American*, p. 27, March 1974.

[11] S. J. Singer, "The Molecular Organization of Membranes," in *Annual Review of Biochemistry*, ed. by E. E. Snell, Annual Reviews, Inc., Palo Alto, Vol. 43, 1974.

[12] R. W. Gurney, *Ionic Processes in Solution*, Chapter 5, McGraw-Hill Book Co., Inc., N. Y. (1953).

[13] P. Sheng, "Hard Rod Model of the Nematic-Isotropic Phase Transition," Chapter 5.

Appendix

Table 1—Thermodynamic Properties of Representative Liquid Crystal Materials from Several Different Classes

Class	Chemical name	Acronym	Transition temperatures and enthalpies*
Schiff's bases or anils[a]	4-Methoxybenzylidene-4'-n-butylaniline	MBBA	X(21,3.2)N(48,0.1) I
	4-Ethoxybenzylidene-4'-n-butylaniline	EBBA	X(36,5.8)N(79,0.2) I
	4-Butoxybenzylidene-4'-ethylaniline		X(51,1.9)N(65,3,0.1) I
	4-Octyloxybenzylidene-4'-n-heptylaniline		X(36,16)S(76,1.7)S(88,1.5) I
	4-Cyanobenzylidene-4'-n-octyloxyaniline	CBOOA	X(72,9.1)S(81,0.012[b])N(106,0.2) I
	Anisylidene-p-aminophenylacetate	APAPA	X(82,6.0)N(110,0.3) I
	Anisylidene-p-aminophenyl-3-methylvalerate		X(36,4.6)N(81,0.1–0.2) I
Biphenyls[c]	4-n-Pentyl-4'-cyanobiphenyl	PCB	X(22.5,4.1)N(35,0.1–0.2) I
	4-n-Nonyl-4'-cyanobiphenyl		X(40.5,8.0)S(44.5,0.1–0.3)N(47.5,0.1–0.3) I
	4-n-Pentoxy-4'-cyanobiphenyl		X(48,6.9)N(67.5,0.1) I
	4-n-Octyloxy-4'-cyanobiphenyl		X(54.5,5.9)S(67,—)N(80,0.6) I
Azoxybenzenes[d]	4-4'-Dimethoxyazoxybenzene	PAA	X(118.2,7.1)N(135.5,0.1) I
	4-4'-Diethoxyazoxybenzene	PAP	X(136.6,6.4)N(167.5,0.3) I
	4-4'-Di-n-heptyloxyazoxybenzene		X(74.4,9.8)S(95.4,0.4)N(124.2,0.2–0.3) I
	4-4'-Di-n-dodecyloxyazoxybenzene		X(81.7,10.1)S(122.0,2.9) I

*The first number in parentheses is the transition temperature in °C, the second number is the enthalpy of transition in kcal/mol; X = crystal, S = smectic, N = nematic, and I = isotropic.
[a]Measurements made at RCA Laboratories by A. W. Levine and M. T. McCaffery.
[b]See D. Djurek, J. Baturic-Rubcic, and K. Franulovic, *Phys. Rev. Letters* 33, 1126 (1974).
[c]Melting point data taken from D. S. Hulme, E. P. Raynes, and K. J. Harrison, *J. Chem. Soc. Chem. Commun.* 1974, p. 98. Data for 4-n-nonyl-4'-cyanabiphenyl obtained from G. W. Gray, private communication.
[d]Measurements made at RCA Laboratories by M. T. McCaffery.

Table 2—Physical Properties of 4-Methoxybenzylidine-4'-n-butylanaline and 4-n-Pentyl-4'-cyanobiphenyl at Room Temperature

Parameter	MBBA		PC B	
$\epsilon_\parallel$	4.7 (a)		19.7 (b)	
$\epsilon_\perp$	5.4 (a)		6.4 (b)	
$\Delta\chi$ [cgs units]	0.97×10^{-7} (c)		—	
n_0 (5145 Å)	1.5616		1.5442	
n_0 (6328 Å)	1.5443	(d)	1.5309	(d)
n_e (5145 Å)	1.8062		1.7360	
n_e (6328 Å)	1.7582		1.7063	
K_{11} [dynes]	6×10^{-7}		6.4×10^{-7} (b)	
K_{22} [dynes]	4×10^{-7}	(e)	—	
K_{33} [dynes]	7.5×10^{-7}		10×10^{-7} (b)	
ρ [g cm^{-3}]	1.088 (f)		—	
$S(T_c)$	0.34 (h)		0.49 (g)	
S (R.T.)	0.64 (h)		0.65 (g)	

aD. Diguet, F. Rondalez, and G. Durand, *C. R. Acad. Sci.* **B271**, 924 (1970).
bMeasurements made at RCA Laboratories by D. Meyerhofer.
cH. Gasparoux and J. Proust, *J. Phys. (Paris)* **32**, 953 (1971).
dMeasurements made at Harvard University by Shen Jen and A. E. Bell.
eI. Haller, *J. Chem. Phys.* **57**, 1400 (1972).
fI. Haller, H. A. Huggins, and M. J. Freiser, *Mol. Cryst. Liq. Cryst.* **16**, 53 (1972).
gMeasurements made at RCA Laboratories by A. E. Bell and E. B. Priestley.
hShen Jen, Noel A. Clark, P. S. Pershan, and E. B. Priestley, *Phys. Rev. Letters* **31**, 1552 (1973).

Index

Additives, 224
Addressing,
  beam scanning, 261, 270
  dual frequency, 266, 267
  matrix, 261, 304
  television, 268
  thermal, 273
  transistor, 268, 269
Alignment,
  effect of grooves, 225, 226
  homeotropic, 222, 291
  homogeneous, 222
  perpendicular, 222, 223, 243, 249, 255
  random parallel, 222
  rubbing, 224, 226, 227, 244
  sloped evaporation, 229
  uniform parallel, 222, 255

Bend, 106
Birefringence, 242
  voltage controlled, 299, 303, 306
Bragg scattering, cholesterics, 6, 12, 204, 205
Bubbles in cells, 239
Bundling effect, 70

Cheverons, 253, 303, 305, 310, 312
Chiral nematic materials, 7, 203
Cholesteric order, 5
Cholesterics,
  pitch, 6, 204
  propagation of light in, 205
  waveguide regime, 212
Classification of liquid crystals, 4
Clearing temperature, 38, 167

Coatings,
  conductive, 220
  indium–tin oxide, 220
  tin oxide, 226
Coherence length, 104, 112
  electric, 115
  magnetic, 112
  magnitude, 114
Conduction regime, 140, 253
Contact angle, 229
Correlation length, 172, 178
Cotton–Mouton coefficient, 195

Density discontinuity, 70
Devices,
  electrochemical reactions in, 328
  photoconductor control of, 271, 272, 314
Diamagnetic anisotropy, 74
Dielectric anisotropy, 24, 242, 298, 306, 309
  effects of substitution, 24
  related to devices, 25
Dielectric constants, 108, 109
Dielectric force, 131
Dielectric regime, 253
Dielectric relaxation, 312, 313, 326
Diffraction by domains, 308
Director, 5, 16, 32, 47, 84, 104, 204, 207, 222, 312
Director field, 104
Disinclination lines, 112, 120
Dissolved gases in cells, 239
Disordered crystals, 2
Dispersion relations, 284

INDEX

Display devices, 25
  effects of pressure, 235
  materials for, 24
  matrix addressed, 261, 304
Display life, 259
Distortion parameters, 105
Distribution function,
  orientational, 35, 51, 71, 72, 73, 76, 84
  smectic-A, 86
  translational, 85
Domains, 129, 300, 301, 303, 310, 312
  threshold for formation, 131, 139
Dynamic scattering, 129, 251, 256, 293, 312, 328, 330

Elastic constants, 106, 158, 159, 160
  magnitude, 107
Electric field effect, 104, 242, 292, 303, 385
Electro-optic transfer function, 297, 304, 305
Electrochemical reactions in devices, 328
Electrochemistry of nematic solvents, 319
Electrohydrodynamic flow, 129, 251
Embryo, 183
  critical radius, 184
  stability, 184, 185
Entropy,
  change on transition, 40, 94, 98, 167
  cratic, 345, 348, 349
  nematic phase, 38, 40, 51
  orientational, 59, 60
  smectic-A phase, 93
  translational, 59, 60
  unitary, 345, 348, 349
Equivalent conductance, 320, 322
Euler–Lagrange equation, 65, 66, 111
Evaporation, sloped, 229
Excluded volume, 60, 63, 64, 65, 68

Field effects, 104, 242, 292, 303, 305
  cholesteric pitch variation, 122
  cholesteric–nematic transition, 120
Field-induced birefringence, 242
Fluctuations, 41, 168
  amplitudes, 174, 180, 181, 182
  heterophase, 182
  homophase, 169
Fluidity, 3, 7, 8, 319, 323, 328
Frank elastic constants, 106
Free energy,
  classical gas of hard rods, 66, 68

Free energy (*cont'd*)
  cratic, 345
  dielectric, 136
  distortion, 104, 105, 106
  elastic, 136
  electric, 162
  external field, 108
  Landau, 146
  magnetic, 162
  nematic phase, 38, 51
  smectic-A phase, 93, 97
  surface, 105
  unitary, 345
Fréedericksz transition, 115
  critical field, 118
Frequency,
  critical cut-off, 253, 258, 259, 261
  transit time, 255

Gas bubbles in cells, 239
Gases dissolved in cells, 239
Grooves, 225, 226
Guest–host effect, 248, 299, 306

Hydrodynamic effects, 308
Hydrodynamic force, 132
Hydrodynamic instability, 129

Index of refraction, anisotropy, 244
Interactions,
  amphiphilic, 335
  electrolyte–solvent, 320
  hydrophilic, 335, 346
  hydrophobic, 335, 344
  intermolecular, 34, 46, 52, 71, 88
  lipophilic, 335
  steric, 59, 63, 70
Ionic mobility, 325

Landau expansion, 153, 157, 164
Landau free energy, 146
Landau parameters, values, 161, 168
Lateral attractions, 20
Light scattering, 189, 287, 291
Local anisotropy, 107
Lyotropic materials, 3, 334, 338

Magnetic birefringence, 193
Matrix array, 262
Mean field approximation, 35, 47, 89
Mesogens, 10, 18
  broadening substituents, 21

Mesogens (cont'd)
  central linkages, 10, 11, 18
  terminal groups, 10, 11, 19
Mesophases, 1, 12
  molecular geometry, 10, 16
  ordered structures, 16
  structural requirements, 17
  thermal stability, 19
Micelles, 336
Misalignment, alkali-induced, 231
Mixtures,
  calculated nematic range, 27
  equilibrium properties, 320
  equivalent conductance, 320
  for display devices, 24
  temperature ranges, 26
Mobility, ionic, 325
Molecular elongation, 10
Molecular order, 4
Molecular rigidity, 10, 11
Multiplexing, 261

Neat phase, 339
Nematic order, 4, 5
  experimental determination, 77, 78, 79
  macroscopic, 74
  microscopic, 72, 73
Nematic phase, 4, 32
  effect of broadening, 21
  effect of chirality, 6, 7, 16
  electrochemistry, 319
  homologous series, 22
  symmetry, 4, 5, 32, 104
  terminal attractions, 20

Odd–even effect, 24
Onsager equations, 65, 66
Optical measurements, 299
Optical properties, cholesterics, 203
Optical rotatory power, cholesterics, 205
Optical waveguides, 212, 281
Order parameter, 71, 75, 154, 166
  discontinuity, 39, 70
  fluctuations, 168
  macroscopic, 75
  nematic, 32, 106
  numerical values, 39
  smectic-A, 86, 87
  temperature dependence, 37, 39
  tensor, 75, 154
Ordered fluid mesophases, 12
Ordered fluids, 2

Orientational distribution function, 35, 51, 71, 72, 73, 76, 84

Packaging, 219
Paranematic susceptibility, 194
Partition function, 36, 51, 61, 62, 63, 64, 65, 67, 72, 92, 97, 165
Phase diagram, 69
  lyotropics, 340
  smectic-A, 87, 95
Phase transitions, 1, 38, 59, 93, 98, 99, 120, 124, 249
Photoconductor–liquid crystal device, 271, 272, 314
Pitch of cholesterics, 6, 204
Plastic crystals, 2
Pleochroic dye, 248, 304, 306
Polar effects, 19
Polymorphous materials, 9
Potential, mean field, 35, 47, 71, 89, 91
Pressure effects in cells, 235
Pretransitional behavior, 41, 168, 169, 189, 195

Rayleigh ratio, 191
Rayleigh scattering, 288, 292
Reverse tilt, 247
Reverse twist, 247
Rubbing, unidirectional, 224, 226, 227, 244

Short-range order, 57, 70
Silane coupling agents, 224
Smectic order, 7
Smectic phase,
  effect of broadening, 21
  homologous series, 20, 22
  lateral attractions, 20
  symmetry, 7, 8, 84
Solutions,
  equilibrium properties, 319, 320
  equivalent conductance, 320, 322
Splay, 106
Storage mode, 255, 299, 314
Structure,
  biological membranes, 342
  cholesteric phase. 5, 6, 204, 205
  lyotropics, 338
  mesogens, 18
  mesophases, 16
  nematic phase, 4, 32, 74
  smectic phases, 7, 84
Supercooling, 156, 184, 185, 186, 189

Surface energy, 188, 189, 227
Surface tension, critical, 228
Susceptibility, 106
Symmetry,
  cholesteric phase, 5
  nematic phase, 4, 32, 104
  smectic-A phase, 7, 84
  smectic-B phase, 8
  smectic-C phase, 7

Temperature, effect on display devices, 236, 237
Terminal attractions, 20
Texture,
  Grandjean, 222, 249, 256
  scattering, 249, 256
Thermal expansion effects, 236
Thermal stability, 17, 19
  effect of polarity, 19
  effect of polarizability, 20

Thermal stability (*cont'd*)
  off-axis substituents, 22
  size of substituents, 20
Thermotropic materials, 3
Thermotropic mesogens, 10
Transient response, 258
Transition temperature, 19, 20, 21, 22, 38, 167
Twist, 106
  director field, 109
  distribution, 111
  right- and left-handed, 112, 122
Twisted nematic cell, 104, 110, 212, 299
Twisted nematic effect, 245, 306

Viscous forces, 133, 136

Walden's rule, 322, 324
Wein effect, 325
Williams' domains, 129, 252